Dynamic Modeling and Simulation for Control Systems, 2nd Edition

Dynamic Modeling and Simulation for Control Systems, 2nd Edition

Editors

Adrian Olaru
Gabriel Frumusanu
Catalin Alexandru

Basel • Beijing • Wuhan • Barcelona • Belgrade • Novi Sad • Cluj • Manchester

Editors

Adrian Olaru
National University of
Science and Technology
Politehnica Bucharest
Bucharest
Romania

Gabriel Frumusanu
"Dunarea de Jos" University
of Galati
Galati
Romania

Catalin Alexandru
Transilvania University of Brasov
Brasov
Romania

Editorial Office
MDPI AG
Grosspeteranlage 5
4052 Basel, Switzerland

This is a reprint of articles from the Special Issue published online in the open access journal *Mathematics* (ISSN 2227-7390) (available at: https://www.mdpi.com/si/mathematics/Dyn_Modeling_Simul_Control_Syst_II).

For citation purposes, cite each article independently as indicated on the article page online and as indicated below:

Lastname, A.A.; Lastname, B.B. Article Title. *Journal Name* **Year**, *Volume Number*, Page Range.

ISBN 978-3-7258-1813-6 (Hbk)
ISBN 978-3-7258-1814-3 (PDF)
doi.org/10.3390/books978-3-7258-1814-3

© 2024 by the authors. Articles in this book are Open Access and distributed under the Creative Commons Attribution (CC BY) license. The book as a whole is distributed by MDPI under the terms and conditions of the Creative Commons Attribution-NonCommercial-NoDerivs (CC BY-NC-ND) license.

Contents

About the Editors vii

Leonardo Carvalho, Jonathan M. Palma, Cecília F. Morais, Bayu. Jayawardhana and O. L. V. Costa
Gain Scheduled Fault Detection Filter for Markovian Jump Linear System with Nonhomogeneous Markov Chain
Reprinted from: *Mathematics* **2023**, *11*, 1713, doi:10.3390/math11071713 1

Samir Aberkane and Vasile Dragan
A Deterministic Setting for the Numerical Computation of the Stabilizing Solutions to Stochastic Game-Theoretic Riccati Equations
Reprinted from: *Mathematics* **2023**, *11*, 2068, doi:10.3390/math11092068 22

Humam Al-Baidhani, Abdullah Sahib and Marian K. Kazimierczuk
State Feedback with Integral Control Circuit Design of DC-DC Buck-Boost Converter
Reprinted from: *Mathematics* **2023**, *11*, 2139, doi:10.3390/math11092139 38

Jiacheng Wang, Yunmei Fang and Juntao Fei
Adaptive Super-Twisting Sliding Mode Control of Active Power Filter Using Interval Type-2-Fuzzy Neural Networks
Reprinted from: *Mathematics* **2023**, *11*, 2785, doi:10.3390/math11122785 56

Dian Jiang, Yunmei Fang and Juntao Fei
An Extended-State Observer Based on Smooth Super-Twisting Sliding-Mode Controller for DC-DC Buck Converters
Reprinted from: *Mathematics* **2023**, *11*, 2835, doi:10.3390/math11132835 76

Karim El Moutaouakil, Abdellatif El Ouissari, Vasile Palade, Anas Charroud, Adrian Olaru, Hicham Baïzri, et al.
Multi-Objective Optimization for Controlling the Dynamics of the Diabetic Population
Reprinted from: *Mathematics* **2023**, *11*, 2957, doi:10.3390/math11132957 95

Rui Jiao, Wei Liu and Yijun Hu
The Optimal Consumption, Investment and Life Insurance for Wage Earners under Inside Information and Inflation
Reprinted from: *Mathematics* **2023**, *11*, 3415, doi:10.3390/math11153415 123

Claudiu Bisu, Adrian Olaru, Serban Olaru, Adrian Alexei, Niculae Mihai and Haleema Ushaq
Monitoring the Wear Trends in Wind Turbines by Tracking Fourier Vibration Spectra and Density Based Support Vector Machines
Reprinted from: *Mathematics* **2024**, *12*, 1307, doi:10.3390/math12091307 141

Maysoon Ghandour, Subhi Jleilaty, Naima Ait Oufroukh, Serban Olaru and Samer Alfayad
Real-Time EtherCAT-Based Control Architecture for Electro-Hydraulic Humanoid
Reprinted from: *Mathematics* **2023**, *12*, 1405, doi:10.3390/math12091405 164

Djamel Ziane, Samir Zeghlache, Mohamed Fouad Benkhoris and Ali Djerioui
Robust Control Based on Adaptative Fuzzy Control of Double-Star Permanent Synchronous Motor Supplied by PWM Inverters for Electric Propulsion of Ships
Reprinted from: *Mathematics* **2024**, *12*, 1451, doi:10.3390/math12101451 189

About the Editors

Adrian Olaru

Adrian Olaru, Ph.D., is a Full Professor at the University Politehnica of Bucharest, Romania. His research mainly include the modeling and simulation of control systems, including hydraulic power systems, transport systems, electrical and hydraulic servo systems, and the dynamic behavior of industrial robots.

Gabriel Frumusanu

Gabriel Frumusanu, Ph.D. is a Full Professor at "Dunarea de Jos" University of Galati, Romania. His research focuses on the numerical modeling of machining systems, manufacturing process control, dynamics of cutting processes, chaos theory, and computer-assisted design.

Catalin Alexandru

Catalin Alexandru, Ph.D. is a Full Professor at Transilvania University of Brasov, Romania. The main research interests include modeling and simulation for mechanical systems, renewable energy systems, and virtual prototyping.

Article

Gain Scheduled Fault Detection Filter for Markovian Jump Linear System with Nonhomogeneous Markov Chain

Leonardo Carvalho [1,*], Jonathan M. Palma [2], Cecília F. Morais [3], Bayu Jayawardhana [4] and Oswaldo L. V. Costa [1]

[1] Departamento de Engenharia de Telecomunicacções e Controle, Escola Politécnica na Universidade de São Paulo, São Paulo 05508-010, SP, Brazil
[2] Facultad de Ingeniería, Universidad de Talca, Curico 3340000, Maule, Chile
[3] Pontifical Catholic University of Campinas (PUC-Campinas), Center for Exact, Environmental and Technological Sciences (CEATEC), Postgraduate Program in Telecommunication Networks Management, Campinas 13086-900, SP, Brazil
[4] Engineering and Technology Institute Groningen, Faculty of Science and Engineering, Rijksuniversiteit Groningen, 9747 AG Groningen, The Netherlands
* Correspondence: carvalho.lp@usp.br

Abstract: In a networked control system scenario, the packet dropout is usually modeled by a time-invariant (homogeneous) Markov chain (MC) process. However, from a practical point of view, the probabilities of packet loss can vary in time and/or probability parameter dependency. Therefore, to design a fault detection filter (FDF) implemented in a semi-reliable communication network, it is important to consider the variation in time of the network parameters, by assuming the more accurate scenario provided by a nonhomogeneous jump system. Such a premise can be properly taken into account within the linear parameter varying (LPV) framework. In this sense, this paper proposes a new design method of $\mathcal{H}_\infty$ gain-scheduled FDF for Markov jump linear systems under the assumption of a nonhomogeneous MC. To illustrate the applicability of the theoretical solution, a numerical simulation is presented.

Keywords: fault-detection filter; Markovian jump linear system; $\mathcal{H}_\infty$ norm; LMI relaxations; nonhomogeneous Markov chains

MSC: 93E10

1. Introduction

In order to keep a manufacturing process as lean as possible, there are several aspects that must be considered. The monitoring capability is one of the features receiving the major spotlight in industrial operations since it is crucial to guarantee that the process is safe for the personnel involved. Among the procedures that constitute the monitoring systems, one that is worth mentioning is the fault-detection (FD) process [1,2].

A fault can be seen as the first indication of more harsh problems. It is any type of unwanted minor behavior that was not expected from the system. It can be caused, for instance, by extended wear due to long periods of time without maintenance. As a consequence of inadequately fixed wear, malfunctions or failures can cause a breakage [3].

In this sense, FD is a model-based process in which any abnormal/unexpected behavior is detected by a two-step procedure. The first step in the FD process is the residue generation, performed by an observer. The second step is the evaluation process, where the residue signal, generated by the observer, is treated by an evaluation function and compared with a predetermined threshold. We assume that a fault has occurred if the evaluation function surpasses the threshold; otherwise, we consider that the system is working as intended [4].

Currently, an important assumption that must be taken into account in FD systems is that the communication between the components is made via semireliable networks, which are associated with occasional packet dropout. These dropouts are caused by different sources as package collision due to high network congestion [5]. The distinction between a dropout and a fault is an essential aspect of the FD process since it makes easier to locate a fault. A viable way to model a packet dropout in a network is to employ a Markov jump linear system (MJLS) framework. This representation is appropriate to handle systems whose dynamic behavior is subject to random abrupt changes, like those caused by network packet dropouts. In this scenario, a Markov chain (MC) is used to model the jumping between the modes of operation of the system [6].

In the literature, there are plenty of examples of FD approaches that consider network behavior in their design. For instance, in [7,8] linear matrix inequality (LMI)-based constraints are provided to design fault detection filters (FDF) by using the $\mathcal{H}_\infty$ norm as a performance index. In [9], the authors have developed an FD approach for underactuated manipulators modeled by MJLS. In [10], an FD method for networked control systems (NCS) under the assumption of the existence of a variable delay between the signals received by the system components is tackled. In [11], a fault detection filter under the MJLS formulation was applied to a control moment gyroscope. In [12], a fault-detection filtering problem is tackled under the Markov switching memristive neural networks. In [13], an observed-based sliding mode control problem based on the event-triggered protocol under the Markovian jump systems framework was presented. In [14], a fault-detection filter for discrete time Markov jump Lur'e systems with bounded sector condition. Observing all the above examples, one fundamental premise in the MJLS context is that the Markov chain (MC) is considered to be homogeneous [15], which means that it does not vary in time. However, since the packet dropout sources (collision, congestion, networked-induced delay) change in time, we consider that a fixed transition probability between the Markovian operation modes does not properly model the network behavior. A way to handle the particularity of a time-varying MC was presented in [16], where the author has proposed new LMI constraints to evaluate the stability of MJLS governed by a nonhomogeneous MC. A particular case of the proposal presented by [16], which allows for designing FDF for MJLS systems affected by nonhomogeneous MCs, consists of using a linear parameter varying (LPV)-based representation for the time-varying transition probability matrix [17,18]. There are several works in the literature that deal with the problem of control (or filter) synthesis for nonlinear systems by using different approaches. For example, regarding the design of fault-tolerant controllers, there are strategies based on fuzzy systems [19] capable of modeling system nonlinearities by using Takagi–Sugeno models, so that if the probability of actuator failure is small, the control mode is normal, and if the probability is high, the control is changed to fault-tolerant mode. Another strategy to deal with nonlinearities that can be found in the literature arises in the context of sliding mode control [20]. In this case, the class of discrete-time nonlinear systems with delays and uncertainties that is considered is the conic type, where the nonlinear terms satisfy the constraint that lies in a known hypersphere with an uncertain center. However, the proposed approach, in addition to considering the loss of packets in the communication network via the Markov chain, deals with the nonlinearity of the systems by using a different strategy from those previously discussed, in which the modes of operation are considered linear but depend on time-varying parameters. Such modeling allows the use of convex optimization methods and LMI-based tools to solve the filtering problem without adding extra levels of complexity.

In view of the above works, the main contribution of the present work are

- the proposition of a new design technique of gain-scheduled FDF for MJLS with nonhomogeneous MC, and
- the numerical simulation to reinforce the usability of the proposed theoretical solution.

The proposed approach describes the nonhomogeneous MC using linear time-varying parameters to model those variations, assuming that these parameters are known or at least measurable. Another important assumption made is that the probability varies in time arbitrarily. Hence, the probability parameter for the following instant $(k+1)$ does not depend on present instant k, which grants the ability to disassociate the Lyapunov function in two distinct simplexes. Based on this assumption, we propose the design of a gain-scheduled fault-detection filter where the scheduling parameter implemented is the one that dictates the variation of the MC. One advantage of the proposed approach, when compared with others found in the literature, is that the design conditions assure the system stability for the entire parameter-varying range since the FDF is scheduled in terms of time-varying parameters modeling the network variation. The major novelty of the proposed technique is the higher level of fidelity in the representation of the network influence in the system model. Since FD is a model-based approach, a more accurate representation of the system can lead to better performance in practice.

The paper is organized as follows. Sections 2 and 4 present the necessary theoretical fundamentals. Section 3 shows how to model the nonhomogeneous Markov Chain by using LPV. Section 5 introduces the problem formulation and the main contributions. Section 6 illustrates the feasibility of applying the proposed technique, by means of a numerical simulation, and Section 7 concludes the paper with some final remarks.

Notation

The real Euclidean space is denoted by $\mathbb{R}^n$ where n represents its dimension and a real matrix with n rows and m columns is represented by $\mathbb{R}^{n \times m}$. The symbol I_n stands for an $n \times n$ identity matrix (or, for simplicity, just I, with an appropriate dimension, whenever no confusion arises) and the symbol $(\cdot)'$ denotes the transpose of a matrix. The operator $\text{Her}(\cdot)$ is used to express the symmetric sum as in $\text{Her}(X) = X + X'$, while the operator $\text{diag}(\cdot)$ represents a diagonal matrix. The symbol $\bullet$ denotes a symmetric block in a partitioned symmetric matrix. The expected value operator is represented by $\mathbb{E}(\cdot)$ and the conditional expected operator is denoted by $\mathbb{E}(\cdot|\cdot)$. The fundamental probability space is described by $(\Omega, \mathcal{F}, \{\mathcal{F}_k\}, \Pr(\cdot))$. The space $\mathcal{L}^2$ is the Hilbert space formed by $\mathcal{F}_k$-measurable random sequences $\{z_k\}_{k=0}^{\infty}$ such that $\|z\|_2 \triangleq \left[\sum_{k=0}^{\infty} \mathbb{E}\{|z(k)|^2\}\right]^{1/2} < \infty$.

2. Preliminaries

A generic discrete-time MJLS is given by

$$\mathcal{G} \equiv \begin{cases} x(k+1) = A_{\theta_k} x(k) + J_{\theta_k} w(k) \\ z(k) = C_{\theta_k} x(k) + D_{\theta_k} w(k) \end{cases} \tag{1}$$

where $x(k) \in \mathbb{R}^{n_x}$ is for the state vector, $w(k) \in \mathbb{R}^{n_w}$ is the exogenous input vector, and $z(k) \in \mathbb{R}^{n_z}$ is the output signal. The state-space matrices of system (1) depend on the index θ_k, which represents a discrete-time Markov chain belonging to a finite set of modes $\mathbb{K} = \{1, \ldots, \sigma\}$, whose switching is ruled by a time-varying transition probability matrix

$$\mathbb{P}(k) = \begin{bmatrix} \rho_{11}(k) & \cdots & \rho_{1\sigma}(k) \\ & \ddots & \\ \rho_{\sigma 1}(k) & \cdots & \rho_{\sigma\sigma}(k) \end{bmatrix}. \tag{2}$$

The entries $\rho_{ij}(k)$ of $\mathbb{P}(k)$ are such that $\rho_{ij}(k) = \Pr(\theta_{k+1} = j | \theta_k = i)$, $\forall k \geq 0$, $\rho_{ij}(k) \geq 0$, and $\sum_{j=1}^{\sigma} \rho_{ij}(k) = 1$. We recall that whenever the transition matrix is time-invariant, that is, $\mathbb{P}(k) = \mathbb{P}$, the associated Markov chain is said to be homogeneous; otherwise, it is called nonhomogeneous (meaning that the probabilities vary in time) [15,21]. It is assumed that $\rho_{ij}(k)$ varies within the following interval: $0 \leq \underline{\rho}_{ij} \leq \rho_{ij}(k) \leq \overline{\rho}_{ij} \leq 1$, where $\underline{\rho}_{ij}$ represents lower bound and $\overline{\rho}_{ij}$ denotes the upper bound. Another important assumption is that the upper and lower bounds of the transition probability are known, and the transition probability variation is instantly measurable. Therefore, all the parameters in (2) may vary in a known range with $\rho_{ij}(k) \in [\underline{\rho}_{ij}, \overline{\rho}_{ij}]$. There are several ways to determine the values of the upper ($\overline{\rho}_{ij}$) and lower ($\underline{\rho}_{ij}$) bounds of $\rho_{ij}(k)$. Those values can be obtained via mathematical modeling, observation, estimation, simulation, or based on the a priori knowledge of the system, such that the estimate can vary among systems and depends on the type of variation to which the system is subjected.

From the constraints $\sum_{j=1}^{\sigma} \rho_{ij}(k) = 1$ and $0 \leq \underline{\rho}_{ij} \leq \rho_{ij}(k) \leq \overline{\rho}_{ij} \leq 1$, the transition matrix (2) can be described by N polytopic intervals, where N depends on the number of transition probabilities that are time-varying. From these polytopic intervals, some techniques can be applied to obtain a gain-scheduled FDF.

In order to exemplify these N polytopic intervals and how to define a time-varying transition matrix by using LPV, let us assume that $\sigma = 5$ and the parameters $\rho_{14}(k)$ and $\rho_{15}(k)$ vary in time; hence, the first row of the transition matrix (2) can be written as

$$\begin{bmatrix} \rho_{11} & \rho_{12} & \rho_{13} & \rho_{14}(k) & \rho_{15}(k) \end{bmatrix}, \tag{3}$$

and from this row, two polytopic intervals ($N = 2$) are obtained:

$$\begin{bmatrix} \rho_{11} & \rho_{12} & \rho_{13} & \underline{\rho}_{14} & \overline{\rho}_{15} \end{bmatrix}, \begin{bmatrix} \rho_{11} & \rho_{12} & \rho_{13} & \overline{\rho}_{14} & \underline{\rho}_{15} \end{bmatrix}. \tag{4}$$

The polytopic intervals obey the constraints $\sum_{j=1}^{\sigma} \rho_{ij}(k) = 1$ and $0 \leq \underline{\rho}_{ij} \leq \rho_{ij}(k) \leq \overline{\rho}_{ij} \leq 1$ simultaneously. The following notation will be used to represent a time-varying row of $\mathbb{P}(k)$ as in (3) with the lower and upper bounds as in (4):

$$\begin{bmatrix} \rho_{11} & \rho_{12} & \rho_{13} & [\underline{\rho}_{14}, \overline{\rho}_{14}] & [\underline{\rho}_{15}, \overline{\rho}_{15}] \end{bmatrix}. \tag{5}$$

The main novelty in this paper is the usage of the same time-varying parameters that coordinate the nonhomogeneous MC variation as gain-scheduled parameters for the design and implementation of the FDF. This concept will be carefully described in Section 3.

Although the time variation that affects the probability matrix is generally represented by modeling $\mathbb{P}(k)$ as belonging to a polytope, in this paper we choose to use another approach, which describes each time-varying row of $\mathbb{P}(k)$ in terms of a linear time-varying parameter vector $\alpha_r(k)$ belonging to a distinct unit simplex Λ_{N_r}, $r = 1, 2, \ldots, m$. The definition of the unit simplex is given by

$$\Lambda_{N_r} \equiv \left\{ \zeta \in \mathbb{R}^{N_r} \,\middle|\, \sum_{i=1}^{N_r} \zeta_i = 1, \zeta_i \geq 0, i = 0, 1, \ldots, N_r \right\}, \tag{6}$$

where m is the number of time-varying rows in the probability matrix. In order to group up all the time-varying parameters of $\mathbb{P}(k)$ in a single domain, we perform a Cartesian product of m simplexes, each one of dimension N_r, in a single domain called multisimplex, and represent it by $\Lambda_N = \Lambda_{N_1} \times \Lambda_{N_2} \times \cdots \times \Lambda_{N_m}$, with the index N given by $N = (N_1, \ldots, N_m)$. For ease of notation $\mathbb{R}^N$ represents the space $\mathbb{R}^{N_1 + N_2 + \cdots + N_m}$. In this sense, a given element $\alpha(k) \in \Lambda_N$ is a vector belonging to $\mathbb{R}^N$ and can be decomposed as $(\alpha_1(k), \alpha_2(k), \ldots, \alpha_m(k))$ according to the structure of Λ_N. Subsequently, each $\alpha_r(k) \in \Lambda_r \subset \mathbb{R}^{N_r}$, $r = 1, \ldots, m$, is decomposed in the form $(\alpha_{r1}, \alpha_{r2}, \ldots, \alpha_{rN_r})$. This approach follows the one adopted in [22].

Hereafter, the transition probability will be denoted by $\rho_{ij}(\alpha(k))$, where the term $\alpha(k) \in \Lambda_N$ represents the time-varying parameter responsible to model the probability of the nonhomogeneous Markov chain at time k.

3. Modeling the Nonhomogeneous Markov Chain by Using the Linear Parameter Varying Approach

We next present the definition of a matrix $O_i(\iota_k)$, where i represents the MC mode, and $\iota_k = (\iota_1(k), \ldots, \iota_m(k))$ denotes a generic LPV parameter. It is assumed that $O_i(\iota_k)$ is affinely dependent on the time-varying parameters $\iota_j(k)$, as described below:

$$O_i(\iota_k) = O_{i_{l_0}} + \sum_{j=1}^{m} \iota_j(k) O_{i_{l_j}}. \tag{7}$$

The matrix in the affine form (7) can be interpreted in the following manner: matrix $O_{i_{l_0}}$ represents the time-invariant part of the filter dynamics. The remaining matrices $O_{i_{l_j}}, j = 1, \ldots m$ denote the time-varying dynamic that depends on the parameters $\iota_j(k)$. To illustrate this particular structure, consider the example presented below, for an MJLS with three operation modes, whose time-varying probability matrix is given by

$$\mathbb{P}(k) = \begin{bmatrix} 0.5 & [0.1, 0.3] & [0.2, 0.4] \\ [0, 0.4] & [0.5, 0.9] & 0.1 \\ 0.2 & 0.6 & 0.2 \end{bmatrix}, \tag{8}$$

where elements $\rho_{12}(k)$, $\rho_{13}(k)$, $\rho_{21}(k)$ and $\rho_{22}(k)$ vary in a known interval $[\underline{\rho}_{ij}, \overline{\rho}_{ij}]$. Since each uncertain row of $\mathbb{P}(k)$ can be represented by a polytopic interval, the representation of the first row is given by

$$\begin{bmatrix} 0.5 & 0.1 & 0.4 \end{bmatrix} \alpha_{11}(k) + \begin{bmatrix} 0.5 & 0.3 & 0.2 \end{bmatrix} \alpha_{12}(k) \tag{9}$$

and the second row is

$$\begin{bmatrix} 0 & 0.9 & 0.1 \end{bmatrix} \alpha_{21}(k) + \begin{bmatrix} 0.4 & 0.5 & 0.1 \end{bmatrix} \alpha_{22}(k) \tag{10}$$

with $\alpha_1(k) = (\alpha_{11}(k), \alpha_{12}(k)) \in \Lambda_2$, $\alpha_2(k) = (\alpha_{21}(k), \alpha_{22}(k)) \in \Lambda_2$, and $\alpha(k) = (\alpha_1(k), \alpha_2(k)) \in \Lambda_2 \times \Lambda_2$. On the other hand, the representation of $\mathbb{P}(k)$, in terms of parameter $\iota(k)$ used in the affine structure, can be done as follows,

$$\mathbb{P}(k) = \underbrace{\begin{bmatrix} 0.5 & 0.2 & 0.3 \\ 0.2 & 0.7 & 0.1 \\ 0.2 & 0.6 & 0.2 \end{bmatrix}}_{\mathbb{P}_{\iota_0}} + \underbrace{\begin{bmatrix} 0 & 1 & -1 \\ 0 & 0 & 0 \\ 0 & 0 & 0 \end{bmatrix}}_{\mathbb{P}_{\iota_1}} \iota_1(k) + \underbrace{\begin{bmatrix} 0 & 0 & 0 \\ 1 & -1 & 0 \\ 0 & 0 & 0 \end{bmatrix}}_{\mathbb{P}_{\iota_2}} \iota_2(k), \tag{11}$$

where $\iota_1(k) \in [-0.1, 0.1]$ and $\iota_2(k) \in [-0.2, 0.2]$. Although the modeling seems to be different, note that a simple change of variables can recover the multisimplex modeling from the affine representation, since

$$\iota_r(k) = \underline{\iota}_r \alpha_{r1}(k) + \overline{\iota}_r \alpha_{r2}(k), \tag{12}$$

where $\iota_r(k) \in [\underline{\iota}_r, \overline{\iota}_r]$, $\alpha_r(k) = (\alpha_{r1}(k), \alpha_{r2}(k)) \in \Lambda_2$, $r = 1, 2$.

In order to clarify how to write a time-varying matrix in the affine form, consider the following affine matrix as

$$\underbrace{\begin{bmatrix} 5 & 0.3 + \iota_1(k) \\ 12 & -2 + 0.5\iota_2(k) \end{bmatrix}}_{O_i(\iota_j)} = \underbrace{\begin{bmatrix} 5 & 0.3 \\ 12 & -2 \end{bmatrix}}_{O_{i_{l_0}}} + \underbrace{\begin{bmatrix} 0 & 1 \\ 0 & 0 \end{bmatrix}}_{O_{i_{l_1}}} \iota_1(k) + \underbrace{\begin{bmatrix} 0 & 0 \\ 0 & 0.5 \end{bmatrix}}_{O_{i_{l_2}}} \iota_2(k), \tag{13}$$

where $\iota_1(k) \in [\underline{\iota}_1, \overline{\iota}_1]$ and $\iota_2(k) \in [\underline{\iota}_2, \overline{\iota}_2]$. By using the multisimplex formulation, $\iota_r(k) = \underline{\iota}_r \alpha_{r1}(k) + \overline{\iota}_r \alpha_{r2}(k)$ for $r = 1, 2$, we recover the representation with $\alpha(k) \in \Lambda_N = \Lambda_2 \times \Lambda_2$, where $N = (2, 2)$. This procedure can be extended for all matrices throughout this paper. Bearing this in mind, in what follows, whenever we write $P_i(\alpha(k))$ for $\alpha(k) \in \Lambda_N$, we mean a representation, as in (7), in terms of generic LPV parameters or, equivalently, in terms the multisimplex parameter $\alpha(k)$.

4. Bounded Real Lemma

The concept of stability for the nonhomogeneous Markov chain is different from its homogeneous counterpart. This discrepancy is caused by the arbitrary variation of the transition probability. Therefore, an upper bound of the $\mathcal{H}_\infty$ norm can only be obtained if system (1) under the assumption of $w(k) \equiv 0$ is exponentially stable in the mean square sense with conditioning of type I (ESMS-CI). This concept was first introduced in [23] and is also presented in [16]. In this sense, before introducing the main results of this paper, some fundamental definitions are presented next.

Definition 1 ([16]). *Assuming that system (1) is ESMS-CI, and $x(0) = 0$, its $\mathcal{H}_\infty$ norm is given by*

$$\|\mathcal{G}\|_\infty = \sup_{w \in \mathcal{L}, \|w\|_2 \neq 0} \frac{\|z\|_2}{\|w\|_2}. \tag{14}$$

Next, we present a sufficient condition version of the bounded real lemma adapted from [16], which allows us to deal with nonhomogeneous MJLS with arbitrarily fast time-varying parameters, where the parameters are modeled by using the multisimplex domain Λ_N. For that, it is assumed that the condition H_1 in Proposition 1 in [16] is satisfied; that is, $\Pr(\theta_k = i) > 0$ for all $i \in \mathbb{K}$ and $k \geq 0$.

Remark 1. *In order to draw the results presented in Lemma 1, it is necessary to consider the assumption that the variation of the probabilities $\rho_{ij}(k)$ is arbitrarily fast. Under this assumption, there is no need to bound the variation limit.*

Lemma 1. *System (1) is ESMS-CI and satisfies $\|\mathcal{G}\|_\infty < \gamma$ if there exist symmetric positive definite matrices $P_i(\alpha(k))$, such that, for each $i \in \mathbb{K}$ and for all $\alpha(k), \alpha(k+1) \in \Lambda_N$, the parameter-dependent LMIs*

$$\underbrace{\begin{bmatrix} A_i & J_i \\ C_i & D_i \end{bmatrix}' \begin{bmatrix} \mathbb{E}_i(P)(\alpha(k),\alpha(k+1)) & 0 \\ 0 & I \end{bmatrix} \begin{bmatrix} A_i & J_i \\ C_i & D_i \end{bmatrix} - \begin{bmatrix} P_i(\alpha(k)) & 0 \\ 0 & \gamma^2 I \end{bmatrix}}_{\Omega_i(\alpha(k))} < 0 \tag{15}$$

are satisfied, where $\mathbb{E}_i(P)(\alpha(k), \alpha(k+1)) = \sum_{j=1}^{\sigma} \rho_{ij}(\alpha(k))P_j(\alpha(k+1))$.

Proof. Here is a sketch of the proof for Lemma 1. Assuming that there exist $P_i(\alpha(k)) = P_i'(\alpha(k))$ such that Equation (15) holds, we have, from Proposition 1 in [16], that system (1) is ESMSC1. Define the cost function as

$$\mathcal{J}_\tau^\gamma = \sum_{k=0}^{\tau} E\left[z(k)'z(k) - \gamma^2 w(k)'w(k)\right]. \tag{16}$$

Observe that $\|\mathcal{G}\|_\infty < \gamma \iff \mathcal{J}_\infty^\gamma \leq -e^2 \|w\|_2^2, \forall \|w\|_2 \neq 0$ and for some $e \neq 0$, where $\mathcal{J}_\infty^\gamma$ represent the cost function for $\tau \to \infty$. Considering the Lyapunov function $\mathcal{V}_{\theta_k}(k, x(k)) \triangleq x(k)' P_{\theta_k}(\alpha(k)) x(k)$, one has

$$\mathcal{J}_\tau^\gamma = \sum_{k=0}^{\tau} E\left[z(k)'z(k) - \gamma^2 w(k)'w(k) - \mathcal{V}_{\theta_k}(k, x(k)) + \mathcal{V}_{\theta_{k+1}}(k+1, x(k+1))\right]$$

$$+ \sum_{k=0}^{\tau} E\left[\mathcal{V}_{\theta_k}(k, x(k)) - \mathcal{V}_{\theta_{k+1}}(k+1, x(k+1))\right]$$

$$= \sum_{k=0}^{\tau} E\Big[z(k)'z(k) - \gamma^2 w(k)'w(k) - \mathcal{V}_{\theta_k}(k, x(k))\Big]$$

$$+ \sum_{k=0}^{\tau} E\Big[E\Big[\mathcal{V}_{\theta_{k+1}}(k+1, x(k+1))|\mathcal{F}_k\Big]\Big]$$

$$- E\Big[\mathcal{V}_{\tau+1}(\tau+1, x(\tau+1)\Big] + E\Big[\mathcal{V}_{\theta_0}(0, x(0))\Big]$$

$$= \sum_{k=0}^{\tau} E\Big[z(k)'z(k) - \gamma^2 w(k)'w(k) - \mathcal{V}_{\theta_k}(k, x(k))\Big]$$

$$+ \sum_{k=0}^{\tau} E\Big[x(k+1)'E\Big[P_{\theta_{k+1}}(\alpha(k+1))|\mathcal{F}_k\Big]x(k+1)\Big]$$

$$- E\Big[\mathcal{V}_{\tau+1}(\tau+1, x(\tau+1))\Big] + E\Big[\mathcal{V}_{\theta_0}(0, x(0))\Big]$$

$$= \sum_{k=0}^{\tau} E\Big[[x(k)'\ w(k)']\Omega_{\theta_k}(\alpha(k))[x(k)'\ w(k)']'\Big]$$

$$- E\Big[\mathcal{V}_{\tau+1}(\tau+1, x(\tau+1))\Big] + E\Big[\mathcal{V}_{\theta_0}(0, x(0))\Big],$$

where $\Omega_i(\alpha(k))$ is presented in Equation (15). Recalling that $x(0) = 0$ so that $\mathcal{V}_{\theta_0}(0, x(0)) = 0$, we have $\forall k \geq 0$ and some $e \neq 0$ that

$$\mathcal{J}_\tau^\gamma = \sum_{k=0}^{\tau} E\Big[[x(k)'\ w(k)']\Omega_{\theta_k}(\alpha(k)) \bullet \Big] - E[\mathcal{V}_{\theta_{\tau+1}}(\tau+1, x(\tau+1))]$$

$$\leq \sum_{k=0}^{\tau} E\Big[[x(k)'\ w(k)']\Omega_{\theta_k}(\alpha(k)) \bullet \Big] \leq -\sum_{k=0}^{\tau} e^2 E[\|w(k)\|^2]. \quad (17)$$

Inequality (17) yields, as $\tau \to \infty$, that $\mathcal{J}_\infty^\gamma \leq -e^2 \|w\|_2^2\ \forall\ \|w\|_2 \neq 0$, showing the desired result. □

5. Problem Formulation and Main Result

The block diagram of Figure 1 illustrates the FD scheme considered in this paper. Note that there are three elements composing the diagram: the system itself ($\mathcal{G}_{\theta_k}$, which represents the plant subjected to a fault), the controller K_{θ_k}, and the gain-scheduled FDF block $\mathcal{F}_{\theta_k}$.

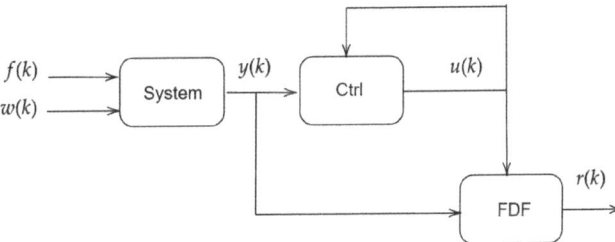

Figure 1. Graphic description of the FD scheme used to design the gain-scheduled fault detection filter.

The main purpose of this section is to design the FDF. Consider that the nonhomogeneous MJLS $\mathcal{G}_{\theta_k}$ is defined as

$$\mathcal{G}_{\theta_k} \equiv \begin{cases} x(k+1) = A_{\theta_k} x(k) + B_{\theta_k} u(k) + J_{\theta_k} w(k) + F_{\theta_k} f(k) \\ y(k) = C_{\theta_k} x(k) + D_{\theta_k} w(k) + E_{\theta_k} f(k) \end{cases}, \quad (18)$$

where $x(k) \in \mathbb{R}^{n_x}$ denotes the system states vector, $u(k) \in \mathbb{R}^{n_u}$ represents the control input, $w(k) \in \mathbb{R}^{n_w}$ is the exogenous/noise input, $y(k) \in \mathbb{R}^{n_y}$ represents the measurement signals, and $f(k) \in \mathbb{R}^{n_f}$ is the fault signal that should be detected. We also assume that $w(k)$, $f(k) \in \mathcal{L}^2$. Finally, when regarding the output-feedback control law, consider the following expression

$$u(k) = \mathcal{K}_{\theta_k} y(k). \quad (19)$$

The controller $\mathcal{K}_{\theta_k}$ is assumed to be designed beforehand.

Remark 2. *Although the formulation presented in this paper considers an output-feedback control law (19) that is mode-dependent (depends on θ_k) but is parameter-independent (does not depend on α_k or the time-varying probabilities $\rho_{ij}(k)$), the synthesis of the FD filter presented next can be extended to deal with a parameter-dependent control law. Regardless of that, for both situations, it is imperative that the controller is designed a priori. The major implementation difference in the latter case is that the controller would be defined as gain-scheduled.*

We define the gain-scheduled FDF under the aforementioned conditions as

$$\mathcal{F}_{\theta_k} \equiv \begin{cases} \eta(k+1) = \mathcal{A}_{\eta\theta_k}(\alpha(k))\eta(k) + \mathcal{M}_{\eta\theta_k}(\alpha(k))u(k) + \dots \\ \qquad\qquad \mathcal{B}_{\eta\theta_k}(\alpha(k))y(k) \\ r(k) = \mathcal{C}_{\eta\theta_k}(\alpha(k))\eta(k) \end{cases}, \quad (20)$$

where $\eta(k) \in \mathbb{R}^{n_\eta}$ is the filter states, $r(k) \in \mathbb{R}^{n_r}$ is the residue signal, $u(k)$ is the control law given by (19), and $y(k)$ represents the measurement output. The FDF is scheduled in terms of the time-varying parameter $\alpha(k)$, which represents the variation in time of the MC. Consequently, it is assumed that the time-varying behavior of the MC is known or at least measurable. The purpose of the gain-scheduled FDF in (20) is to generate a residue signal $r(k)$, which is used to detect the fault.

Remark 3. *All matrices that compose the filter (20) are written in the affine form as in (7), that is*

$$\mathcal{A}_{\eta i}(\beta_k) = \mathcal{A}_{\eta i \beta_0} + \sum_{j=1}^{m} \beta_j(k) \mathcal{A}_{\eta i \beta_j}, \quad (21)$$

and similarly for $\mathcal{M}_{\eta i}(\beta_k)$, $\mathcal{B}_{\eta i}(\beta_k)$, $\mathcal{C}_{\eta i}(\beta_k)$. Recall that the goal of this paper is to design the gain-scheduled FDF (20) where the schedule parameter represents the variation in the nonhomogeneous MC, as explained in Section 3.

We define $e(k) = r(k) - f(k)$ and the augmented system as

$$\mathcal{G}_{aug} \equiv \begin{cases} \tilde{x}(k+1) = \tilde{A}_{\theta_k}(\alpha(k))\tilde{x}(k) + \tilde{J}_{\theta_k}(\alpha(k))\tilde{w}(k) \\ e(k) = \tilde{C}_{\theta_k}(\alpha(k))\tilde{x}(k) + \tilde{D}_{\theta_k}(\alpha(k))\tilde{w}(k) \end{cases}, \quad (22)$$

where $\tilde{x}(k) = [x(k)\ \eta(k)]$, and $\tilde{w}(k) = [u(k)\ w(k)\ f(k)]$. The matrices that compose the augmented system are

$$\tilde{A}_i(\alpha(k)) = \begin{bmatrix} A_i & 0 \\ \mathcal{B}_{\eta i}(\alpha(k)) C_i & \mathcal{A}_{\eta i}(\alpha(k)) \end{bmatrix},$$

$$\tilde{J}_i(\alpha(k)) = \begin{bmatrix} B_i & J_i & F_i \\ \mathcal{M}_{\eta i}(\alpha(k)) & \mathcal{B}_{\eta i}(\alpha(k))D_i & \mathcal{B}_{\eta i}(\alpha(k))E_i \end{bmatrix}, \text{ and}$$

$$\tilde{C}_i(\alpha(k)) = \begin{bmatrix} 0 & C_{\eta i}(\alpha(k)) \end{bmatrix}, \quad \tilde{D}_i(\alpha(k)) = \begin{bmatrix} 0 & 0 & -I \end{bmatrix}. \tag{23}$$

In order to provide an FDF (20), which generates a residue signal $r(k)$ that is robust against noise and associated with a fast response whenever a fault occurs, we set our goal as the synthesis of a proper FD filter by minimizing an upper bound γ for the $\mathcal{H}_\infty$ norm of the augmented system (22), such that

$$\sup_{w(k)\in\mathcal{L}^2,\ \|w(k)\|_2\neq 0} \frac{\|e(k)\|_2}{\|w(k)\|_2} < \gamma. \tag{24}$$

In the following, we present our main result of the paper, which provides parameter-dependent bilinear matrix inequalities (BMI) for the design of an FDF (20) for system (18) with $\mathcal{H}_\infty$ guaranteed cost. For compactness, hereafter, the dependence on time of parameter $\alpha(k)$ and $\alpha(k+1)$ is omitted, such that they will be respectively replaced by α and α^+.

Theorem 1. *If there exist, for all $i \in \mathbb{K}$, symmetric positive definite parameter-dependent matrices $Z_i(\alpha)$, $R_i(\alpha)$, $\tilde{Z}_1(\alpha^+)$, $\bar{R}_i(\alpha^+)$, matrices $W_i(\alpha)$, $\bar{W}_i(\alpha^+)$, $X_i(\alpha)$, $Y_i(\alpha)$ $\mathcal{O}_i(\alpha)$, $\mathcal{B}_{\eta i}(\alpha)$, $\mathcal{M}_{\eta i}(\alpha)$, $\nabla_i(\alpha)$ with appropriate dimensions, and a scalar $\xi \in (0,2)$, such that the BMIs (25) hold for all $\alpha, \alpha^+ \in \Lambda_N$ and $i \in \mathbb{K}$,*

$$\begin{bmatrix} \Pi_{11} & \bullet & \bullet & \bullet & \bullet & \bullet & \bullet & \bullet & \bullet & \bullet \\ \Pi_{21} & \Pi_{22} & \bullet & \bullet & \bullet & \bullet & \bullet & \bullet & \bullet & \bullet \\ 0 & 0 & -\gamma^2 I & \bullet & \bullet & \bullet & \bullet & \bullet & \bullet & \bullet \\ 0 & 0 & 0 & -\gamma^2 I & \bullet & \bullet & \bullet & \bullet & \bullet & \bullet \\ 0 & 0 & 0 & 0 & -\gamma^2 I & \bullet & \bullet & \bullet & \bullet & \bullet \\ \Pi_{61} & \Pi_{62} & 0 & 0 & 0 & \Pi_{66} & \bullet & \bullet & \bullet & \bullet \\ \Pi_{71} & \Pi_{72} & 0 & 0 & 0 & \Pi_{76} & \Pi_{77} & \bullet & \bullet & \bullet \\ \Pi_{81} & \Pi_{82} & Y_i^a(\alpha)'(B_i) & Y_i^a(\alpha)'(J_i) & Y_i^a(\alpha)'(F_i) & \xi\Pi_{81} & \xi\Pi_{82} & -\tilde{Z}(\alpha^+) & \bullet & \bullet \\ \Pi_{91} & \Pi_{92} & Y_i^b(\alpha)'\mathcal{M}_{\eta i}(\alpha) & \Pi_{94} & \Pi_{95} & \xi\Pi_{91} & \xi\Pi_{92} & -\bar{W}(\alpha^+) & -\bar{R}(\alpha^+) & \bullet \\ \nabla_i(\alpha) & \nabla_i(\alpha) & 0 & 0 & -I & \xi\nabla_i(\alpha) & \xi\nabla_i(\alpha) & 0 & 0 & -I \end{bmatrix} < 0, \tag{25}$$

where

$$Y_i^a(\alpha) = [\rho_{i1}^{1/2}(\alpha)I_{n_x} \cdots \rho_{i\sigma}^{1/2}(\alpha)I_{n_x}], \quad Y_i^b(\alpha) = [\rho_{i1}^{1/2}(\alpha)I_{n_\eta} \cdots \rho_{i\sigma}^{1/2}(\alpha)I_{n_\eta}],$$

$$Y_i^c(\alpha) = [\rho_{i1}^{1/2}(\alpha)I_{n_x+n_\eta} \cdots \rho_{i\sigma}^{1/2}(\alpha)I_{n_x+n_\eta}], \quad \Pi_{11} = Z_i(\alpha) - \text{Her}(X_i(\alpha)),$$

$$\Pi_{21} = W_i(\alpha) - Y_i(\alpha) - X_i(\alpha)', \quad \Pi_{61} = Z_i(\alpha) - \xi X_i(\alpha)' - X_i(\alpha),$$

$$\Pi_{71} = W_i(\alpha) - \xi X_i(\alpha)' - Y_i(\alpha), \quad \Pi_{81} = Y_i^a(\alpha)'A_i X_i(\alpha),$$

$$\Pi_{91} = Y_i^b(\alpha)'(\mathcal{B}_{\eta i}(\alpha)C_i X_i(\alpha) + \mathcal{O}_i(\alpha)), \quad \Pi_{22} = R_i(\alpha) - \text{Her}(Y_i(\alpha)),$$

$$\Pi_{62} = W_i(\alpha)' - \xi Y_i(\alpha)' - X_i(\alpha), \quad \Pi_{72} = R_i(\alpha) - \xi Y_i(\alpha)' - Y_i(\alpha),$$

$$\Pi_{82} = \Pi_{81}, \quad \Pi_{92} = \Pi_{91}, \quad \Pi_{94} = Y_i^b(\alpha)'\mathcal{B}_{\eta i}(\alpha)D_i,$$

$$\Pi_{95} = Y_i^b(\alpha)'(\mathcal{B}_{\eta i}(\alpha)E_i), \quad \Pi_{66} = -\xi\,\text{Her}(X_i(\alpha)),$$

$$\Pi_{76} = -\xi(Y_i(\alpha) - X_i(\alpha)'), \quad \Pi_{77} = -\xi\,\text{Her}(Y_i(\alpha)),$$

and

$$\tilde{Z}(\alpha^+) = \text{diag}(\tilde{Z}_1(\alpha^+),\ldots,\tilde{Z}_\sigma(\alpha^+)), \quad \bar{W}^+(\alpha^+) = \text{diag}(\bar{W}_1^+(\alpha^+),\ldots,\bar{W}_\sigma(\alpha^+)),$$
$$\bar{R}_i(\alpha^+) = \text{diag}(\bar{R}_1(\alpha^+),\ldots,\bar{R}_\sigma(\alpha^+)), \tag{26}$$

then γ is an upper bound for the $\mathcal{H}_\infty$ norm of the augmented system (22), where the matrices that compose the FDF in the form of (20) are given by $\mathcal{A}_{\eta i}(\alpha) = \mathcal{O}_i(\alpha)Y_i(\alpha)^{-1}$, $\mathcal{B}_{\eta i}(\alpha)$, $\mathcal{M}_{\eta i}(\alpha)$, $\mathcal{C}_{\eta i}(\alpha) = \nabla_i Y_i(\alpha)^{-1}(\alpha)$ for all $i \in \mathbb{K}$.

Proof of Theorem 1. Consider the augmented matrices as from (23), and define the matrices $P_i(\alpha)$ and $H_i(\alpha)$ according to

$$P_i(\alpha) = \begin{bmatrix} Z_i(\alpha) & \bullet \\ W_i(\alpha) & R_i(\alpha) \end{bmatrix}, \quad H_i(\alpha) = \begin{bmatrix} X_i(\alpha) & X_i(\alpha) \\ Y_i(\alpha) & Y_i(\alpha) \end{bmatrix}. \tag{27}$$

In addition, we define the matrices $\bar{P}_i(\alpha^+)$ and $\bar{P}(\alpha^+)$ as follows:

$$\bar{P}_i(\alpha^+) = \begin{bmatrix} \bar{Z}_i(\alpha^+) & \bullet \\ \bar{W}_i(\alpha^+) & \bar{R}_i(\alpha^+) \end{bmatrix}, \quad \bar{P}(\alpha^+) = \mathrm{diag}(\bar{P}_1(\alpha^+), \ldots, \bar{P}_\sigma(\alpha^+)). \tag{28}$$

Consider the following change of variables: $\mathcal{O}_i(\alpha) = \mathcal{A}_{\eta i}(\alpha) Y_i(\alpha)$, $\nabla_i(\alpha) = \mathcal{C}_{\eta i}(\alpha) Y_i(\alpha)$. Then we obtain the following identities:

$$\begin{aligned}
\tilde{A}_i(\alpha) H_i(\alpha) &= \begin{bmatrix} A_i + B_i K_i(\alpha) C_i & 0 \\ \mathcal{M}_{\eta i}(\alpha) K_i(\alpha) C_i + \mathcal{B}_{\eta i}(\alpha) C_i & \mathcal{A}_{\eta i}(\alpha) \end{bmatrix} \begin{bmatrix} X_i(\alpha) & X_i(\alpha) \\ Y_i(\alpha) & Y_i(\alpha) \end{bmatrix} \\
&= \begin{bmatrix} (A_i + B_i K_i(\alpha) C_i) X_i(\alpha) & (A_i + B_i K_i(\alpha) C_i) X_i(\alpha) \\ \Xi_i(\alpha) + \mathcal{A}_{\eta i}(\alpha) Y_i(\alpha) & \Xi_i(\alpha) + \mathcal{A}_{\eta i}(\alpha) Y_i(\alpha) \end{bmatrix}, \\
\tilde{C}_i(\alpha) H_i(\alpha) &= \begin{bmatrix} 0 & \mathcal{C}_{\eta i}(\alpha) \end{bmatrix} \begin{bmatrix} X_i(\alpha) & X_i(\alpha) \\ Y_i(\alpha) & Y_i(\alpha) \end{bmatrix} \\
&= \begin{bmatrix} \mathcal{C}_{\eta i}(\alpha) Y_i(\alpha) & \mathcal{C}_{\eta i}(\alpha) Y_i(\alpha) \end{bmatrix},
\end{aligned} \tag{29}$$

where $\Xi_i(\alpha) = (\mathcal{M}_{\eta i}(\alpha) K_i C_i + \mathcal{B}_{\eta i}(\alpha) C_i) X_i(\alpha)$. Consequently, the BMI (25) can be rewritten as

$$\begin{bmatrix} P_i(\alpha) - \mathrm{Her}(H_i(\alpha)) & \bullet & \bullet & \bullet & \bullet \\ 0 & -\gamma^2 I & \bullet & \bullet & \bullet \\ P_i(\alpha) - \xi H_i(\alpha)' - H_i(\alpha) & 0 & -\xi\,\mathrm{Her}(H_i(\alpha)) & \bullet & \bullet \\ Y_i^c(\alpha)' \tilde{A}_i(\alpha) H_i(\alpha) & Y_i^c(\alpha)' \tilde{J}_i(\alpha) & \xi Y_i^c(\alpha)' \tilde{A}_i(\alpha) H_i(\alpha) & -\bar{P}(\alpha^+) & \bullet \\ \tilde{C}_i(\alpha) H_i(\alpha) & \tilde{D}_i(\alpha) & \xi \tilde{C}_i(\alpha) H_i(\alpha) & 0 & -I \end{bmatrix} < 0. \tag{30}$$

By using the projection lemma (see [24]), (30) can be rewritten as follows,

$$\mathcal{D} + \mathcal{U}' H_i \mathcal{V} + \mathcal{V} H_i' \mathcal{U} < 0, \tag{31}$$

where

$$\mathcal{D} = \begin{bmatrix} P_i & \bullet & \bullet & \bullet & \bullet \\ 0 & -\gamma^2 I & \bullet & \bullet & \bullet \\ P_i & 0 & 0 & \bullet & \bullet \\ 0 & Y_i^{c'} \tilde{J}_i & 0 & -\bar{P}^+ & \bullet \\ 0 & \tilde{D}_i & 0 & 0 & -I \end{bmatrix}, \quad \mathcal{U}' = \begin{bmatrix} -I \\ 0 \\ -I \\ Y_i^{c'} \tilde{A}_i \\ \tilde{C}_i \end{bmatrix}, \quad \mathcal{V}' = \begin{bmatrix} I \\ 0 \\ \xi I \\ 0 \\ 0 \end{bmatrix}. \tag{32}$$

(Observe that in the remaining of the proof, the time-varying parameter α is omitted for notation simplicity, as well as the dependence on α^+, which will be replaced by the superscript index "+".)

By taking the following basis for the null space of $\mathcal{U}$ and $\mathcal{V}$

$$\mathcal{N}_\mathcal{U} = \begin{bmatrix} I & 0 & 0 & 0 \\ 0 & I & 0 & 0 \\ -I & 0 & \tilde{A}_i' Y_i^c & \tilde{C}_i' \\ 0 & 0 & I & 0 \\ 0 & 0 & 0 & I \end{bmatrix}, \quad \mathcal{N}_\mathcal{V} = \begin{bmatrix} \xi I & 0 & 0 & 0 \\ 0 & I & 0 & 0 \\ -I & 0 & 0 & 0 \\ 0 & 0 & I & 0 \\ 0 & 0 & 0 & I \end{bmatrix} \tag{33}$$

and by applying the equivalence conditions of the projection lemma, we get

$$\mathcal{N}_\mathcal{U}' \mathcal{D} \mathcal{N}_\mathcal{U} = \begin{bmatrix} -P_i & \bullet & \bullet & \bullet \\ 0 & -\gamma^2 I & \bullet & \bullet \\ Y_i^{c'} \tilde{A}_i P_i & Y_i^{c'} \tilde{J}_i & -\bar{P}^+ & \bullet \\ \tilde{C}_i P_i & \tilde{D}_i & 0 & -I \end{bmatrix} < 0, \tag{34}$$

$$\mathcal{N}_\mathcal{V}' \mathcal{D} \mathcal{N}_\mathcal{V} = \begin{bmatrix} (\xi^2 - 2\xi) P_i & \bullet & \bullet & \bullet \\ 0 & -\gamma^2 I & \bullet & \bullet \\ 0 & Y_i^{c'} \tilde{J}_i & -\bar{P}^+ & \bullet \\ 0 & \tilde{D}_i & 0 & -I \end{bmatrix} < 0. \tag{35}$$

Notice that from the first term in (35) we can state that $0 < \xi < 2$. Now, pre- and postmultiplying (34) by $\text{diag}(X_i, I, \bar{\mathcal{X}}^+, I)$, where $X_i = P_i^{-1}$, $\bar{\mathcal{X}}^+ = (\bar{P}^+)^{-1}$, we obtain

$$\begin{bmatrix} -X_i & \bullet & \bullet & \bullet \\ 0 & -\gamma^2 I & \bullet & \bullet \\ \bar{\mathcal{X}}^+ Y_i^{c\prime} \bar{A}_i & \bar{\mathcal{X}}^+ Y_i^{c\prime} \bar{J}_i & -\bar{\mathcal{X}}^+ & \bullet \\ \bar{C}_i & \bar{D}_i & 0 & -I \end{bmatrix} < 0. \tag{36}$$

Notice that $\widehat{\mathcal{X}}^+$ represents a block diagonal matrix given by

$$\widehat{\mathcal{X}}^+ = \text{diag}(\bar{X}_1^+, \ldots, \bar{X}_\sigma^+), \tag{37}$$

with $\bar{X}_i^+ = (\bar{P}_i^+)^{-1}$. By using the Schur complement in (36) and noticing that

$$Y_i^c \bar{\mathcal{X}}^+ Y_i^{c\prime} = \sum_{j=1}^{\sigma} \rho_{ij} \bar{X}_j^+, \tag{38}$$

we have the condition (15) satisfied, so that the result follows from the bounded real lemma, presented in Lemma 1, for the nonhomogenous MJLS. □

Remark 4. *Observe that the conditions of Theorem 1 constitute of infinite-dimensional problems, which can be solved by using homogeneous polynomial approximations for the optimization variables (LMI relaxations) and then testing the positivity of the polynomial matrix inequalities by means of a finite set of LMIs. For this purpose, the authors strongly recommend the use of the toolbox robust LMI parser (ROLMIP), whose tutorial can be found in [25].*

Remark 5. *Theorem 1 can be adapted to handle the FDF synthesis problem for homogeneous MJLS with constant or uncertain but time-invariant probability matrix by simply making $\alpha(k+1) = \alpha(k) = \alpha$.*

The constraints presented in Theorem 1 are BMIs, which means that is necessary to use appropriate tools in order to solve them. Among a number of techniques available in the literature [26–28], we employ the coordinate descend algorithm (CDA), since it is a well-known and widely used tool to solve such issues. Accordingly, an iterative procedure based in CDA is given below to solve Theorem 1.

In Algorithm 1, ϕ represents the stop criteria and $t_{\max}$ is the maximum number of iterations allowed. Observe that if a solution is found in the first iteration of CDA, the iterative procedure will converge to an optimized solution or at least keep the same solution found in the first iteration. The CDA is better detailed in [26,27].

Algorithm 1 Coordinate descent algorithm.

Coordinate descent algorithm (CDA):
Input: $\mathcal{B}_{\eta i}$, γ, $t_{\max}$, ϕ.
Output: $\mathcal{A}_{\eta i}$, $\mathcal{B}_{\eta i}$, $\mathcal{M}_{\eta i}$, $\mathcal{C}_{\eta i}$.
Initialization:
While: $\frac{\gamma^{t-1} - \gamma^t}{\gamma^{t-1}} \leq \eta$ or $t \leq t_{\max}$ **do:**
Step 1: Find a solution for the LMI constraint (25) obtaining the values of X using as an input $\mathcal{B}_i$, which can be obtained by using any method, for example, the one in Theorem 1 in [7].
Step 2: Now find a solution for the same LMI constraint (25) to obtain $\mathcal{A}_{\eta i}$, $\mathcal{B}_{\eta i}$, $\mathcal{M}_{\eta i}$, $\mathcal{C}_{\eta i}$, but this time by using X as an input. Also obtain the value of γ.

6. Numerical Example

To illustrate the applicability of the theoretical results, a numerical example of a coupled tanks model (see Figure 2) is presented next. In this example, we assume that the plant itself is time-invariant, nevertheless, there are some time-varying parameters associated with the model of the communication network, which is represented by a nonhomogeneous Markov chain.

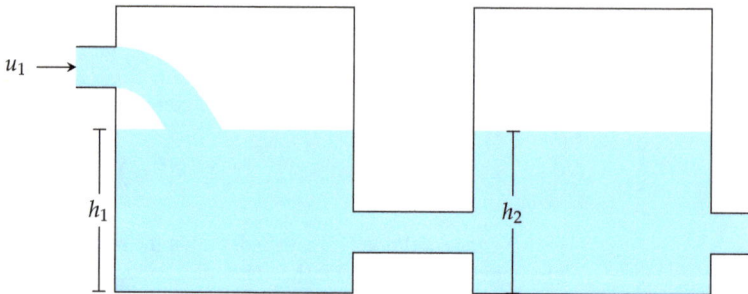

Figure 2. Coupled tanks model considered in the numerical example. The level on each tank h_1, h_2, measured independently from each other, are the system states, while the control input is the inlet flow u_1 into the first tank.

6.1. Simulation Setup

The parameters and modeling of the coupled tanks system were extracted from [29], such that the continuous-time state-space matrices are given by

$$A = \begin{bmatrix} -0.0239 & -0.0127 \\ 0.0127 & -0.0285 \end{bmatrix}, B = \begin{bmatrix} 0.71 \\ 0 \end{bmatrix},$$

$$J = \begin{bmatrix} 0.0071 \\ 0 \end{bmatrix}, F = \begin{bmatrix} 0.071 \\ 0 \end{bmatrix}, K = \begin{bmatrix} -1.03 & -0.33 \end{bmatrix}. \tag{39}$$

The sampling time used is $T_s = 1$s. Note that in order to represent the fault as an abnormal input on the first tank, the matrix associated with the fault signal F is 10% of the control input matrix B.

Regarding the network modeling, we assume that each tank is far away from the other; therefore, data gathered from each sensor is transmitted via two distinct networks. Network 1 transmits the measurement of the first tank, and Network 2 transmits the measurements of the second tank. The transmission of the measurement signals through a semireliable communication network is modeled by using a simplified Gilbert–Elliot model, as done in [30] while the packet dropout is represented by the zero-input approach from [31], meaning that when a packet loss occurs, we assume that the value of the received signal is null. Hence, the complete network behavior is represented by four distinct operation modes, as illustrated by Figure 3. The first one is that where all the measurements are correctly transmitted (called "Ok Ok" in Figure 3); the second one considers that the measurement on the first tank is successfully transmitted, but occurs a packet dropout of the measurement from the second tank (called "Ok Drop" in Figure 3); the third case is the opposite of the second one (called "Drop Ok" in Figure 3); and the last mode represents the case where all measurements were lost (called "Drop Drop" in Figure 3).

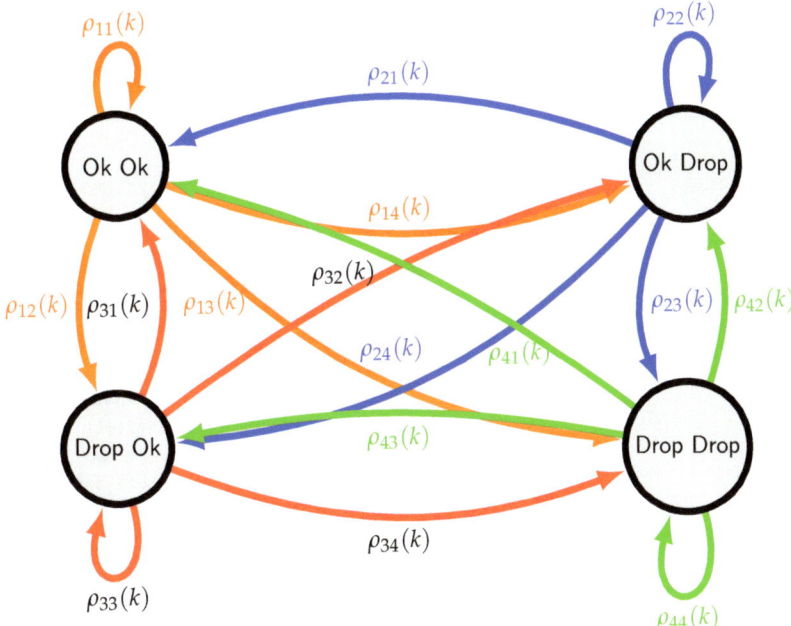

Figure 3. Graphic representation of the network states modeled as a Markov chain where the mode Ok Ok denotes that the networks responsible for transmitting the measurement of both sensors are operating on the nominal state, the mode Ok Drop represents the situation where there is a problem in the network in charge of transmitting the measurements of the second tank, the mode Drop Ok represents a failed transmission of the first tank measurements, and the mode Drop Drop stands for the case when both networks present some issues.

Remark 6. *It is essential to point out that the proposed solution allows modeling the network packet dropout rate varying in time. However, the feasibility of the proposed solution is dictated by three main features, the system dynamic, the number of Markov modes, and range variation. Therefore, it is important to keep in mind that an overly complex Markov chain with a high range of variation will request more computational effort, and in some situations, a feasible solution may not be achieved.*

From the previous network description, we can write the matrices that represent the measurement signal $y(k)$ are denoted by (the subindex is associated with the network operation mode)

$$C_1 = \begin{bmatrix} 1 & 0 \\ 0 & 1 \end{bmatrix}, \quad C_2 = \begin{bmatrix} 1 & 0 \\ 0 & 0 \end{bmatrix}, \quad C_3 = \begin{bmatrix} 0 & 0 \\ 0 & 1 \end{bmatrix}, \quad C_4 = \begin{bmatrix} 0 & 0 \\ 0 & 0 \end{bmatrix},$$
$$D_1 = \begin{bmatrix} 0.01 \\ 0.01 \end{bmatrix}, \quad D_2 = \begin{bmatrix} 0.01 \\ 0 \end{bmatrix}, \quad D_3 = \begin{bmatrix} 0 \\ 0.01 \end{bmatrix}, \quad D_4 = \begin{bmatrix} 0 \\ 0 \end{bmatrix}, \quad (40)$$
$$E_{1,2,3,4} = \begin{bmatrix} 0 \\ 0 \end{bmatrix}$$

Observe that we imposed E equal to zero in all modes because despite the fault in the example representing an abnormal input on the first tank, the sensors are assumed to be healthy throughout the simulation.

To illustrate the flexibility of the proposed approach, a particular scenario was tested. Assume that we solely know the boundary of the transition probability matrix that governs the jumps among the four network modes. Another important assumption considered in the design process is that in the fourth mode, Drop Drop, where there is no sensor information, the scheduling parameter is not accessible, implying that the matrices that compose the FDF for this mode are designed in the robust form instead of the affine form.

The time-varying transition probability matrix is given by

$$\mathbb{P}(k) = \underbrace{\begin{bmatrix} 0.3 & 0.2 & 0.1 & 0.4 \\ 0.1 & 0.4 & 0.3 & 0.2 \\ 0.1 & 0.2 & 0.4 & 0.3 \\ 0.1 & 0.2 & 0.1 & 0.6 \end{bmatrix}}_{\mathbb{P}_{\beta_0}} + \underbrace{\begin{bmatrix} 0 & 0 & 0 & 0 \\ 1 & -1 & 0 & 0 \\ 0 & 0 & 0 & 0 \\ 0 & 0 & 0 & 0 \end{bmatrix} \beta_1(k)}_{\mathbb{P}_{\beta_1}(k)} + \underbrace{\begin{bmatrix} 0 & 0 & 0 & 0 \\ 0 & 0 & 0 & 0 \\ 0 & 1 & -1 & 0 \\ 0 & 0 & 0 & 0 \end{bmatrix} \beta_2(k)}_{\mathbb{P}_{\beta_2}(k)}, \quad (41)$$

where $\beta_1(k) \in [0, 0.3]$, $\beta_2(k) \in [0, 0.3]$, and by consequence $\beta_r(k) = 0\alpha_{r1}(k) + 0.3\alpha_{r2}(k)$, $r = 1, 2$, where $\alpha(k) \in \Lambda_N = \Lambda_2 \times \Lambda_2$, $N = (2, 2)$.

Additionally, the transition probability among network states represented in Figure 3 by $\rho_{ij}(k)$ corresponds to an entry in the transition probability matrix $\mathbb{P}(k)$. Those transition probabilities depend on time as follows,

$$\rho_{ij}(k) = \rho 0_{ij} + \beta_1(k)\rho 1_{ij} + \beta_2(k)\rho 2_{ij},$$

where $\rho 0_{ij}$, $\rho 1_{ij}$ and $\rho 2_{ij}$ represent the elements of ith row and jth column from the following matrices $\mathbb{P}_{\beta_0}(k)$, $\mathbb{P}_{\beta_1}(k)$ and $\mathbb{P}_{\beta_2}(k)$ of Equation (41). Furthermore, the time evolution of the time-varying parameters $\beta_1(k)$ and $\beta_2(k)$ is illustrated by Figure 4.

By using the above numerical values of the plant and the network, we are now able to apply Theorem 1 to provide a solution for the FDF, so that the filter matrices in (20) (with $\zeta = 0.9$ in Theorem 1) are given by

$$\begin{aligned}
&\mathcal{A}_{\eta 1\beta_0} = \begin{bmatrix} 0.56 & 0.00 \\ -0.00 & -0.00 \end{bmatrix}, \quad \mathcal{A}_{\eta 1\beta_1} = \begin{bmatrix} -0.01 & -0.00 \\ 0.00 & -0.00 \end{bmatrix}, \quad \mathcal{A}_{\eta 1\beta_2} = \begin{bmatrix} -0.01 & -0.00 \\ 0.00 & -0.00 \end{bmatrix}, \\
&\mathcal{A}_{\eta 2\beta_0} = \begin{bmatrix} 0.12 & 0.00 \\ 0.00 & 0.00 \end{bmatrix}, \quad \mathcal{A}_{\eta 2\beta_1} = \begin{bmatrix} -0.01 & -0.00 \\ -0.00 & -0.00 \end{bmatrix}, \quad \mathcal{A}_{\eta 2\beta_2} = \begin{bmatrix} -0.01 & -0.00 \\ -0.00 & -0.00 \end{bmatrix}, \\
&\mathcal{A}_{\eta 3\beta_0} = \begin{bmatrix} 0.00 & -0.02 \\ -0.00 & -0.00 \end{bmatrix}, \quad \mathcal{A}_{\eta 3\beta_1} = \begin{bmatrix} -0.00 & 0.03 \\ 0.00 & 0.00 \end{bmatrix}, \quad \mathcal{A}_{\eta 3\beta_2} = \begin{bmatrix} -0.00 & 0.03 \\ 0.00 & 0.00 \end{bmatrix}, \\
&\mathcal{A}_{\eta 4\beta_0} = \begin{bmatrix} -0.00 & -0.00 \\ 0.00 & -0.00 \end{bmatrix}, \\
&\mathcal{B}_{\eta 1\beta_0} = \begin{bmatrix} 0.04 & 0.00 \\ -0.00 & 0.00 \end{bmatrix}, \quad \mathcal{B}_{\eta 1\beta_1} = \begin{bmatrix} 0.01 & -0.00 \\ -0.00 & 0.00 \end{bmatrix}, \quad \mathcal{B}_{\eta 1\beta_2} = \begin{bmatrix} 0.01 & -0.00 \\ -0.00 & 0.00 \end{bmatrix}, \\
&\mathcal{B}_{\eta 2\beta_0} = \begin{bmatrix} 0.03 & 0.00 \\ 0.00 & 0.00 \end{bmatrix}, \quad \mathcal{B}_{\eta 2\beta_1} = \begin{bmatrix} 0.01 & 0.00 \\ 0.00 & 0.00 \end{bmatrix}, \quad \mathcal{B}_{\eta 2\beta_2} = \begin{bmatrix} 0.01 & 0.00 \\ 0.00 & 0.00 \end{bmatrix}, \\
&\mathcal{B}_{\eta 3\beta_0} = \begin{bmatrix} 0.00 & 0.03 \\ 0.00 & 0.00 \end{bmatrix}, \quad \mathcal{B}_{\eta 3\beta_1} = \begin{bmatrix} 0 & 0.01 \\ 0 & 0.00 \end{bmatrix}, \quad \mathcal{B}_{\eta 3\beta_2} = \begin{bmatrix} 0 & 0.01 \\ 0 & 0.00 \end{bmatrix}, \\
&\mathcal{B}_{\eta 4\beta_0} = \begin{bmatrix} 0.00 & 0.00 \\ 0.00 & 0.00 \end{bmatrix}, \\
&\mathcal{M}_{\eta 1\beta_0} = \begin{bmatrix} 0.04 \\ 0.00 \end{bmatrix}, \quad \mathcal{M}_{\eta 1\beta_1} = \begin{bmatrix} 0.01 \\ 0.00 \end{bmatrix}, \quad \mathcal{M}_{\eta 1\beta_2} = \begin{bmatrix} 0.01 \\ 0.00 \end{bmatrix}, \\
&\mathcal{M}_{\eta 2\beta_0} = \begin{bmatrix} 0.03 \\ 0.00 \end{bmatrix}, \quad \mathcal{M}_{\eta 2\beta_1} = \begin{bmatrix} 0.01 \\ 0.00 \end{bmatrix}, \quad \mathcal{M}_{\eta 2\beta_2} = \begin{bmatrix} 0.01 \\ 0.00 \end{bmatrix}, \\
&\mathcal{M}_{\eta 3\beta_0} = \begin{bmatrix} 0.03 \\ 0.00 \end{bmatrix}, \quad \mathcal{M}_{\eta 3\beta_1} = \begin{bmatrix} 0.01 \\ 0.00 \end{bmatrix}, \quad \mathcal{M}_{\eta 3\beta_2} = \begin{bmatrix} 0.01 \\ 0.00 \end{bmatrix}, \\
&\mathcal{M}_{\eta 4\beta_0} = \begin{bmatrix} 0.01 \\ 0.00 \end{bmatrix}, \\
&\mathcal{C}_{\eta 1\beta_0} = \begin{bmatrix} 1.20 & -0.12 \end{bmatrix}, \quad \mathcal{C}_{\eta 1\beta_1} = \begin{bmatrix} -0.01 & -0.01 \end{bmatrix}, \quad \mathcal{C}_{\eta 1\beta_2} = \begin{bmatrix} -0.01 & -0.01 \end{bmatrix}, \\
&\mathcal{C}_{\eta 2\beta_0} = \begin{bmatrix} 0.32 & 0.01 \end{bmatrix}, \quad \mathcal{C}_{\eta 2\beta_1} = \begin{bmatrix} -0.00 & -0.00 \end{bmatrix}, \quad \mathcal{C}_{\eta 2\beta_2} = \begin{bmatrix} -0.01 & -0.00 \end{bmatrix}, \\
&\mathcal{C}_{\eta 3\beta_0} = \begin{bmatrix} 0.02 & -0.02 \end{bmatrix}, \quad \mathcal{C}_{\eta 3\beta_1} = \begin{bmatrix} -0.00 & 0.01 \end{bmatrix}, \quad \mathcal{C}_{\eta 3\beta_2} = \begin{bmatrix} -0.00 & 0.01 \end{bmatrix}, \\
&\mathcal{C}_{\eta 4\beta_0} = \begin{bmatrix} -0.01 & -0.00 \end{bmatrix}.
\end{aligned} \quad (42)$$

6.2. Simulation Result

In this section, for comparison purposes, we present simulation results with one design of FDF that assumes complete knowledge of the modes and the other design that uses the results from [8], where an $\mathcal{H}_\infty$ guaranteed cost is used as the performance criterion. This comparison is important to show that the consideration of the nonhomogeneous MC impacts the FDF performance. This particular paper was chosen to be compared with the proposed approach since both are based on the MJLS framework and are based on the H_∞ norm. Another critical piece of information that can be gathered from this comparison is the complexity of the problem versus performance gain. We remind that the parameters $\beta_1(k)$, $\beta_2(k)$ are assumed to be instantly measurable. Hence, the parameters $\beta_1(k)$, $\beta_2(k)$ vary during the simulation according to the information provided by Figure 4.

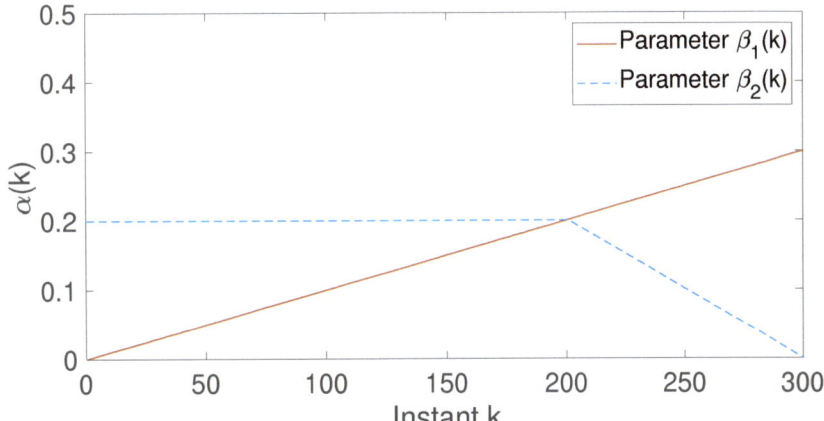

Figure 4. The behavior of the parameters $\beta_1(k)$, $\beta 2(k)$ during each simulation. The parameters $\beta_1(k)$, $\beta_2(k)$ are responsible for representing the time-varying aspect of the transition matrix $\mathbb{P}(k)$ in (41).

There are three faults that the system is subjected to during simulation. Those faults, presented in Figure 5, were applied on the first tank, and both affect the signal multiplying matrix F in (39). A Monte Carlo simulation with 100 interactions was performed, and Figures 6 and 7 illustrate the obtained results.

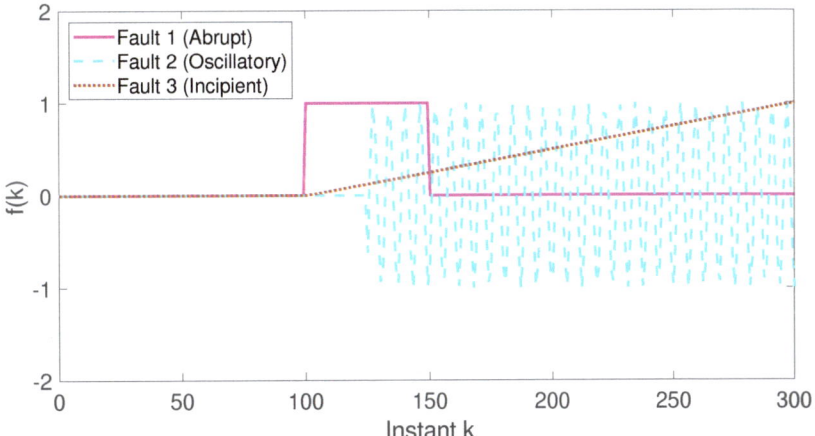

Figure 5. The representation of the three faults that were inflicted in the system during simulations. The magenta curve represents an abrupt fault, the cyan curve denotes an oscillatory fault, and the brown curve is the incipient fault.

Figure 6 presents the system's states when subjected to all three fault cases, and also a case without fault. From Figure 6, it is possible to observe that the faults do not surpass 10% of the nominal state value.

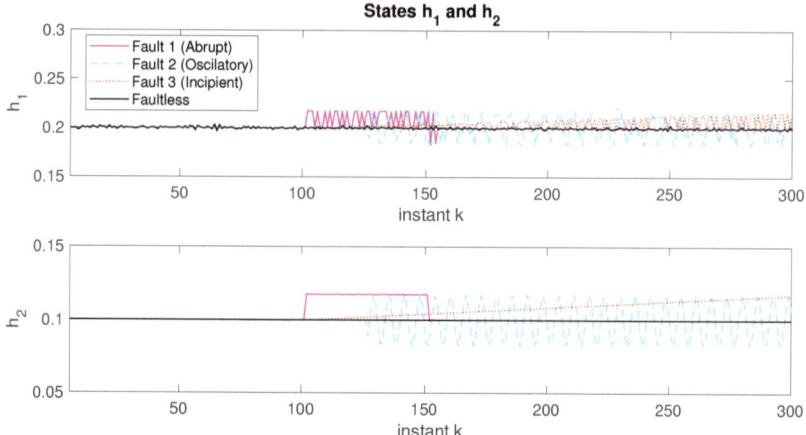

Figure 6. System states when subjected to all three faults, where the magenta curves denote the states with an abrupt fault, the cyan curves represent the states for the oscillatory fault, and brown curves are the states results for the incipient fault. The black curves represent the state without faults.

The residue signal generated by the FDFs for each type of fault, and also the case without fault are presented in Figure 7. We may state that all the FDF worked as expected, have been affected by the faults when it occurs, and that when there is not a fault the residue kept close to zero. Note that the results obtained by the FDF designed with the complete knowledge of the modes show more sensitivity against the fault, which helps to detect the fault faster.

A fault-detection procedure has two major stages: the residue generation and evaluation process. In Figure 7, the residue signals obtained via simulation were presented. To execute the next stage, it is necessary to define two tools: the evaluation function (EVAL(k)) and the threshold (TH). The definition of these two tools are a deep issue that will not be tackled here; therefore, for a more detailed discussion, please refer to [1,4]. Consider the following definition for the evaluation function:

$$\text{EVAL}(k) \triangleq \sqrt{\sum_{i=k-L}^{k} r(i)'r(i)}, \qquad (43)$$

where L represents the evaluation window, which, in this particular simulation is assumed to be $L = 250$. The threshold TH is used to assess the evaluation function EVAL(k). If EVAL(k) > TH there is a fault, and for the opposite case there is no fault. For the simulations, the value of the threshold was arbitrarily set to TH = 5. From the aforementioned, the evaluation functions for all the FDF considering each fault were calculated and presented in Figure 8.

As can be seen in Figure 8, all FDFs are able to detect the different faults; however, there is a clear difference in the time that it takes for each FDF, as indicated by Table 1. Note that the proposed approach presented a faster detection for all three faults. This occurred because the proposed solution considers a more trustworthy model. Thus, an abnormal change can be detected more quickly.

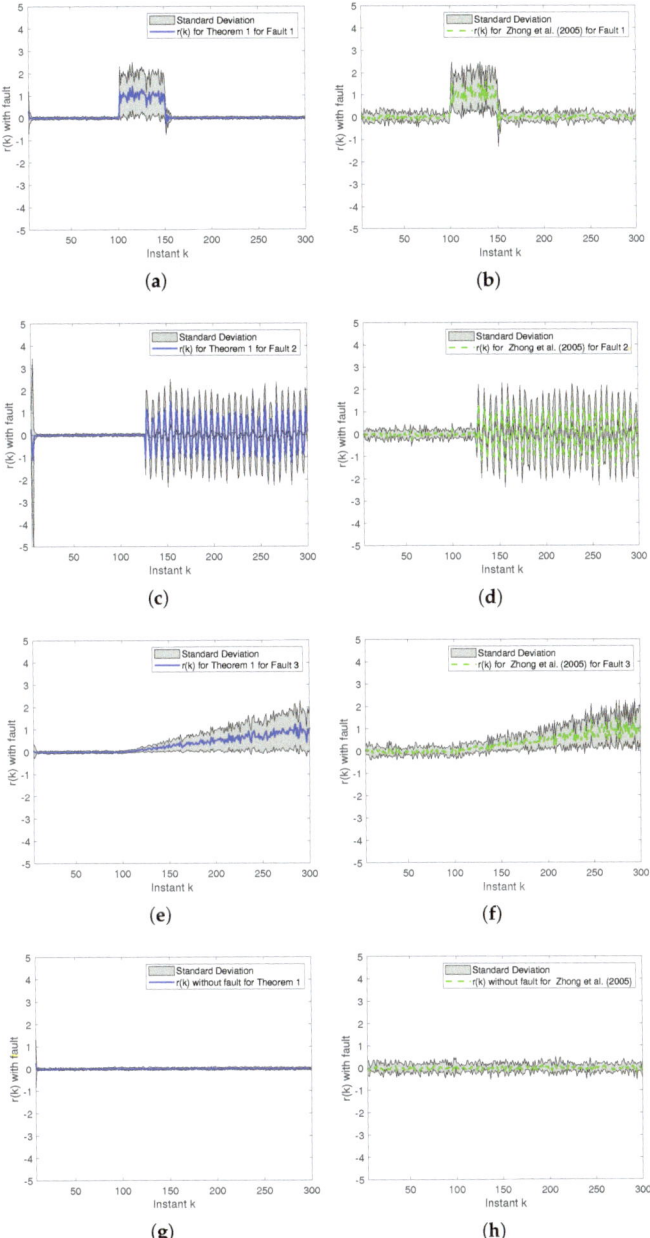

Figure 7. The residue signal obtained for FDF designed by using Theorem 1 under the assumption of knowledge of all modes (to the left), and the FDF provided by [8] (to the right). Both designs are subjected to three type of faults, and also the situation without fault. (**a**) Residue signal for FDF designed via Theorem 1 subjected to Fault 1. (**b**) Residue signal for FDF designed via [8] subjected to Fault 1. (**c**) Residue signal for FDF designed via Theorem 1 subjected to Fault 2. (**d**) Residue signal for FDF designed via [8] subjected to Fault 2. (**e**) Residue signal for FDF designed via Theorem 1 subjected to Fault 3. (**f**) Residue signal for FDF designed via [8] subjected to Fault 3. (**g**) Residue signal for FDF designed via Theorem 1 without fault. (**h**) Residue signal for FDF designed via [8] without fault.

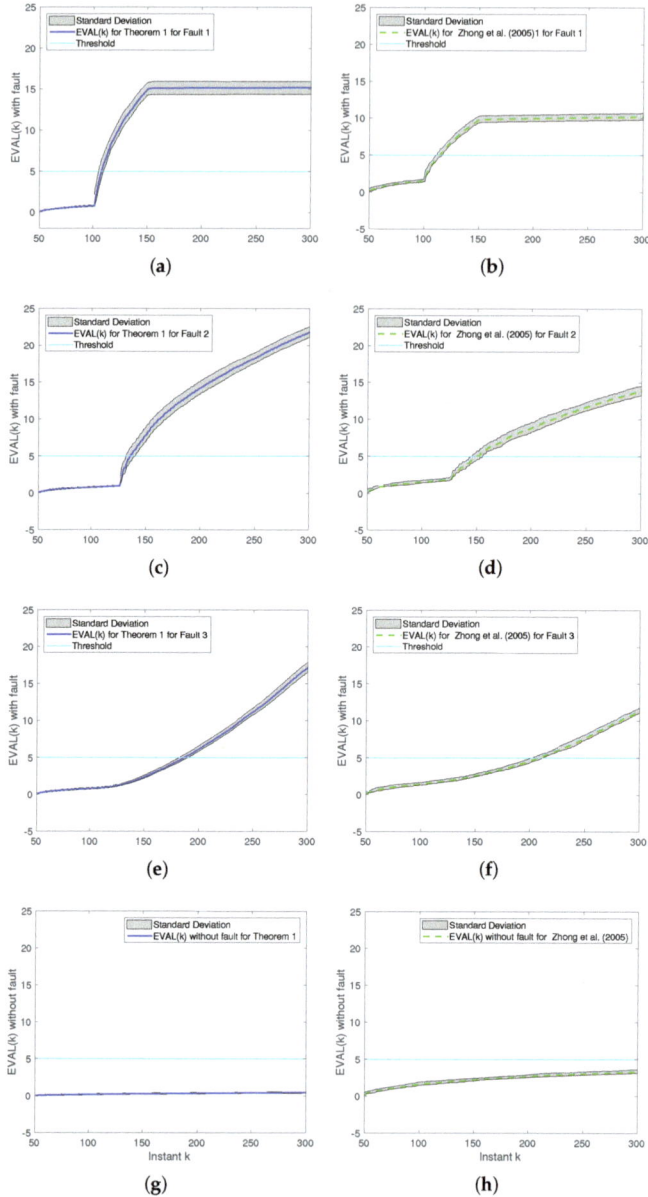

Figure 8. The evaluation function obtained for FDF designed by using Theorem 1 under the assumption of knowledge of all modes (to the left), and the FDF provided by [8] (to the right). Both designs are subjected to three type of faults, and also the situation without fault. (**a**) Evaluation function for FDF designed via Theorem 1 subjected to Fault 1. (**b**) Evaluation function for FDF designed via [8] subjected to Fault 1. (**c**) Evaluation function for FDF designed via Theorem 1 subjected to Fault 2. (**d**) Evaluation function for FDF designed via [8] subjected to Fault 2. (**e**) Evaluation function for FDF designed via Theorem 1 subjected to Fault 3. (**f**) Evaluation function for FDF designed via [8] subjected to Fault 3. (**g**) Evaluation function for FDF designed via Theorem 1 without fault. (**h**) Evaluation function for FDF designed via [8] without fault.

Table 1. The detection window for all FDF considering each fault after the Monte Carlo simulation.

Design Method	Fault 1 (k)	Fault 2 (k)	Fault 3 (k)
Theorem 1 (All)	[105 114]	[133 140]	[183 188]
[8]	[112 117]	[145 156]	[202 212]

Another interesting aspect is that the FDF designed by using [8] (the FDF under the assumption that the MC is homogeneous) is affected by the variation even when there is no fault occurrence. This particularity can be observed in Figure 8h, where the evaluation value drifts away from zero. This is an issue that affects the performance of the FDF, since it may cause false alarms. Note that this phenomenon does not occur on the FDF designed by using the proposed approach, showing that the consideration of the nonhomogeneous MC is useful to increase the performance of the FDF, especially for the situation where a nonhomogeneous MC is used to model characteristics that are inherently time-varying, such as a wireless sensor network. The last piece of information gathered from the simulation is that even though the proposed solution was based on more assumptions and a higher number of LMI constraints, which is clearly more computationally costly compared to the results in [8], the proposed approach provides superior performance.

7. Conclusions

The main contribution of this paper is the development of a new design method for gain-scheduled FDF considering that the plant measurement signals are transmitted through a network whose dropouts are modeled by using the MJLS theory, with the Markov chain being nonhomogeneous. The premise of nonhomogeneous MC is tackled by using the LPV modelling, specifically applied to the probability transition matrix, allowing us to design the FDF under this particular circumstance. The FDF synthesis results are obtained by using parameter-dependent LMI constraints that employ $\mathcal{H}_\infty$ norm as performance index. Since the gain-scheduled FDF varies according to the probability transition matrix variation, the proposed FDF is optimal for all the variation range of the Markov chain. To illustrate this, a comparison between the proposed technique with another method from the literature that does not contemplate the nonhomogeneous MC assumption was made. The simulation results show that the nonhomogeneous assumption positively impacts the FDF performance, allowing the proposed technique to detect the fault in a smaller time window, which can be seen in Table 1. Furthermore the FDF does not present false variations in the evaluation function, that could be identified as faults, when a fault does not occur, as can be seen in Figure 8g,h. A possible next step along this research line would be to include a sensitivity $\mathcal{H}_-$ index in the design to further improve the FDF performance.

Author Contributions: Conceptualization, L.C., J.M.P. and C.F.M.; Methodology, C.F.M. and L.C.; Validation, C.F.M., B.J. and O.L.V.C.; Investigation, L.C. and J.M.P.; Writing—original draft, L.C.; Writing—review & editing, J.M.P., C.F.M., B.J. and O.L.V.C.; Visualization, L.C.; Supervision, B.J. and O.L.V.C.; Project administration, O.L.V.C.; Funding acquisition, J.M.P. All authors have read and agreed to the published version of the manuscript.

Funding: This research was funded by ANID-FONDECYT grant number 11201049 and 1220903. And The APC was funded by ANID-FONDECYT. The first, second, and third authors were funded by the Chilean National Agency for Research and Development, project ANID-FONDECYT Iniciacion finance code N° 11201049 and ANID-FONDECYT 1220903. The fifth author was supported by Conselho Nacional de Desenvolvimento Científico e Tecnológico (CNPq), process No. 304149/2019-5, and by Instituto Nacional de Ciência e Tecnologia para Sistemas Autônomos Cooperativos (INSAC), process CNPq/INCT -465755/2014-3 and FAPESP/INCT-2014/50851-0.

Data Availability Statement: Not applicable.

Conflicts of Interest: The authors declare no conflict of interest.

Abbreviations

The following abbreviations are used in this manuscript:

Fault Detection	FD
Markov Jump Linear Systems	MJLS
Markov Chain	MC
Networked Control System	NCS
Fault Detection Filter	FDF
Linear Parameter Varying	LPV
Exponentially Stable in the Mean Square Sense with Conditioning of Type 1	ESMS-CI
Linear Matrix Inequality	LMI
Coordinate Descend Algorithm	CDA

References

1. Chen, J.; Patton, R.J. *Robust Model-Based Fault Diagnosis for Dynamic Systems*; Springer Science & Business Media: Berlin/Heidelberg, Germany, 2012; Volume 3.
2. Patton, R.J.; Frank, P.M.; Clark, R.N. *Issues of Fault Diagnosis for Dynamic Systems*; Springer Science & Business Media: Berlin/Heidelberg, Germany, 2013.
3. Isermann, R.; Schwarz, R.; Stolzl, S. Fault-tolerant drive-by-wire systems. *IEEE Control Syst.* **2002**, *22*, 64–81.
4. Frank, P.M.; Ding, X. Survey of robust residual generation and evaluation methods in observer-based fault detection systems. *J. Process. Control* **1997**, *7*, 403–424. [CrossRef]
5. Al-Karaki, J.N.; Kamal, A.E. Routing techniques in wireless sensor networks: A survey. *IEEE Wirel. Commun.* **2004**, *11*, 6–28. [CrossRef]
6. Wu, J.; Chen, T. Design of networked control systems with packet dropouts. *IEEE Trans. Autom. Control* **2007**, *52*, 1314–1319. [CrossRef]
7. de Paula Carvalho, L.; de Oliveira, A.M.; do Valle Costa, O.L. Robust fault detection H_∞ filter for Markovian jump linear systems with partial information on the jump parameter. *IFAC-PapersOnLine* **2018**, *51*, 202–207. [CrossRef]
8. Zhong, M.; Ye, H.; Shi, P.; Wang, G. Fault detection for Markovian jump systems. *IEE Proc.-Control Theory Appl.* **2005**, *152*, 397–402. [CrossRef]
9. Wu, L.; Luo, W.; Zeng, Y.; Li, F.; Zheng, Z. Fault detection for underactuated manipulators modeled by Markovian jump systems. *IEEE Trans. Ind. Electron.* **2016**, *63*, 4387–4399. [CrossRef]
10. Zhai, D.; An, L.; Li, J.; Zhang, Q. Fault detection for stochastic parameter-varying Markovian jump systems with application to networked control systems. *Appl. Math. Model.* **2016**, *40*, 2368–2383. [CrossRef]
11. de Paula Carvalho, L.; Toriumi, F.Y.; Angélico, B.A.; do Valle Costa, O.L. Model-based fault detection filter for Markovian jump linear systems applied to a control moment gyroscope. *Eur. J. Control* **2021**, *59*, 99–108. [CrossRef]
12. Lin, A.; Cheng, J.; Park, J.H.; Yan, H.; Qi, W. Fault Detection Filtering of Nonhomogeneous Markov Switching Memristive Neural Networks with Output Quantization. *Inf. Sci.* **2023**, *632*, 715–729. [CrossRef]
13. Zhao, L.; Zhang, H.; Hu, J.; Xu, L. Event-based Sliding Mode Control for Markovian Jump Systems with Time-varying Delays: An Observer Method. *Int. J. Control Autom. Syst.* **2023**, 1–10. [CrossRef]
14. de Paula Carvalho, L.; Jayawardhana, B.; do Valle Costa, O.L. Fault Detection Filter for Discrete-Time Markov Jump Lur'e Systems. In Proceedings of the 2021 European Control Conference (ECC), Delft, The Netherlands, 29 June–2 July 2021; pp. 1826–1831.
15. Ross, S.M. *Introduction to Probability Models*; Academic Press: Cambridge, MA, USA, 2014.
16. Aberkane, S. Bounded real lemma for nonhomogeneous Markovian jump linear systems. *IEEE Trans. Autom. Control* **2012**, *58*, 797–801. [CrossRef]
17. Palma, J.M.; Morais, C.F.; Oliveira, R.C.L.F. $\mathcal{H}_2$ gain-scheduled filtering for discrete-time LPV systems using estimated time-varying parameters. In Proceedings of the 2018 American Control Conference, Milwaukee WI, USA, 27–29 June 2018; pp. 4367–4372. [CrossRef]
18. Palma, J.M.; Morais, C.F.; Oliveira, R.C. A less conservative approach to handle time-varying parameters in discrete-time linear parameter-varying systems with applications in networked control systems. *Int. J. Robust Nonlinear Control* **2020**, *30*, 3521–3546. [CrossRef]
19. Wu, J.; Shi, H.; Jiang, X.; Su, C.; Li, P. Stochastic fuzzy predictive fault-tolerant control for time-delay nonlinear system with actuator fault under a certain probability. *Optim. Control Appl. Methods* **2022**. [CrossRef]
20. He, S.; Lyu, W.; Liu, F. Robust H∞ Sliding Mode Controller Design of a Class of Time-Delayed Discrete Conic-Type Nonlinear Systems. *IEEE Trans. Syst. Man Cybern. Syst.* **2021**, *51*, 885–892. [CrossRef]
21. Iosifescu, M. *Finite Markov Processes and Their Applications*; Courier Corporation: North Chelmsford, MA, USA, 2014.
22. Morais, C.F.; Braga, M.F.; Oliveira, R.C.; Peres, P.L. H_2 control of discrete-time Markov jump linear systems with uncertain transition probability matrix: Improved linear matrix inequality relaxations and multi-simplex modelling. *IET Control Theory Appl.* **2013**, *7*, 1665–1674. [CrossRef]

23. Dragan, V.; Morozan, T.; Stoica, A.M. *Mathematical Methods in Robust Control of Discrete-Time Linear Stochastic Systems*; Springer: Berlin/Heidelberg, Germany, 2010.
24. Boyd, S.; El Ghaoui, L.; Feron, E.; Balakrishnan, V. *Linear Matrix Inequalities in System and Control Theory*; SIAM: Philadelphia, PA, USA, 1994.
25. Agulhari, C.M.; Felipe, A.; Oliveira, R.; Peres, P.L. Manual of "The Robust LMI Parser", Version 3.0. 2019. Available online: https://www.researchgate.net/profile/Mohamed-Mourad-Lafifi/post/Can_anyone_help_me_to_find_the_maximum_number_of_linear_matrix_inequality_LMI_that_can_be_solved_by_using_YALMIP_or_LMI_toolbox/attachment/5cb8ae87cfe4a7df4ae9bc27/AS%3A749084939923456%401555607175083/download/Manual+of+%E2%80%9CThe+Robust+LMI+Parser%E2%80%9D+%E2%80%93+Version+3.0.pdf (accessed on 3 March 2023).
26. Simon, E.; R-Ayerbe, P.; Stoica, C.; Dumur, D.; Wertz, V. LMIs-based coordinate descent method for solving BMIs in control design. *IFAC Proc. Vol.* **2011**, *44*, 10180–10186. [CrossRef]
27. Wang, Y.; Zemouche, A.; Rajamani, R. A sequential LMI approach to design a BMI-based multi-objective nonlinear observer. *Eur. J. Control* **2018**, *44*, 50–57. [CrossRef]
28. de Paula Carvalho, L.; Rosa, T.; Jayawardhana, B.; Costa, O. Fault accommodation controller under Markovian jump linear systems with asynchronous modes. *Int. J. Robust Nonlinear Control* **2020**, *30*, 8503–8520. [CrossRef]
29. Feedback Instruments, Ltd. *FeedBack Coupled Tanks Control Experiments 33-041S (For Use with MATLAB)*, 1st ed.; Feedback Instruments, Ltd.: Crowborough, UK, 2013; pp. 1–49.
30. Gilbert, E.N. Capacity of a burst-noise channel. *Bell Syst. Tech. J.* **1960**, *39*, 1253–1265. [CrossRef]
31. Schenato, L. To zero or to hold control inputs with lossy links? *IEEE Trans. Autom. Control* **2009**, *54*, 1093–1099. [CrossRef]

Disclaimer/Publisher's Note: The statements, opinions and data contained in all publications are solely those of the individual author(s) and contributor(s) and not of MDPI and/or the editor(s). MDPI and/or the editor(s) disclaim responsibility for any injury to people or property resulting from any ideas, methods, instructions or products referred to in the content.

Article

A Deterministic Setting for the Numerical Computation of the Stabilizing Solutions to Stochastic Game-Theoretic Riccati Equations

Samir Aberkane [1,2,*,†] and Vasile Dragan [3,4,†]

1 Campus Sciences, Université de Lorraine, CRAN, UMR 7039, BP 70239, Vandoeuvre-les-Nancy CEDEX, 54506 Nancy, France
2 CNRS, CRAN, UMR 7039, 54500 Nancy, France
3 Institute of Mathematics of the Romanian Academy, P.O. Box 1-764, RO-014700 Bucharest, Romania; vasile.dragan@imar.ro
4 The Academy of the Romanian Scientists, Str. Ilfov, 3, 50044 Bucharest, Romania
* Correspondence: samir.aberkane@univ-lorraine.fr
† These authors contributed equally to this work.

Abstract: In this paper, we are interested in the numerical aspects of the class of generalized Riccati difference equations which are involved in linear quadratic (LQ) stochastic difference games. More specifically, we address the problem of the numerical computation of the stabilizing solutions for this class of nonlinear difference equations. We propose an iterative *deterministic* algorithm for the computation of such a global solution. The performances of the proposed algorithm are illustrated with some numerical examples.

Keywords: stochastic Riccati equations; stochastic control; iterative computation; deterministic approach

MSC: 93E03; 93C05; 93E20; 60J27; 60J10; 91A05; 91A15

Citation: Aberkane, S.; Dragan, V. A Deterministic Setting for the Numerical Computation of the Stabilizing Solutions to Stochastic Game-Theoretic Riccati Equations. *Mathematics* **2023**, *11*, 2068. https://doi.org/10.3390/math11092068

Academic Editors: Adrian Olaru, Gabriel Frumusanu and Catalin Alexandru

Received: 25 February 2023
Revised: 21 April 2023
Accepted: 23 April 2023
Published: 27 April 2023

Copyright: © 2023 by the authors. Licensee MDPI, Basel, Switzerland. This article is an open access article distributed under the terms and conditions of the Creative Commons Attribution (CC BY) license (https://creativecommons.org/licenses/by/4.0/).

1. Introduction

In this paper, we address the problem of the numerical computation of the stabilizing solutions of a class of generalized Riccati difference equations. The considered nonlinear matrix equation occurs in connection with zero-sum linear LQ stochastic difference game control problems (see [1] for more precision regarding this aspect). One of the particularities of such equations lies in the *sign indefiniteness* of their quadratic terms. This sign indefiniteness makes the characterization (as well as the numerical computation) of global solutions to such nonlinear matrix difference equations far more challenging when compared with the sign-definite counterpart. Even though some interesting results have already been reported in the literature (see [2,3] and the references therein), there are still substantial open problems in this field.

In [1], we addressed some theoretical aspects related to the nonlinear difference equations under consideration. The present paper can be viewed as the numerical counterpart of [1]. We propose a globally convergent iterative algorithm for the computation of the stabilizing solutions to this class of Riccati equations. To the best of the authors' knowledge, the numerical algorithms developed in the literature for the computation of the solutions to stochastic Riccati equations are mainly based on *stochastic* approaches consisting of transformation of the original problem into the problem of solving a sequence of *coupled stochastic* Riccati equations (see [4,5] and the references therein) that rely again on some iterative procedures for their numerical resolution. One of the most remarkable features of our proposed algorithm is its *deterministic* nature, in the sense that one has to solve at each main iteration a system of *uncoupled deterministic* Riccati equations. This allows us to use direct methods (invariant or deflating subspace-based methods; see [6]) for the numerical solutions to such deterministic equations. We believe that such a fundamental difference in the construction of what we called above deterministic and stochastic algorithms will have an important impact from the computation-time point of view. This will be illustrated via numerical experiments.

We mention here that in [7], we proposed a deterministic iterative algorithm for the numerical computation of the stabilizing solutions to a class of generalized Riccati equations related to the so-called continuous-time, full-information stochastic $\mathcal{H}_\infty$ control. The discrete-time counterpart of this type of Riccati equation is a particular case of the more general class of Riccati equations considered in the present paper. We have recently shown (see [1]) that the proof of the existence and uniqueness of the stabilizing solution for this more general class of Riccati equations presents substantial differences when compared with the full-information $\mathcal{H}_\infty$-type Riccati equations, even though we followed a similar philosophy in the proof procedure. We believe that we have a similar situation from the numerical computation point of view. The results reported in the present paper are more general and contain substantial differences when compared with [7].

This paper is organized as follows. In Section 2, we describe the problem that we address. In Section 3, we introduce the main results of the paper. Some numerical experiments are included in Section 4.

Notations: $\mathfrak{N} = \{1, 2, ..., N\}$, where $N \geq 1$ is a fixed natural number. A^T stands for the transpose of the matrix A, and $Tr[A]$ denotes the trace of a matrix A. The notation $X \geq Y$ ($X > Y$), where X and Y are symmetric matrices, means that $X - Y$ is positive semi-definite (positive definite). In block matrices, $\star$ indicates symmetric terms, where $\begin{pmatrix} A & B \\ B^T & C \end{pmatrix} = \begin{pmatrix} A & \star \\ B^T & C \end{pmatrix} = \begin{pmatrix} A & B \\ \star & C \end{pmatrix}$. The expression $MN\star$ is equivalent to MNM^T, while $M\star$ is equivalent to MM^T. Consider the following space of matrices: $\mathcal{M}_{n,m}^N = \mathbb{R}^{n \times m} \times \cdots \times \mathbb{R}^{n \times m}$. In the case where $n = m$, we shall write $\mathcal{M}_n^N$ instead of $\mathcal{M}_{n,n}^N$.

We introduce the following convention of notations:

- If $\mathbb{B} = (\ B(1),\ \cdots,\ B(N)\ ) \in \mathcal{M}_{n,m}^N$ and $\mathbb{D} = (\ D(1),\ \cdots,\ D(N)\ ) \in \mathcal{M}_{m,p}^N$, then $\mathbb{C} = \mathbb{B}\mathbb{D} \in \mathcal{M}_{n,p}^N$, where $\mathbb{C} = (\ C(1),\ \cdots,\ C(N)\ ), C(i) = B(i)D(i), 1 \leq i \leq N$.
- $\mathbb{B}^T = (\ B^T(1),\ \cdots,\ B^T(N)\ ) \in \mathcal{M}_{m,n}^N$.
- If $\mathbb{A} = (\ A(1),\ \cdots,\ A(N)\ ) \in \mathcal{M}_n^N$ with $\det(A(i)) \neq 0, 1 \leq i \leq N$, then $\mathbb{A}^{-1} = (\ A^{-1}(1),\ \cdots,\ A^{-1}(N)\ )$.

As usual, $\mathcal{S}_n \in \mathbb{R}^{n \times n}$ denotes the subspace of symmetric matrices of a size $n \times n$, and $\mathcal{S}_n^N = \mathcal{S}_n \times \cdots \times \mathcal{S}_n$. $\mathcal{S}_n^N$ is a finite, dimensional real Hilbert space with respect to the inner product:

$$\langle \mathbb{X}, \mathbb{Y} \rangle = \sum_{i=1}^N Tr[X(i)Y(i)] \qquad (1)$$

for all $\mathbb{X} = (X(1), X(2), ..., X(N)), \mathbb{Y} = (Y(1), Y(2), ..., Y(N)) \in \mathcal{S}_n^N$. Throughout this paper, $\mathbb{E}[\cdot]$ stands for the mathematical expectation and $\mathbb{E}[\cdot|\theta_t = i]$ denotes the conditional expectation with respect to the event $\{\theta_t = i\}$.

2. Problem Setting

2.1. Problem Description

Consider the following nonlinear difference equation in the space $\mathcal{S}_n^N$:

$$\mathbb{X}(t) = \Pi_1(t)[\mathbb{X}(t+1)] + \mathbb{M}(t) - \left[\Pi_2(t)[\mathbb{X}(t+1))] + \mathbb{L}(t)\right]\left[\mathbb{R}(t) + \Pi_3(t)[\mathbb{X}(t+1)]\right]^{-1} \star \qquad (2)$$

where $t \in \mathbb{Z}_+ = \{0, 1, 2, \ldots\}$ with an unknown function $\mathbb{X}(t) = (\ X(t,1),\ \cdots,\ X(t,N)\ )$. Here, $\Pi_k(t)[\mathbb{X}] = (\ \Pi_k(t)[\mathbb{X}](1),\ \cdots,\ \Pi_k(t)[\mathbb{X}](N)\ )$ $(1 \leq k \leq 3)$ are defined by

$$\begin{cases} \Pi_1(t)[\mathbb{X}](i) = \sum_{j=0}^r A_j^T(t,i) \Xi(t)[\mathbb{X}](i) A_j(t,i) \\ \Pi_2(t)[\mathbb{X}](i) = \sum_{j=0}^r A_j^T(t,i) \Xi(t)[\mathbb{X}](i) B_j(t,i) \\ \Pi_3(t)[\mathbb{X}](i) = \sum_{j=0}^r B_j^T(t,i) \Xi(t)[\mathbb{X}](i) B_j(t,i) \\ \Xi(t)[\mathbb{X}](i) = \sum_{j=1}^N p_t(i,j) X(j) \end{cases} \quad (3)$$

where $1 \leq i \leq N$ for all $\mathbb{X} = (\ X(1),\ \cdots,\ X(N)\ ) \in \mathcal{S}_n^N$. In (2) $\mathbb{M}(t) = (M(t,1), \ldots, M(t,N)) \in \mathcal{S}_n^N$, $\mathbb{R}(t) = (R(t,1), \ldots, R(t,N)) \in \mathcal{S}_m^N$, and $\mathbb{L}(t) = (L(t,1), \ldots, L(t,N)) \in \mathcal{M}_{n,m}^N$. Regarding the coefficients of Equation (2), we make the following assumption:

(H1) (a) $\{A_j(t,i)\}_{t \geq 0} \subset \mathbb{R}^{n \times n}$, $\{B_j(t,i)\}_{t \geq 0} \subset \mathbb{R}^{n \times m}$ $(0 \leq j \leq r)$, $\{M(t,i)\}_{t \geq 0} \subset \mathcal{S}_n$, $\{L(t,i)\}_{t \geq 0} \subset \mathbb{R}^{n \times m}$, and $\{R(t,i)\}_{t \geq 0} \subset \mathcal{S}_m$ for all $i \in \mathfrak{N}$ are periodic matrix-valued sequences of a period $\mathfrak{p}$. $\{P_t\}_{t \geq 0}$ with $P_t := (p_t(i,j))_{(i,j) \in \mathfrak{N} \times \mathfrak{N}}$ is also assumed to be a periodic matrix-valued sequence of a period $\mathfrak{p}$.

(b) For each $t \geq 0$, P_t is a *strong nondegenerate stochastic matrix* (i.e., $p_t(i,j) \geq 0$, $\sum_{k=1}^N p_t(i,k) = 1$, $p_t(i,i) > 0$ for all $i, j \in \mathfrak{N}$).

The discrete-time backward nonlinear equation (Equation (2)) will be called a **generalized discrete-time Riccati equation (GDTRE)** in the rest of this paper.

We consider the following partitions of the coefficients of Equation (2):

$$B_j(t,i) = (\ B_{j1}(t,i)\ \ B_{j2}(t,i)\ ),\ B_{jk}(t,i) \in \mathbb{R}^{n \times m_k},\ 0 \leq j \leq r, \quad (4)$$
$$L(t,i) = (L_1(t,i)\ \ L_2(t,i)),\ L_k(t,i) \in \mathbb{R}^{n \times m_k}, k = 1, 2$$

and

$$R(t,i) = \begin{pmatrix} R_{11}(t,i) & R_{12}(t,i) \\ \star & R_{22}(t,i) \end{pmatrix},\ R_{lj}(t,i) \in \mathbb{R}^{m_l \times m_j},\ l, j = 1, 2. \quad (5)$$

Consider the following partitions corresponding to Equations (4) and (5):

$$\begin{cases} \Pi_2(t)[\mathbb{X}](i) = (\ \Pi_{21}(t)[\mathbb{X}](i)\ \ \Pi_{22}(t)[\mathbb{X}](i)\ ) \\ \Pi_3(t)[\mathbb{X}](i) = \begin{pmatrix} \Pi_{311}(t)[\mathbb{X}](i) & \Pi_{312}(t)[\mathbb{X}](i) \\ \star & \Pi_{322}(t)[\mathbb{X}](i) \end{pmatrix} \end{cases} \quad (6)$$

with

$$\begin{cases} \Pi_{2k}(t)[\mathbb{X}](i) = \sum_{j=1}^r A_j^T(t,i) \Xi(t)[\mathbb{X}](i) B_{jk}(t,i) \\ \Pi_{3lk}(t)[\mathbb{X}](i) = \sum_{j=1}^r B_{jl}^T(t,i) \Xi(t)[\mathbb{X}](i) B_{jk}(t,i) \end{cases} ; k, l = 1, 2.$$

The GDTRE (Equation (2)) plays a key role in the solution of a zero-sum LQ stochastic difference game control problem described by the controlled system

$$\begin{cases} x(t+1) = A_0(t, \theta_t) x(t) + B_{01}(t, \theta_t) u_1(t) + B_{02}(t, \theta_t) u_2(t) + \sum_{k=1}^r \big[A_k(t, \theta_t) x(t) \\ \quad + B_{k1}(t, \theta_t) u_1(t) + B_{k2}(t, \theta_t) u_2(t) \big] w_k(t) \\ x(t_0) = x_0 \end{cases} \quad (7)$$

and the quadratic performance criterion

$$\mathcal{J}(x_0, u_1(\cdot), u_2(\cdot)) = \mathbb{E}\left[\sum_{t_0}^{\infty} \begin{pmatrix} x_u(t) \\ u_1(t) \\ u_2(t) \end{pmatrix}^T \begin{pmatrix} M(t, \theta_t) & L_1(t, \theta_t) & L_2(t, \theta_t) \\ \star & R_{11}(t, \theta_t) & R_{12}(t, \theta_t) \\ \star & \star & R_{22}(t, \theta_t) \end{pmatrix} \star \right] \quad (8)$$

where $x_u(t)$ is the solution to the initial value problem (IVP) (Equation (7)), $t \geq t_0 \geq 0$, and $u(\cdot) = \begin{pmatrix} u_1^T(\cdot) & u_2^T(\cdot) \end{pmatrix}^T$. In the first equation (Equation (7)), $\{w_t\}_{t \geq 0}$, $\left(w_t = (w_1(t), \cdots, w_r(t))^T\right)$ is a sequence of independent random vectors, and the triple $(\{\theta_t\}_{t \geq 0}, \{P_t\}_{t \geq 0}, \mathfrak{N})$ is a time non-homogeneous Markov chain defined in a given probability space $(\Omega, \mathcal{F}, \mathrm{P})$ with the finite states set $\mathfrak{N} = \{1, \cdots, N\}$ and the sequence of transition probability matrices $\{P_t\}_{t \geq 0}$. Regarding processes $\{\theta_t\}_{t \geq 0}$ and $\{w_t\}_{t \geq 0}$, the following assumptions are made:

(H2) $\{w_t\}_{t \geq 0}$ is a sequence of independent random vectors with the following properties: $\mathbb{E}[w(t)] = 0$, $\mathbb{E}[w(t)w^T(t)] = I_r$, and $t \geq 0$, with I_r being the identity matrix of a size r.

(H3) (a) For each $t \geq 0$, the σ algebra $\mathcal{F}_t$ is independent of the σ algebra $\mathcal{G}_t$, where $\mathcal{F}_t = \sigma(w(s); 0 \leq s \leq t)$ and $\mathcal{G}_t = \sigma(\theta_s; 0 \leq s \leq t)$.

(b) $\pi_0(i) := \mathcal{P}\{\theta_0 = i\} > 0$ for all $i \in \mathfrak{N}$.

The following assumption regarding the weight matrices $M(t, i)$, $R(t, i)$ and $L(t, i)$ is made:

(H4) For each $(t, i) \in \mathbb{Z}_+ \times \mathfrak{N}$, we have

$$R_{22}(t, i) \geq \rho_2 I_{m_2} \tag{9}$$

$$M(t, i) - L_2(t, i) R_{22}^{-1}(t, i) L_2^T(t, i) \geq 0 \tag{10}$$

$$R_{11}(t, i) - R_{12}(t, i) R_{22}^{-1}(t, i) R_{12}^T(t, i) \leq -\rho_1 I_{m_1} \tag{11}$$

with $\rho_j > 0$ and $j = 1, 2$, given constant scalars.

Let

$$\mathcal{R}(t, \mathbb{X}(t+1), i) := R(t, i) + \Pi_3(t)[\mathbb{X}(t+1)](i). \tag{12}$$

In [1], we considered two different types of admissible strategies, namely the full-state feedback and full-information feedback strategies. We succeeded in showing that for both strategies, the solution to the LQ game relies on the unique bounded and stabilizing solution to the GDTRE (Equation (2)) satisfying a sign condition of the form

$$\mathcal{R}_{22}^{\sharp}(t, \mathbb{X}(t+1), i) = R_{11}(t, i) + \Pi_{311}(t)[\mathbb{X}(t+1)](i) - \left[R_{12}(t, i) + \Pi_{312}(t)[\mathbb{X}(t+1)](i)\right]$$
$$\times \left[R_{22}(t, i) + \Pi_{322}(t)[\mathbb{X}(t+1)](i)\right]^{-1} \star \leq -\delta_1 I_{m_1} \tag{13}$$

$$\mathcal{R}_{22}(t, \mathbb{X}(t+1), i) = R_{22}(t, i) + \Pi_{322}(t)[\mathbb{X}(t+1)](i) \geq \delta_2 I_{m_2} \tag{14}$$

for all $t \in \mathcal{I}$, $1 \leq i \leq N$, $\delta_k > 0$, and $k = 1, 2$ being constants.

The sign conditions in Equations (13) and (14) mean that the quadratic part of the GDTRE (Equation (2)) is of an *indefinite sign*. This sign indefiniteness makes the characterization and the numerical computation of the global solutions to the GDTRE (Equation (2)) much more intricate than in the sign-definite case.

Remark 1.

(i) The solutions $\{\mathbb{X}(t)\}_{t \in \mathcal{I}}$ to the GDTRE (Equation (2)) satisfying the conditions in Equations (13) and (14) will be called admissible solutions.

(ii) If $\mathbb{X}(\cdot) : \mathcal{I} \to \mathcal{S}_n^N$ is an admissible solution to the GDTRE (Equation (2)), then we have the following factorization:

$$R(t,i) + \Pi_3(t)[\mathbb{X}(t+1)](i) = \begin{pmatrix} V_{11}(t)[\mathbb{X}(t+1)](i) & 0 \\ V_{21}(t)[\mathbb{X}(t+1)](i) & V_{22}(t)[\mathbb{X}(t+1)](i) \end{pmatrix}^T \\ \times \begin{pmatrix} -I_{m_1} & 0 \\ 0 & I_{m_2} \end{pmatrix} \star \quad (15)$$

where $V_{kk}(t)[\mathbb{X}(t+1)](i) \geq c_k I_{m_k}$, $k = 1, 2$, $i \in \mathfrak{N}$, and $t \in \mathcal{I}$.

(iii) For a precise definition of the stabilizing solution to the GDTRE (Equation (2)), one can refer to [1].

We derived in [1] the conditions for the existence and uniqueness of the stabilizing solution to Equation (2). In the present paper, we are interested in the numerical aspects of the GDTRE (Equation (2)). Our objective here is to propose a globally convergent algorithm for the computation of the unique stabilizing solution to Equation (2) with the sign (indefinite) conditions in Equations (13) and (14). We will propose an iterative *deterministic* algorithm which is based on the numerical computation of the bounded and stabilizing solutions of a sequence of Riccati difference equations arising in the deterministic framework. In order to accomplish this, we consider the following sequence of uncoupled Riccati difference equations (which are specific to the deterministic framework):

$$X^k(t,i) = \bar{A}_0^T(t,i) X^k(t+1,i) \bar{A}_0(t,i) + M_i^k(t) \\ - \left(\bar{A}_0^T(t,i) X^k(t+1,i) \bar{B}_0(t,i) + L_i^k(t) \right) \left(R_i^k(t) + \bar{B}_0^T(t,i) X^k(t+1,i) \bar{B}_0(t,i) \right)^{-1} \star \quad (16)$$

where

$$\begin{cases} \bar{A}_0(t,i) = \sqrt{p_t(i,i)} A_0(t,i) \\ \bar{B}_0(t,i) = \sqrt{p_t(i,i)} B_0(t,i) \\ M_i^k(t) = \tilde{\Pi}_1(t)[\mathbb{X}^{k-1}(t+1)](i) + A_0^T(t,i) \bar{\Xi}(t)[\mathbb{X}^{k-1}(t+1)](i) A_0(t,i) + M(t,i) \\ \tilde{\Pi}_1(t)[\mathbb{X}^{k-1}(t+1)](i) = \sum_{j=1}^r A_j^T(t,i) \Xi(t)[\mathbb{X}^{k-1}(t+1)](i) A_j(t,i) \\ \bar{\Xi}(t)[\mathbb{X}^{k-1}(t+1)](i) = \sum_{\substack{j=1 \\ j \neq i}}^N p_t(i,j) X^{k-1}(t+1,j) \\ L_i^k(t) = \tilde{\Pi}_2(t)[\mathbb{X}^{k-1}(t+1)](i) + A_0^T(t,i) \bar{\Xi}(t)[\mathbb{X}^{k-1}(t+1)](i) B_0(t,i) + L(t,i) \\ \tilde{\Pi}_2(t)[\mathbb{X}^{k-1}(t+1)](i) = \sum_{j=1}^r A_j^T(t,i) \Xi(t)[\mathbb{X}^{k-1}(t+1)](i) B_j(t,i) \\ R_i^k(t) = \tilde{\Pi}_3(t)[\mathbb{X}^{k-1}(t+1)](i) + B_0^T(t,i) \bar{\Xi}(t)[\mathbb{X}^{k-1}(t+1)](i) B_0(t,i) + R(t,i) \\ \tilde{\Pi}_3(t)[\mathbb{X}^{k-1}(t+1)](i) = \sum_{j=1}^r B_j^T(t,i) \Xi(t)[\mathbb{X}^{k-1}(t+1)](i) B_j(t,i) \end{cases} \quad (17)$$

By taking $X_i^0(t) = 0$, $1 \leq i \leq N$, $t \in \mathbb{Z}_+$, we may construct the inductive sequences $\{X_i^k(t)\}_{k \geq 1}$, $1 \leq i \leq N$, $X_i^k(\cdot)$, which are the unique bounded and stabilizing solution to the Riccati difference equation (Equation (16)). The aim of this study is to provide a set of conditions which guarantee that $X_i^k(\cdot)$ is well defined for all $k \geq 1$ and $\lim_{k \to \infty} X_i^k(t) = X_s(t,i)$ for all $1 \leq i \leq N$ and $t \in \mathbb{Z}_+$.

Remark 2. *Note that*

$$\hat{\Pi}(t)[\mathbb{X}](i) = \begin{pmatrix} \Theta_1(t)[\mathbb{X}](i) & \Theta_2(t)[\mathbb{X}](i) \\ \star & \Theta_3(t)[\mathbb{X}](i) \end{pmatrix} \geq 0 \quad (18)$$

if $\mathbb{X}$ is such that $X(i) \geq 0$, where $\Theta_1(t)[\mathbb{X}](i) = \tilde{\Pi}_1(t)[\mathbb{X}](i) + A_0^T(t,i) \bar{\Xi}(t)[\mathbb{X}](i) A_0(t,i)$, $\Theta_2(t)[\mathbb{X}](i) = \tilde{\Pi}_2(t)[\mathbb{X}](i) + A_0^T(t,i) \bar{\Xi}(t)[\mathbb{X}](i) B_0(t,i)$, and $\Theta_3(t)[\mathbb{X}](i) = \tilde{\Pi}_3(t)[\mathbb{X}](i) +$

$B_0^T(t,i)\bar{\Xi}(t)[\mathbb{X}](i)B_0(t,i)$, $1 \leq i \leq N$. This follows by noticing that Equation (18) could be rewritten as

$$\hat{\Pi}(t)[\mathbb{X}](i) = \begin{pmatrix} A_0^T(t,i) \\ B_0^T(t,i) \end{pmatrix} \bar{\Xi}(t)[\mathbb{X}](i) \star + \sum_{j=1}^{r} \begin{pmatrix} A_j^T(t,i) \\ B_j^T(t,i) \end{pmatrix} \Xi(t)[\mathbb{X}](i) \star. \quad (19)$$

Remark 3. *In the Numerical Experiments section, we will clarify the deterministic nature of the proposed algorithm and highlight the contribution of such a paradigm.*

2.2. Some Intermediate Results

Let us formally set $u_2(t) \equiv u_2^{\mathbb{KW}}(t) = K(t,\theta_t)x(t) + W(t,\theta_t)u_1(t)$. Hence, Equations (7) and (8) are rewritten as follows:

$$x(t+1) = A_{0\mathbb{K}}(t,\theta_t)x(t) + B_{0\mathbb{W}}(t,\theta_t)u_1(t) + \sum_{k=1}^{r} w_k(t)(A_{k\mathbb{K}}(t,\theta_t)x(t) + B_{k\mathbb{W}}(t,\theta_t)u_1(t)) \quad (20)$$

$$\mathcal{J}_{\mathbb{KW}}(t_0, x_0, u_1) = \mathbb{E}\left[\sum_{t=t_0}^{\infty} \begin{pmatrix} x_{u_1}(t) \\ u_1(t) \end{pmatrix}^T \begin{pmatrix} M_{\mathbb{K}}(t,\theta_t) & L_{\mathbb{KW}}(t,\theta_t) \\ \star & R_{\mathbb{W}}(t,\theta_t) \end{pmatrix} \star \right] \quad (21)$$

where $x_{u_1}(t)$ is the solution to Equation (20) corresponding to $u_1(t)$ and

$$\begin{cases} A_{k\mathbb{K}}(t,i) = A_k(t,i) + B_{k2}(t,i)K(t,i) \\ B_{k\mathbb{W}}(t,i) = B_{k1}(t,i) + B_{k2}(t,i)W(t,i) \\ M_{\mathbb{K}}(t,i) = M(t,i) + L_2(t,i)K(t,i) + K^T(t,i)L_2^T(t,i) + K^T(t,i)R_{22}(t,i)K(t,i) \\ L_{\mathbb{KW}}(t,i) = L_1(t,i) + K^T(t,i)R_{12}^T(t,i) + (L_2(t,i) + K^T(t,i)R_{22}(t,i))W(t,i) \\ R_{\mathbb{W}}(t,i) = \begin{pmatrix} I_{m_1} \\ W(t,i) \end{pmatrix}^T R(t,i) \begin{pmatrix} I_{m_1} \\ W(t,i) \end{pmatrix} \end{cases} \quad (22)$$

With the above system (Equation (20)) and the corresponding quadratic functional in Equation (21), we associate the following Riccati-type difference equation of the type in Equation (2):

$$X(t,i) = \Pi_{\mathbb{K}}(t)[\mathbb{X}(t+1)](i) + M_{\mathbb{K}}(t,i) - \left(\Pi_{\mathbb{KW}}(t)[\mathbb{X}(t+1)](i) + L_{\mathbb{KW}}(t,i)\right)$$
$$\times \left(R_{\mathbb{W}}(t,i) + \Pi_{\mathbb{W}}(t)[\mathbb{X}(t+1)](i)\right)^{-1} \star \quad (23)$$

where

$$\begin{cases} \Pi_{\mathbb{K}}(t)[\mathbb{X}](i) = \sum_{j=0}^{r} A_{j\mathbb{K}}^T(t,i)\Xi(t)[\mathbb{X}](i)A_{j\mathbb{K}}(t,i) \\ \Pi_{\mathbb{KW}}(t)[\mathbb{X}](i) = \sum_{j=0}^{r} A_{j\mathbb{K}}^T(t,i)\Xi(t)[\mathbb{X}](i)B_{j\mathbb{W}}(t,i) \\ \Pi_{\mathbb{W}}(t)[\mathbb{X}](i) = \sum_{j=0}^{r} B_{j\mathbb{W}}^T(t,i)\Xi(t)[\mathbb{X}](i)B_{j\mathbb{W}}(t,i) \end{cases} \quad (24)$$

for all $\mathbb{X} \in \mathcal{S}_n^N$.

In the following, we associate to the GDTRE (Equation (2)) the set $\mathcal{A}^{\mathbb{KW}}$, which consists of all pairs of feedback gains $(\mathbb{K}(\cdot), \mathbb{W}(\cdot))$, where $t \to \mathbb{K}(t) = \begin{pmatrix} K(t,1), & \cdots, & K(t,N) \end{pmatrix}$: $\mathbb{Z}_+ \to \mathcal{M}_{m_2,n}^N$ and $t \to \mathbb{W}(t) = \begin{pmatrix} W(t,1), & \cdots, & W(t,N) \end{pmatrix}$: $\mathbb{Z}_+ \to \mathcal{M}_{m_2,m_1}^N$ are p-periodic matrix-valued sequences having the following properties:

(i) The zero solution of the stochastic linear system

$$x(t+1) = A_{0\mathbb{K}}(t,\theta_t)x(t) + \sum_{k=1}^{r} w_k(t)A_{k\mathbb{K}}(t,\theta_t)x(t) \quad (25)$$

is exponentially stable in the mean square sense (ESMS) (see Definition 3.1 from [8] for details).

(ii) The corresponding GRDE (Equation (23)) has a unique bounded and stabilizing solution $\tilde{\mathbb{X}}_{\mathbb{KW}}(\cdot)$ satisfying the sign condition

$$R_{\mathbb{W}}(t,i) + \Pi_{\mathbb{W}}(t)[\tilde{\mathbb{X}}_{\mathbb{KW}}(t+1)](i) \leq -\xi \mathbb{I} \qquad (26)$$

for some positive scalar ξ (which may depend upon $(\mathbb{K}(\cdot), \mathbb{W}(\cdot))$) where $(t,i) \in \mathbb{Z}_+ \times \mathfrak{N}$.

The following result gives a necessary and sufficient condition which helps us to decide if the set $\mathcal{A}^{\overline{\mathbb{KW}}}$ is empty or not:

Proposition 1. *Under the considered assumptions, the following two assertions are equivalent:*
(i) $\mathcal{A}^{\mathbb{KW}}$ *is not empty;*
(ii) *There exist* $\mathfrak{p}$*-periodic sequences* $t \to \mathbb{Z}(t) : \mathbb{Z}_+ \to \mathcal{S}_n^N$, $t \to \mathbb{K}(t) : \mathbb{Z}_+ \to \mathcal{M}_{m_2,n}^N$ *and* $t \to \mathbb{W}(t) : \mathbb{Z}_+ \to \mathcal{M}_{m_2,m_1}^N$ *solving the following matrix inequalities*

$$\begin{pmatrix} \Pi_{\mathbb{K}}(t)[\mathbb{Z}(t+1)](i) + M_{\mathbb{K}}(t,i) - \mathbb{Z}(t,i) & \Pi_{\mathbb{KW}}(t)[\mathbb{Z}(t+1)](i) + L_{\mathbb{KW}}(t,i) \\ \star & R_{\mathbb{W}}(t,i) + \Pi_{\mathbb{W}}(t)[\mathbb{Z}(t+1)](i) \end{pmatrix} < 0 \qquad (27)$$

Proof. One can apply Theorem 5.6 in [8] to the Riccati difference equation

$$Y(t,i) = \Pi_{\mathbb{K}}(t)[\mathbb{Y}(t+1)](i) - M_{\mathbb{K}}(t,i) - \left(\Pi_{\mathbb{KW}}(t)[\mathbb{Y}(t+1)](i) - L_{\mathbb{W}}(t,i)\right)$$
$$\times \left(-R_{\mathbb{W}}(t,i) + \Pi_{\mathbb{W}}(t)[\mathbb{Y}(t+1)](i)\right)^{-1} \star \qquad (28)$$

obtained from Equation (23) by taking $Y(t,i) = -X(t,i)$, $(t,i) \in \mathbb{Z}_+ \times \mathfrak{N}$. □

We end this section by giving the existence conditions for the unique bounded and stabilizing solution to Equation (2). To this end, we introduce the following auxiliary system:

$$\begin{cases} x(t+1) = \check{A}_0(t,\theta_t)x(t) + \sum_{j=1}^r w_j(t)\check{A}_j(t,\theta_t)x(t) \\ y(t) = \check{C}(t,\theta_t)x(t) \end{cases} \qquad (29)$$

where

$$\check{A}_j(t,i) = A_j(t,i) - B_{j2}(t,i)R_{22}^{-1}(t,i)L_2^T(t,i), \quad 0 \leq j \leq r \qquad (30)$$

and $\check{C}(t,i)$ is obtained from the factorization $M(t,i) - L_2(t,i)R_{22}^{-1}(t,i)L_2^T(t,i) = \check{C}^T(t,i)\check{C}(t,i)$ for all $i \in \mathfrak{N}$, $t \geq 0$.

Theorem 1. *Assume the following:*
(a) *Assumptions (**H1**–**H4**) are fulfilled;*
(b) *The set $\mathcal{A}^{\mathbb{KW}}$ is not empty;*
(c) *The auxiliary system in Equation (29) is exactly detectable at a time instant $t_0 = 0$;*

Then, $\tilde{\mathbb{X}}(\cdot)$, defined as $\tilde{X}(t,i) = \lim_{\tau \to \infty} X_\tau(t,i)$, coincides with the unique admissible stabilizing and $\mathfrak{p}$*-periodic solution $\mathbb{X}_s(\cdot)$ to Equation (2), where for each $\tau > 0$, $\mathbb{X}_\tau(t) = (X_\tau(t,1), \cdots X_\tau(t,N))$ is the solution to Equation (2) satisfying the conditions $X_\tau(\tau+1,i) = 0$ and $1 \leq i \leq N$.*

Remark 4. *For the definition of the notion of exact detectability at the time instant $t_0 = 0$, one can refer to [1].*

Remark 5. *Note that the above theorem was proven in [1] under the assumption of stochastic detectability of the system in Equation (29) instead of exact detectability at the time instant $t_0 = 0$. One can show that the concept of exact detectability at the time instant $t_0 = 0$ is wider than the stochastic detectability one. Hence, the above result can be applied to a larger class of stochastic systems than the one reported in [1]. From the technical point of view, the improvement reported in this paper consists of the modification of Lemma 4.7 from [1], which is proven here under exact* detectability at the assumption at the time instant $t_0 = 0$. For the reader's convenience, we include a sketch of the proof of this Lemma in Appendix A.

3. Main Results

For each $k \geq 1$, $1 \leq i \leq N$, the Riccati difference equation (Equation (16)) may be regarded as a special case of Equation (2). Hence, the Riccati difference equation (Equation (16)) is related to the deterministic LQ control problem described by the controlled system

$$x(t+1) = \bar{A}_0(t,i)x(t) + \bar{B}_0(t,i)u(t) \tag{31}$$

where $t \geq 0$, $x(0) = x_0$, as well as the cost functional

$$\mathcal{J}_i^k(x_0, u) = \sum_{t=0}^{\infty} \begin{pmatrix} x_u(t) \\ u(t) \end{pmatrix}^T \mathcal{M}_i^k(t) \star \tag{32}$$

where $x_u(t)$ is the solution to the IVP described by the controlled system in Equation (31), $t \geq 0$, $x(0) = x_0$, and

$$\mathcal{M}_i^k(t) = \begin{pmatrix} M_i^k(t) & L_i^k(t) \\ \star & R_i^k(t) \end{pmatrix} \tag{33}$$

with $M_i^k(t)$, $L_i^k(t)$, and $R_i^k(t)$ being defined in Equation (17).

We formally set $u_2(t) \equiv u_{2,i}^{KW}(t) = K(t,i)x(t) + W(t,i)u_1(t)$. Hence, Equations (31) and (32) are rewritten as follows:

$$x(t+1) = \bar{A}_{0K}(t,i)x(t) + \bar{B}_{0W}(t,i)u_1(t) \tag{34}$$

$$\mathcal{J}_{KW}^{k,i}(x_0, u_1) = \sum_{t=0}^{\infty} \begin{pmatrix} x_{u_1}(t) \\ u_1(t) \end{pmatrix}^T \begin{pmatrix} M_K^k(t,i) & L_{KW}^k(t,i) \\ \star & R_W^k(t,i) \end{pmatrix} \star \tag{35}$$

where $x_{u_1}(t)$ is the solution to Equation (34) corresponding to $u_1(t)$ and

$$\begin{cases} \bar{A}_{0K}(t,i) = \bar{A}_0(t,i) + \bar{B}_{02}(t,i)K(t,i) \\ \bar{B}_{0W}(t,i) = \bar{B}_{01}(t,i) + \bar{B}_{02}(t,i)W(t,i) \\ M_K^k(t,i) = M^k(t,i) + L_2^k(t,i)K(t,i) + K^T(t,i)(L_2^k)^T(t,i) + K^T(t,i)R_{22}^k(t,i)K(t,i) \\ L_{KW}^k(t,i) = L_1^k(t,i) + K^T(t,i)R_{12}^{k^T}(t,i) + \left(L_2^k(t,i) + K^T(t,i)R_{22}^k(t,i)\right)W(t,i) \\ R_W^k(t,i) = \begin{pmatrix} I_{m_1} \\ W(t,i) \end{pmatrix}^T R^k(t,i) \star \end{cases} \tag{36}$$

With the above system (Equation (34)) and the corresponding quadratic functional in Equation (35), we associate the following Riccati difference equation:

$$X^k(t,i) = \bar{A}_{0K}^T(t,i)X^k(t+1,i)\bar{A}_{0K}(t,i) + M_K^k(t,i) - \left(\bar{A}_{0K}^T(t,i)X^k(t+1,i)\bar{B}_{0W}(t,i)\right.$$
$$\left. + L_{KW}^k(t,i)\right)\left(R_W^k(t) + \bar{B}_{0W}^T(t,i)X^k(t+1,i)\bar{B}_{0W}(t,i)\right)^{-1} \star \tag{37}$$

The notion of a stabilizing solution for Equation (37) is defined in the same way as for Equation (2).

In the following, we denote with $\mathcal{A}_{k,i}^{KW}$ the set of all pairs of feedback gains $(K_i(\cdot), W_i(\cdot))$, where $K_i(\cdot) : \mathbb{Z}_+ \to \mathbb{R}^{m_2 \times n}$ and $W_i(t) : \mathbb{Z}_+ \to \mathbb{R}^{m_2 \times m_1}$ are $\mathfrak{p}$-periodic matrix-valued sequences having the following properties:

(i) The zero solution of the closed-loop system

$$x(t+1) = \bar{A}_{0K}(t,i)x(t) \tag{38}$$

is exponentially stable.

(ii) The corresponding GRDE (Equation (37)) has a unique stabilizing and $\mathfrak{p}$-periodic solution $\tilde{\mathbb{X}}_{KW}(\cdot)$ satisfying the sign condition

$$R_W^k(t,i) + \bar{B}_{0W}^T(t,i) X^k(t+1,i) \bar{B}_{0W}(t,i) \leq -\zeta \mathbb{I} \tag{39}$$

for some positive scalar ζ (which may depend upon $(\mathbb{K}(\cdot), \mathbb{W}(\cdot))$), and $(t,i) \in \mathbb{Z}_+ \times \mathfrak{N}$.

Following similar arguments to those in the proof of Proposition 1, the following result is deduced:

Proposition 2. *Under the considered assumptions, the following two assertions are equivalent:*

(i) $\mathcal{A}_{k,i}^{\mathbb{KW}}$ *is not empty;*

(ii) *There exist $\mathfrak{p}$-periodic sequences $t \to Z(t,i) : \mathbb{Z}_+ \to \mathcal{S}^n$, $t \to K(t,i) : \mathbb{Z}_+ \to \mathbb{R}^{m_2 \times n}$, and $t \to W(t,i) : \mathbb{Z}_+ \to \mathbb{R}^{m_2 \times m_1}$ solving the following matrix inequalities:*

$$\begin{pmatrix} \bar{A}_{0K}^T(t,i) Z(t+1,i) \star + M_K^k(t,i) - Z(t,i) & \bar{A}_{0K}^T(t,i) Z(t+1,i) \bar{B}_{0W}(t,i) + L_{KW}^k(t,i) \\ \star & R_W^k(t,i) + \bar{B}_{0W}^T(t,i) Z(t+1,i) \star \end{pmatrix} < 0 \tag{40}$$

We are now in position to prove the main result of this paper. To this end, we introduce the following auxiliary system:

$$\begin{cases} x(t+1) = \check{A}_0^k(t,i) x(t) \\ y(t) = \check{C}^k(t,i) x(t) \end{cases} \tag{41}$$

where

$$\check{A}_0^k(t,i) = \bar{A}_0(t,i) - \bar{B}_{02}(t,i) (R_{22}^k)^{-1}(t,i) (L_2^k)^T(t,i), \tag{42}$$

and $\check{C}^k(t,i)$ is obtained from the factorization

$$M^k(t,i) - L_2^k(t,i) (R_{22}^k)^{-1}(t,i) (L_2^k)^T(t,i) = (\check{C}^k)^T(t,i) \check{C}^k(t,i)$$

for all $i \in \mathfrak{N}, t \geq 0$.

Theorem 2. *Assume the following:*

(a) *Assumptions* (**H$_1$**–**H$_4$**) *are fulfilled;*

(b) *The set $\mathcal{A}^{\mathbb{KW}}$ is not empty;*

(c) *The auxiliary system in Equation (29) is stochastically detectable.*

Under these conditions, if we take $X_i^0(t) \equiv 0$, where $1 \leq i \leq N$, then for each $k \geq 1$, $X_i^k(\cdot)$ is well defined as the unique minimal and positive semi-definite solution to the Riccati difference equation (Equation (16)), and we have the following:

(i) $X_i^k(\cdot)$ *is a periodic sequence of a period $\mathfrak{p}$ and satisfies the sign conditions of the types in Equations (13) and (14);*

(ii) $0 = X_i^0(t) \leq X_i^1(t) \leq \cdots \leq X_i^k(t) \leq \cdots \leq \mathbb{X}_s(t,i)$ *for all $(t,i) \in \mathbb{Z}_+ \times \mathfrak{N}$ and $\mathbb{X}_s(t) = (\mathbb{X}_s(t,1), \cdots, \mathbb{X}_s(t,N))$ as the unique stabilizing and $\mathfrak{p}$-periodic solution to Equation (2);*

(iii) *If the auxiliary system in Equation (41) is detectable, then $X_i^k(\cdot)$ is just the stabilizing solution of the Riccati difference equation (Equation (16));*

(iv) $\lim\limits_{k \to \infty} X_i^k(t) = \mathbb{X}_s(t,i)$ *for all $(t,i) \in \mathbb{Z}_+ \times \mathfrak{N}$.*

Proof. Since $\mathcal{A}^{\mathbb{KW}}$ is not empty, it follows from Proposition 1 that there exist $\mathfrak{p}$-periodic sequences $t \to \mathbb{Z}(t) : \mathbb{Z}_+ \to \mathcal{S}_n^N$, $t \to \mathbb{K}(t) : \mathbb{Z}_+ \to \mathcal{M}_{m_2,n}^N$, and $t \to \mathbb{W}(t) : \mathbb{Z}_+ \to$

$\mathcal{M}_{m_2,m_1}^N$ solving the matrix inequalities in Equation(27). Note that Equation (27) could be rewritten as

$$\begin{pmatrix} \bar{A}_{0K}^T(t,i)Z(t+1,i)\star + M_K^1(t,i) - Z(t,i) & \bar{A}_{0K}^T(t,i)Z(t+1,i)\bar{B}_{0W}(t,i) + L_{KW}^1(t,i) \\ \star & R_W^1(t) + \bar{B}_{0W}^T(t,i)Z(t+1,i)\star \end{pmatrix}$$
$$+ \hat{\Pi}_{KW}[Z(t+1)](i) < 0 \qquad (43)$$

where

$$\hat{\Pi}_{KW}[Z(t+1)](i) = \begin{pmatrix} A_{0K}^T(t,i) \\ B_{0W}^T(t,i) \end{pmatrix} \bar{\Xi}[Z(t+1)](i)\star$$
$$+ \sum_{k=1}^r \begin{pmatrix} A_{kK}^T(t,i) \\ B_{kW}^T(t,i) \end{pmatrix} \Xi[Z(t+1)](i)\star \geq 0 \qquad (44)$$

because $Z(t,i) \geq 0$ for all $(t,i) \in \mathbb{Z}_+ \times \mathfrak{N}$. This allows us to deduce that

$$\begin{pmatrix} \bar{A}_{0K}^T(t,i)Z(t+1,i)\star + M_K^1(t,i) - Z(t,i) & \bar{A}_{0K}^T(t,i)Z(t+1,i)\bar{B}_{0W}(t,i) + L_{KW}^1(t,i) \\ \star & R_W^1(t) + \bar{B}_{0W}^T(t,i)Z(t+1,i)\star \end{pmatrix} < 0 \qquad (45)$$

Hence, under Proposition 2, it follows that $\mathcal{A}_{1,i}^{KW}$ is not empty. If $k = 1$, then the Riccati difference equation (Equation (16)) reduces to

$$X(t,i) = \bar{A}_0^T(t,i)X(t+1,i)\bar{A}_0(t,i) + M(t,i)$$
$$- \left(\bar{A}_0^T(t,i)X(t+1,i)\bar{B}_0(t,i) + L(t,i)\right)\left(R(t) + \bar{B}_0^T(t,i)X(t+1,i)\bar{B}_0(t,i)\right)^{-1}\star \qquad (46)$$

where $1 \leq i \leq N$. From Proposition 4.4 in [1], we deduce that if $X_{i\tau}(\cdot)$ is the solution to Equation (46) which satisfies $X_{i\tau}(\tau) = 0$, then it is well defined for all $0 \leq t \leq \tau$ and $\tau > 0$, where $1 \leq i \leq N$, and for each $1 \leq i \leq N$, with $X_i^1(\cdot)$ defined by

$$X_i^1(t) = \lim_{\tau \to \infty} X_{i\tau}(t) \qquad (47)$$

This is the unique minimal positive semi-definite solution to Equation (46). Moreover, $t \to X_i^1(t)$ is a periodic sequence of a period $\mathfrak{p}$.

Let us notice that the Riccati difference equation (Equation (2)) satisfied by its stabilizing solution $\mathbb{X}_s(\cdot)$ may be rewritten as

$$X_s(t,i) = \bar{A}_0^T(t,i)X_s(t+1,i)\bar{A}_0(t,i) + M_{s,i}(t)$$
$$- \left(\bar{A}_0^T(t,i)X_s(t+1,i)\bar{B}_0(t,i) + L_{s,i}(t)\right)\left(R_{s,i}(t) + \bar{B}_0^T(t,i)X_s(t+1,i)\bar{B}_0(t,i)\right)^{-1}\star \qquad (48)$$

where

$$\begin{pmatrix} M_{s,i}(t) & L_{s,i}(t) \\ \star & R_{s,i}(t) \end{pmatrix} = \begin{pmatrix} M(t,i) & L(t,i) \\ \star & R(t,i) \end{pmatrix} + \hat{\Pi}(t)[\mathbb{X}_s(t)](i). \qquad (49)$$

Since $X_s(t,i) \geq 0$, we deduce from Equation (18) that

$$\begin{pmatrix} M_{s,i}(t) & L_{s,i}(t) \\ \star & R_{s,i}(t) \end{pmatrix} \geq \begin{pmatrix} M(t,i) & L(t,i) \\ \star & R(t,i) \end{pmatrix}.$$

Hence, by applying Theorem 4.2 in [1] in the special case of the Riccati difference equation (Equations (46) and (48)), we may infer that $X_{i\tau}(t) \leq X_s(t,i)$ for all $0 \leq t \leq \tau$, $\tau > 0$, and $1 \leq i \leq N$. By taking the limit for $\tau \to \infty$, we obtain $0 \leq X_i^1(t) \leq X_s(t,i)$ for all $(t,i) \in \mathbb{R}_+ \times \mathfrak{N}$. From the matrix inequality

$$R(t,i) + \Pi_3(t)\left[X_i^1(t)\right](i) \leq R(t,i) + \Pi_3(t)[X_s(t)](i)$$

we deduce, via Lemma 4.5 in [9], that for each $1 \leq i \leq N$, $X_i^1(\cdot)$ satisfies the sign conditions in Equations (13) and (14). Thus, assertions (i) and (ii) from the statement are fulfilled for $k = 1$.

By using the Lyapunov-type characterization of the stochastic detectability of linear stochastic systems (see, for example, Chapter 4 in [8]), one can show that the stochastic detectability of the auxiliary system (Equation (29)) implies the detectability of the deterministic system

$$\begin{cases} x(t+1) = \check{A}_0(t,i)x(t) \\ y(t) = \check{C}(t,i)x_t \end{cases} \quad (50)$$

for each $1 \leq i \leq N$, where $\check{A}_0(t,i) = \tilde{A}_0(t,i) - \tilde{B}_{02}(t,i)R_{22}^{-1}(t,i)L_2^T(t,i)$. Therefore, under assumption (c) in the statement, it follows that $X_i^1(\cdot)$ is just the bounded and stabilizing solution of the Riccati difference equation (Equation (16)) in the special case $k = 1$, which confirms the validity of assertion (iii) from the statement for $k = 1$.

Let us assume that for $k \geq 2$ and for any $1 \leq l \leq k - 1$ and $1 \leq i \leq N$, the functions $X_i^l(\cdot)$ are well defined as unique minimal and positive semi-definite solutions of the Riccati difference equation (Equation (16)) (written for k and replaced by l) and have properties (i–iii) from the statement. We now show that for $l = k$ and $1 \leq i \leq N$, the Riccati difference equation (Equation (16)) has a minimal solution $X_i^k(\cdot)$ which is positive semi-definite, and it is a $\mathfrak{p}$-periodic sequence satisfying the sign conditions in Equations (13) and (14). Moreover, we have

$$0 \leq X_i^1(t) \leq \cdots \leq X_i^l(t) \leq \cdots \leq X_i^{k-1}(t) \leq X_i^k(t) \leq \cdots \leq X_s(t,i) \quad (51)$$

$(t,i) \in \mathbb{R}_+ \times \mathfrak{N}$.

If $(\mathbb{K}(\cdot), \mathbb{W}(\cdot)) \in \mathcal{A}^{\mathrm{KW}}$, then we rewrite Equation (27) in the form

$$\begin{pmatrix} \tilde{A}_{0K}^T(t,i)Z(t+1,i)\star + M_K^k(t,i) - Z(t,i) & \tilde{A}_{0K}^T(t,i)Z(t+1,i)\tilde{B}_{0W}(t,i) + L_{KW}^k(t,i) \\ \star & R_W^k(t) + \tilde{B}_{0W}^T(t,i)Z(t+1,i)\star \end{pmatrix}$$
$$+ \hat{\Pi}_{\mathrm{KW}}[\mathbb{Z}(t+1) - \mathbb{X}^{k-1}(t+1)](i) < 0 \quad (52)$$

in which $(t,i) \in \mathbb{Z}_+ \times \mathfrak{N}$, where $\mathbb{X}^{k-1}(t) = \begin{pmatrix} X_1^{k-1}(t), & \cdots, & X_N^{k-1}(t) \end{pmatrix}$ and $\hat{\Pi}_{\mathrm{KW}}[\mathbb{Z}(t+1) - \mathbb{X}^{k-1}(t+1)](i)$ is computed as in Equation (44) with $\mathbb{Z}(t+1)$ replaced by $\mathbb{Z}(t+1) - \mathbb{X}^{k-1}(t+1)$.

Recalling that stochastic detectability implies exact detectability at time instant $t_0 = 0$ (see Remark 5), it follows from Proposition 4.4 in [1] and Theorem 1 that $X_s(t,i) \leq \tilde{X}_{\mathrm{KW}}(t,i)$ for all $(t,i) \in \mathbb{Z}_+ \times \mathfrak{N}$. Note also that by using similar arguments to those in Chapter 5 from [8], one can show that $\tilde{X}_{\mathrm{KW}}(t,i) \leq Z(t,i)$ for all $(t,i) \in \mathbb{Z}_+ \times \mathfrak{N}$. Hence, we deduce that $X_i^{k-1} \leq X_s(t,i) \leq Z(t,i)$ for all $(t,i) \in \mathbb{Z}_+ \times \mathfrak{N}$. Thus, $\hat{\Pi}_{\mathrm{KW}}[\mathbb{Z}(t+1) - \mathbb{X}^{k-1}(t+1)](i) \geq 0$. This allows us to conclude that the matrix-valued sequences $Z_i(\cdot) = Z(\cdot,i)$ satisfy

$$\begin{pmatrix} \tilde{A}_{0K}^T(t,i)Z(t+1,i)\star + M_K^k(t,i) - Z(t,i) & \tilde{A}_{0K}^T(t,i)Z(t+1,i)\tilde{B}_{0W}(t,i) + L_{KW}^k(t,i) \\ \star & R_W^k(t) + \tilde{B}_{0W}^T(t,i)Z(t+1,i)\star \end{pmatrix} < 0 \quad (53)$$

Therefore, we may conclude that $\mathcal{A}_{k,i}^{\mathrm{KW}}$ is not empty for all $1 \leq i \leq N$ if $\mathcal{A}^{\mathrm{KW}}$ is not empty. Thus, we deduce that the solutions $X_{i\tau}^k(\cdot)$ to the difference equation (Equation (16)) which satisfy the condition $X_{i\tau}^k(\tau) = 0$ are well defined for all $0 \leq t \leq \tau$, $\forall \tau > 0$, and $i \in \mathfrak{N}$. By applying Proposition 4.4 from [1] in the special case of the Riccati difference equation (Equation (16)), we infer that $X_i^k(\cdot)$, defined by $X_i^k(t) = \lim_{\tau \to \infty} X_{i\tau}^k(t)$, is the minimal positive semi-definite and $\mathfrak{p}$-periodic solution of the Riccati difference equation (Equation (16)).

From Equations (17) and (49), we obtain

$$\begin{pmatrix} M_{s,i}(t) & L_{s,i}(t) \\ \star & R_{s,i}(t) \end{pmatrix} - \begin{pmatrix} M_i^k(t) & L_i^k(t) \\ \star & R_i^k(t) \end{pmatrix} = \hat{\Pi}(t)\left[\mathbb{X}_s(t) - \mathbb{X}^{k-1}(t)\right](i) \quad (54)$$

By again invoking the inequalities $X_i^{k-1}(t) \leq X_s(t,i)$ and $\forall (t,i) \in \mathbb{K}_+ \times \mathfrak{N}$, we may obtain $\hat{\Pi}(t)\left[\mathbb{X}_s(t+1) - \mathbb{X}^{k-1}(t+1)\right](i) \geq 0$. By applying Theorem 4.2 in [1] in the special case of Equations (16) and (48), we deduce that $X_{i\tau}^k(t) \leq X_s(t,i)$ for all $0 \leq t \leq \tau$, $\tau > 0$, and $1 \leq i \leq N$. By taking the limit for $\tau \to \infty$, we deduce that

$$X_i^k(t) \leq X_s(t,i) \tag{55}$$

for all $(t,i) \in \mathbb{Z}_+ \times \mathfrak{N}$. On the other hand, Equation (17) yields

$$\begin{pmatrix} M_i^k(t) & L_i^k(t) \\ \star & R_i^k(t) \end{pmatrix} - \begin{pmatrix} M_i^{k-1}(t) & L_i^{k-1}(t) \\ \star & R_i^{k-1}(t) \end{pmatrix} = \hat{\Pi}(t)\left[\mathbb{X}^{k-1}(t) - \mathbb{X}^{k-2}(t)\right](i)$$

Since $X_i^{k-2}(t) \leq X_i^{k-1}(t)$ and $(t,i) \in \mathbb{K}_+ \times \mathfrak{N}$, one obtains $\hat{\Pi}(t)\left[\mathbb{X}^{k-1}(t+1) - \mathbb{X}^{k-2}(t+1)\right]$, where $(i) \geq 0$. This allows to us apply Theorem 4.2 from [1] in the special case of the Riccati difference equation (Equation (16)) to deduce that $X_{i\tau}^{k-1}(t) \leq X_{i\tau}^k(t)$, $\forall\, 0 \leq t \leq \tau$, $\tau > 0$, and $1 \leq i \leq N$. By letting $\tau \to \infty$, we obtain

$$X_i^{k-1}(t) \leq X_i^k(t) \tag{56}$$

$\forall (t,i) \in \mathbb{Z}_+ \times \mathfrak{N}$. Thus, Equations (55) and (56) confirm the validity of Equation (51).

Furthermore, Equation (55) yields

$$R(t,i) + \Pi_3(t)[\mathbb{X}^k(t)](i) \leq R(t,i) + \Pi_3(t)[\mathbb{X}_s(t)](i)$$

$\forall (t,i) \in \mathbb{Z}_+ \times N$. These matrix inequalities, together with Lemma 4.5 from [9], allow us to conclude that $X_i^k(\cdot)$ satisfies the sign conditions in Equations (13) and (14).

Finally, let us remark that if the auxiliary system in Equation (41) is detectable, then the minimal solution $X_i^k(\cdot)$ coincides with the bounded and stabilizing solution of Equation (16) for any $1 \leq i \leq N$. Thus, we have shown inductively that $X_i^k(\cdot)$ can be constructed for any $k \geq 1$ and $1 \leq i \leq N$ which satisfies properties (i–iii) from the statement. Now we remark that Equation (51) allows us to conclude that the sequences $\{X_i^k(t)\}_{k \geq 1}$, $1 \leq i \leq N$, and $t \geq 0$ are convergent. Let $Y(t,i) = \lim_{k \to \infty} X_i^k(t)$, $(t,i) \in \mathbb{R}_+ \times \mathfrak{N}$. By taking the limit for $k \to \infty$ in Equation (16), we obtain that $\{Y(t)\}_{t \in \mathbb{Z}}$ is a positive semi-definite and p-periodic solution of Equation (2). Based on the minimality property of the stabilizing solution of the Riccati equation (Equation (2)), we deduce that $X_s(t) \leq Y(t)$, $t \in \mathbb{Z}$, and hence

$$Y(t) = X_s(t). \tag{57}$$

Thus, the proof is complete. □

4. Numerical Experiments

The time-invariant case will be considered in this section. We will refer to the algorithm proposed here as **Algo_Deter**. In this example, and in order to evaluate the performance of **Algo_Deter**, we will compare it with an algorithm that belongs to the class of *stochastic algorithms* (see Section 1 for a description of this class of algorithms). We propose using here a stochastic algorithm that we adapted from [10] to our setting. This algorithm is referred to as **Algo_Stoch**. We recall here that for solving the deterministic Riccati equations appearing in **Algo_Deter**, one can use direct methods (invariant or deflating subspace-based methods). We refer the reader interested in direct methods to [6,11,12] and the references therein. We also recall that at each main iteration of **Algo_Stoch**, one has to use iterative methods. We will show, from the computation time point of view, the superiority of **Algo_Deter** when compared with **Algo_Stoch**, which is due to the *direct or iterative* method opposition.

We will use the following simulation protocol:

1. Set the example numbers **n_good** = 0, **n_Deter** = 0, and **n_Stoch** = 0, where **n_good** represents the number of examples for which both **Algo_Deter** and **Algo_Stoch** converge, **n_Deter** is the number of examples for which **Algo_Deter** converges but not **Algo_Stoch**, and **n_Stoch** is the number of examples for which **Algo_Stoch** converges but not **Algo_Deter**;
2. Choose n, m_1, and m_2 randomly and uniformly among the integers from 1 to 10 and fix $N = 3$;
3. Generate randomly the corresponding system matrices;
4. If the assumptions in Theorem 2 are not verified, then go back to step 2;
5. Use **Algo_Deter** and **Algo_Stoch** to solve the corresponding generalized Riccati equation. Let the stabilizing solution obtained using **Algo_Deter** be $\mathbb{X}_1$ and the solution obtained using **Algo_Stoch** be $\mathbb{X}_2$, with CPU_time_1 and CPU_time_2 being the respective CPU running times;
    (a) If neither algorithms converge, then go back to step 2;
    (b) If **Algo_Deter** converges but not **Algo_Stoch**, then set **n_Deter** = **n_Deter** + 1 and go back to step 2;
    (c) If **Algo_Deter** does not converge but **Algo_Stoch** does, then set **n_Stoch** = **n_Stoch** + 1 and go back to step 2;
    (d) If both algorithms converge, then set **n_good** = **n_good** + 1 and compute the error $R_i = \frac{1}{N} \sum_{j=1}^{N} \|X_1(j) - X_2(j)\|$ and the coefficient $\rho_i = \frac{\text{CPU\_time\_2}}{\text{CPU\_time\_1}}$;
6. Repeat steps 2–6 until **n_good** = 100.

We generated random test samples with a specified level of accuracy $\epsilon = 10^{-8}$ for both algorithms.

The obtained results are listed in Table 1 and Figure 1. In Table 1, $O(R_i)$ is the order of magnitude of R_i, and "**Number of Examples**" indicates the number of examples corresponding to the same order of magnitude of R_i. It follows from the obtained results that when **Algo_Deter** and **Algo_Stoch** converged, the obtained stabilizing solutions were computed with comparable accuracies.

As expected, and thanks to the use of direct resolution methods instead of iterative ones, one can see clearly from Figure 1 the improvement brought about by **Algo_Deter** from the computation time point of view.

During this experiment, we also obtained the following results: **n_deter** = 36 and **n_Stoch** = 0. This shows that **Algo_Deter** still worked well in cases where **Algo_Stoch** failed. We believe that this was due partly to the fact that in **Algo_Stoch**, the computation of the sequence of approximations of the stabilizing solution relies on the computation of a vanishing matrix sequence $\{Z^{(k)}(t)\}_{k \geq 0}$, while in **Algo_Deter**, one directly computes the sequence of approximations $\{X^{(k)}(t)\}_{k \geq 0}$. The vanishing nature of the matrix sequence $\{Z^{(k)}(t)\}_{k \geq 0}$ could induce ill conditioning in its computation.

Table 1. Accuracy comparison for 100 random examples.

$O(R_i)$	Number of Examples
10^{-9}	66
ine 10^{-10}	34

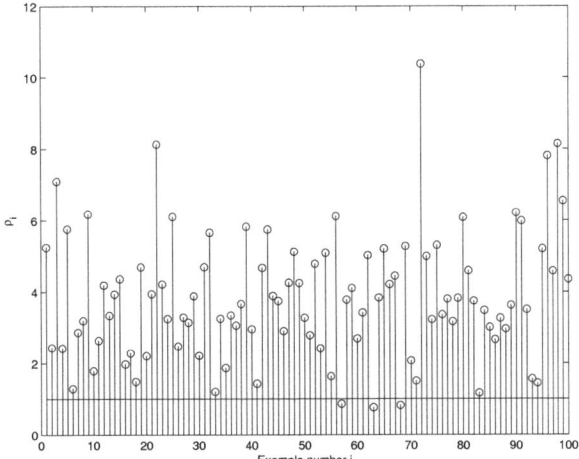

Figure 1. Plot of the quantity $\rho_i = \frac{\text{CPU\_time\_2}}{\text{CPU\_time\_1}}$.

5. Conclusions

In this paper, we addressed the problem of the numerical computation of the stabilizing solution for a class of generalized Riccati difference equations. We proposed an iterative *deterministic* algorithm for the computation of such a global solution. The performances of the proposed algorithm were illustrated via a comparison with existing algorithms in the literature. Our ongoing efforts are twofold. On one side, we are interested in the numerical computation of some global solutions to Riccati equations arising in stochastic Nash and Stackelberg games. The degree of maturity of numerical methods for such an aim is very weak when compared with its deterministic analogue. On the other side, we are also interested in generalized Riccati equations arising in mean field LQ games. Such equations present a coupling that makes this problem very challenging.

Author Contributions: Conceptualization, S.A. and V.D.; Methodology, S.A. and V.D.; Software, S.A.; Validation, S.A. and V.D.; Formal analysis, S.A. and V.D.; Investigation, S.A. and V.D. All authors have read and agreed to the published version of the manuscript.

Funding: This research received no external funding.

Data Availability Statement: Not acceptable.

Conflicts of Interest: The authors declare no conflict of interest.

Appendix A

Lemma A1. *Assume that the assumptions of Theorem 1 hold. If $\mathbb{X}(\cdot)$ is bounded on $\mathbb{Z}_+$, a positive semi-definite solution to Equation (2), then the system*

$$x(t+1) = [A_0(t,\theta_t) + B_{02}(t,\theta_t)V_{22}^{-1}(t,\theta_t)V_2(t,\theta_t)F(t,\theta_t)$$
$$+ \sum_{k=1}^{r} w_k(t)(A_k(t,\theta_t) + B_{k2}(t,\theta_t)V_{22}^{-1}(t,\theta_t)V_2(t,\theta_t)F(t,\theta_t))]x(t) \quad \text{(A1)}$$

is ESMS, where $F(t,i)$ is defined as in Lemma 4.7 from [1], $V_2(t,\theta_t) = \begin{bmatrix} V_{21}(t,\theta_t) & V_{22}(t,\theta_t) \end{bmatrix}$, and $V_{jk}(t,\theta_t) = V_{jk}(t)[\mathbb{X}(t+1)](\theta_t)$, as introduced in Remark 1.

Proof. Using similar arguments to those in [1], one can show that Equation (2) can be rewritten as

$$X(t,i) = \sum_{k=0}^{r}(A_k(t,i) + B_{2k}(t,i)\Gamma(t,i))^T \Xi(t)[\mathbb{X}(t+1)](i)\star$$
$$+ F_1^T(t,i)V_{11}^T(t,i)V_{11}(t,i)F_1(t,i) + \check{C}^T(t,i)\check{C}(t,i)$$
$$+ \left[L_2(t,i) + \Gamma^T(t,i)R_{22}(t,i)\right]R_{22}^{-1}(t,i)\star \tag{A2}$$

where $\Gamma(t,i) = F_2(t,i) + V_{22}^{-1}(t,i)V_{21}(t,i)F_1(t,i)$ and $(t,i) \in \mathbb{Z}_+ \times \mathfrak{N}$.

Let us associate with Equation (A2) the system

$$\begin{cases} x(t+1) = [A_0(t,\theta_t) + B_{02}(t,\theta_t)\Gamma(t,\theta_t) + \sum_{k=1}^{r} w_k(t)(A_k(t,\theta_t) + B_{k2}(t,\theta_t)\Gamma(t,\theta_t))]x(t) \\ y(t) = \begin{pmatrix} \check{C}(t,\theta_t) \\ V_{11}(t,\theta_t)F_1(t,\theta_t) \\ R_{22}^{-\frac{1}{2}}(t,\theta_t)(L_2^T(t,\theta_t) + R_{22}(t,\theta_t)\Gamma(t,\theta_t)) \end{pmatrix} x(t) \end{cases} \tag{A3}$$

Note that the first equation in Equation (A3) is simply Equation (A1). Hence, the conclusion may be obtained by applying Theorem 3.2 from [13] in the case of the system in Equation (A3). To this end, we have to show that the system in Equation (A3) is exactly detectable at the time instant $t_0 = 0$.

Let $x(t; 0, x_0)$ be a solution to the system in Equation (A3) with the property that the corresponding output $y(t; 0, x_0)$ satisfies

$$y(t; 0, x_0) = 0 \quad \text{a.s.} \quad \forall t \geq 0. \tag{A4}$$

This means that
$$\check{C}(t)x(t; 0, x_0) = 0 \tag{A5}$$

$$V_{11}(t,\theta_t)F_1(t,\theta_t)x(t; 0, x_0) = 0 \tag{A6}$$

and
$$R_{22}^{-\frac{1}{2}}(t,\theta_t)\left(L_2^T(t,\theta_t) + R_{22}(t,\theta_t)\Gamma(t,\theta_t)\right)x(t; 0, x_0) = 0, \quad \text{a.s.} \quad \forall t \geq 0. \tag{A7}$$

Since $R_{22}(t,\theta_t) > 0$ and $V_{11}(t,\theta_t) > 0$, Equations (A6) and (A7) yield

$$F_1(t,\theta_t)x(t; 0, x_0) = 0 \tag{A8}$$

and
$$F_2(t,\theta_t)x(t; 0, x_0) = -R_{22}^{-1}(t,\theta_t)L_2^T(t,\theta_t)x(t; 0, x_0) \quad \text{a.s.} \quad \forall t \geq 0. \tag{A9}$$

By substituting Equations (A8) and (A9) in the first equation from Equation (A3), written for $x(t)$ and replaced by $x(t; 0, x_0)$, we obtain that $x(\cdot; 0, x_0)$ is a solution to Equation (29). From Equation (A5), together with the exact detectability at the time instant $t_0 = 0$ of the system in Equation (29), we deduce that

$$\lim_{t \to \infty} \mathbb{E}\left[|x(t; 0, x_0)|^2\right] = 0. \tag{A10}$$

Finally, Equations (A4) and (A10) allow us to conclude that Equation (A3) is exactly detectable at the time instant $t_0 = 0$. Finally, by using the result from Theorem 3.2 in [13], the proof is completed. □

References

1. Aberkane, S.; Dragan, V. On the existence of the stabilizing solution of generalized Riccati equations arising in zero-sum stochastic difference games: The time-varying case. *J. Differ. Equ. Appl.* **2020**, *26*, 913–951. [CrossRef]
2. McAsey, M.; Mou, L. Generalized Riccati equations arising in stochastic games. *Linear Algebra Its Appl.* **2006**, *416*, 710–723. [CrossRef]
3. Yu, Z. An Optimal Feedback Control-Strategy Pair for Zero-sum Linear-Quadratic Stochastic Differential Game: The RIccati Equation Approach. *Siam J. Control. Optim.* **2015**, *53*, 2141–2167. [CrossRef]
4. Feng, Y.; Anderson, B. An iterative algorithm to solve state-perturbed stochastic algebraic Riccati equations in LQ zero-sum games. *Syst. Control. Lett.* **2010**, *59*, 50–56. [CrossRef]
5. Ivanov, I.G. Iterations for solving a rational Riccati equation arising in stochastic control. *Comput. Math. Appl.* **2007**, *53*, 977–988. [CrossRef]
6. Bini, D.A.; Iannazzo, B.; Meini, B. *Numerical Solution of Algebraic Riccati Equations*; Society for Industrial and Applied Mathematics: Philadelphia, PA, USA, 2012.
7. Dragan, V.; Aberkane, S. Computing The Stabilizing Solution of a Large Class of Stochastic Game Theoretic Riccati Differential Equations: A Deterministic Approximation. *SIAM J. Control. Optim.* **2017**, *55*, 650–670. [CrossRef]
8. Dragan, V.; Morozan, T.; Stoica, A.M. *Mathematical Methods in Robust Control of Discrete-Time Linear Stochastic Systems*; Springer: New York, NY, USA, 2010.
9. Freiling, G.; Hochhaus, A. Properties of the solutions of rational matrix difference equations. *Adv. Differ. Equ. IV Comput. Math. Appl.* **2003**, *45*, 1137–1154. [CrossRef]
10. Dragan, V.; Aberkane, S.; Ivanov, I. On computing the stabilizing solution of a class of discrete-time periodic Riccati equations. *Int. J. Robust Nonlinear Control* **2015**, *25*, 1066–1093. [CrossRef]
11. Mehrmann, V. *The Autonomous Linear Quadratic Control Problem. Theory and Numerical Solution*; Series Lecture Notes in Control andInformation Sciences; Springer: Berlin, Germany, 1991; Volume 163.
12. Sima, V. *Algorithms for Linear-Quadratic Optimization*; Series Pure and Applied Mathematics: A Series of Monographs and Textbooks; Marcel Dekker, Inc.: New York, NY, USA, 1996; Volume 200.
13. Dragan, V.; Costa, E.F.; Popa, I.L.; Aberkane, S. Exact detectability: Application to generalized Lyapunov and Riccati equations. *Syst. Control. Lett.* **2021**, *157*, 105032. [CrossRef]

Disclaimer/Publisher's Note: The statements, opinions and data contained in all publications are solely those of the individual author(s) and contributor(s) and not of MDPI and/or the editor(s). MDPI and/or the editor(s) disclaim responsibility for any injury to people or property resulting from any ideas, methods, instructions or products referred to in the content.

Article

State Feedback with Integral Control Circuit Design of DC-DC Buck-Boost Converter

Humam Al-Baidhani [1,2], Abdullah Sahib [3] and Marian K. Kazimierczuk [1,*]

[1] Department of Electrical Engineering, Wright State University, Dayton, OH 45435, USA
[2] Department of Computer Techniques Engineering, Faculty of Information Technology, Imam Ja'afar Al-Sadiq University, Baghdad 10011, Iraq
[3] Department of Electronic and Communication Technologies, Technical Institute, Al-Furat Al-Awsat Technical University, Najaf 54003, Iraq; abdward780@atu.edu.iq
* Correspondence: marian.kazimierczuk@wright.edu

Abstract: The pulse-with modulated (PWM) dc-dc buck-boost converter is a non-minimum phase system, which requires a proper control scheme to improve the transient response and provide constant output voltage during line and load variations. The pole placement technique has been proposed in the literature to control this type of power converter and achieve the desired response. However, the systematic design procedure of such control law using a low-cost electronic circuit has not been discussed. In this paper, the pole placement via state-feedback with an integral control scheme of inverting the PWM dc-dc buck-boost converter is introduced. The control law is developed based on the linearized power converter model in continuous conduction mode. A detailed design procedure is given to represent the control equation using a simple electronic circuit that is suitable for low-cost commercial applications. The mathematical model of the closed-loop power converter circuit is built and simulated using SIMULINK and Simscape Electrical in MATLAB. The closed-loop dc-dc buck-boost converter is tested under various operating conditions. It is confirmed that the proposed control scheme improves the power converter dynamics, tracks the reference signal, and maintains regulated output voltage during abrupt changes in input voltage and load current. The simulation results show that the line variation of 5 V and load variation of 2 A around the nominal operating point are rejected with a maximum percentage overshoot of 3.5% and a settling time of 5.5 ms.

Keywords: analog control circuit; dc-dc converter; pole placement; pulse-width modulated; state feedback with integral control

MSC: 37M05

1. Introduction

The PWM dc-dc converters are utilized in modern aircraft power systems and portable communication devices due to their high efficiency, small size, and low cost. Portable electronic devices such as cell phones and laptops require a well-regulated dc supply voltage to operate properly. However, the dc-dc converters encounter line and load variations during their normal operation, which fluctuate the load voltage. Therefore, a controller is required to provide a constant voltage and improve the transient response of the power converter. Modern control techniques have been applied to control the power converter dynamics due to their robustness against large disturbances. In [1], neural inverse optimal control (NIOC) for a regenerative braking system in an electric vehicle (EV). A neural identifier has been trained with an extended Kalman filter (EKF) to estimate the dc-dc buck-boost power converter dynamics. An artificial neural network-based controller has also been developed for a bidirectional power flow management system that comprises a dual-source low-voltage buck-boost converter [2]. However, the practical implementation of the control schemes in [1,2] is complicated.

Other research efforts have proposed model predictive control (MPC) and adaptive control techniques as alternatives for artificial neural network-based controllers. For instance, the MPC of the buck-boost converter has been introduced in [3], in which a switching algorithm is proposed to minimize the error for the power converter. In [4], a centralized model predictive control has been developed to stabilize the DC microgrid with versatile buck-boost converters. A direct model reference adaptive control [5] and an optimal adaptive control [6] have also been presented for boost and voltage source converters, respectively. A nonlinear control based on the Lyapunov function has been developed in [7] for power management systems, whereas an inverse-system decoupling control method has been presented in [8] for a dc-dc buck-boost converter. Despite the robust control performance, the previous control strategies require tedious mathematical computations and high-cost for practical implementation.

Feedback linearization methods have been discussed in [9–13] for dc-dc power converters. A feedback control law based on full feedback linearization has been introduced for the buck, boost, and buck-boost converters [9]. Feedback linearization has been presented to control the buck-boost power converter [10,11], boost converter [12], and modular multilevel converter-bidirectional dc-dc power converter [13]. However, the aforementioned research efforts fall short of introducing systematic design procedures for the practical implementation of feedback linearization control law. Other research endeavors have proposed full-state feedback control via a pole placement technique [14–18]. In contrast to the classical voltage-mode controllers, all the state variables of the power converter are fed back through constant gains. Such a feature allows the state feedback control law to place the closed-loop poles arbitrarily in the left-half-plane (LHP). Thus, the closed-loop system response can be shaped such that the desired specifications are achieved.

The state feedback control based on the normalized linear state-space average model has been presented in [14] to regulate the output voltage of the dc-dc converters. In [15], the state feedback control is applied to the dc-dc converters and compared with different methods, such as fuzzy logic and neural network controllers. Furthermore, moving unstable poles to the LHP based on a digital state feedback control has been presented in [16]. Such control methods have been presented to regulate the system state variables and achieve the desired transient response. However, the steady-state error elimination has not been discussed. Other methods, such as a power smoothing control using sliding-mode control, a pole placement criterion [17], and a minimum degree pole placement-based digital adaptive control [18], have been proposed for power converters. State feedback with integral control of a PWM push-pull dc-dc power converter has also been reported in [19].

Recently, a pole placement and sensitivity function shaping technique has been applied to the dc-dc buck converter [20]. The control system has been validated using MATLAB/SIMULINK. Experimental validation has been performed on a dc-dc buck converter with a constant power load (CPL) using a hardware-in-the-loop (HIL) system, where dSPACE DS1104 has been utilized to implement the control law. In [21], a state feedback control via pole placement is designed on the basis of a nonlinear model of a fuel cell interleaved buck-boost converter. The aforementioned control systems yield robust control performance, mitigate the non-minimum phase issue, and improve the transient response of the power converter. However, design complexity and high-cost implementation have been noticed. In addition, the systematic design procedure and realization of such a control scheme using a simple analog circuit have not been reported. The comparison among previous control methods is presented in Table 1.

Table 1. Modern control techniques of dc-dc power converters.

Control Technique	Advantages		Disadvantages		References
Neural inverse optimal control (NIOC)	i.	Robustness against large disturbances.	i.	Complexity of practical control system design.	[1]
Artificial neural network-based control	ii.	Estimating converter dynamics.	ii.	High-cost control system implementation.	[2]
Model predictive control (MPC)	i. ii.	Fast dynamical response. Accurate tracking performance.		Practical implementation has not been discussed.	[3]
Centralized MPC	i. ii.	Fast dynamical response. Less computational efforts than traditional MPC.		High-cost control system implementation.	[4]
Direct model reference adaptive control		Robustness against voltage and frequency variations.		Complexity of control system implementation.	[5]
Optimal adaptive control		Estimation of uncertainties and disturbances		High-cost control system implementation (dSPACE).	[6]
Lyapunov-based nonlinear control		Robustness against load variations.		Practical implementation has not been covered.	[7]
Inverse-system decoupling control		Disturbance rejection capability.		Design procedure of control circuit has not been provided.	[8]
Feedback linearization control		Mitigation of CPL and zero dynamics.		Design procedure of control circuit has not been provided.	[9–13]
State-feedback control via pole placement		Placement of closed-loop poles at desired locations.		Steady-state error issue. Design procedure of control circuit has not been provided.	[14–18]
State-feedback with integral control		State variables regulation and steady-state error elimination.		Design procedure of control circuit has not been introduced.	[19]
pole placement control with sensitivity function		Mitigation of CPL and non-minimum phase issue.		High-cost control system implementation (dSPACE).	[20]

Motivated by the control design approach in [22,23], the pole placement via state-feedback with integral control of an inverting PWM dc-dc buck-boost converter in continuous conduction mode (CCM) is introduced. The contributions of this research work are listed below:

- The state-feedback with integral control law is designed based on an ideal small-signal model and tested with a nonlinear power converter model that includes all parasitic components;
- The realization of the proposed control circuit has been introduced using op-amps, resistors, and a capacitor;
- The closed-loop SIMULINK model and the corresponding closed-loop Simscape power converter circuit have been simulated in MATLAB to validate the design approach;
- The transient characteristics, tracking performance, and disturbance rejection capability of the proposed control circuit have been investigated.

The control scheme is designed to track the desired trajectory and improve the transient response of the power converter. The control system parameters are selected to place the closed-loop poles at the desired location and guarantee the system's stability.

The rest of the paper is organized as follows. Section 2 introduces the mathematical model of the power converter in CCM. Section 3 discusses the state feedback with integral control design. Section 4 presents the realization of the analog control circuit. In Section 5, the control design procedure flowchart is introduced. The results and discussion are given in Section 6, and Section 7 covers the conclusions.

2. Mathematical Model of Inverting DC-DC Buck-Boost Converter

2.1. Nonlinear Model

The topology of inverting the dc-dc buck-boost converter is depicted in Figure 1a. The power converter is highly nonlinear because of the switching network presented by the MOSFET S and the diode D_1. The inductor L and the capacitor C represent the energy storage components in the circuit. The switching elements S and D_1 operate alternatively in CCM, which give two possible structures for the dc-dc converter [17]. The non-ideal equivalent circuit of the power converter is given in Figure 1b. As shown in Figure 1b, the equivalent series resistances (ESRs) of L and C are r_L and r_C, respectively. Moreover, r_F, V_F, and r_{DS} represent the parasitic components of the diode D_1 and switch S, respectively.

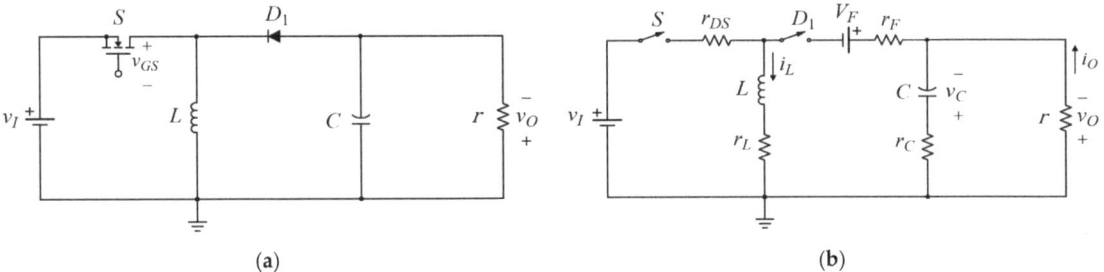

Figure 1. (**a**) The inverting dc-dc buck-boost converter circuit. (**b**) The equivalent circuit of the non-ideal buck-boost converter in CCM.

Based on the averaging theory, the large-signal averaged model of the dc-dc buck-boost converter is derived in [24] using Kirchhoff's voltage and current laws. The nonlinear dynamics and output voltage v_O are expressed as

$$\begin{cases} \frac{di_L}{dt} = \frac{1}{L}[(v_I - r_{DS}i_L)d_T + (v_O - V_F - r_F i_L)\bar{d}_T - r_L i_L] \\ \frac{dv_C}{dt} = -\frac{1}{C}[i_L \bar{d}_T + i_O], \end{cases} \quad (1)$$

and

$$v_O = v_C - r_C(i_L \bar{d}_T + i_O). \quad (2)$$

In (1) and (2), the input voltage v_I, load resistor r, load current i_O, inductor current i_L, output voltage v_O, and capacitor voltage v_C are represented as large-signal quantities. In addition, d_I is the large-signal quantity of the time interval at which S is ON, whereas $\bar{d}_T$ is the large-signal quantity of the time interval at which S is OFF. The duty cycle d_T is defined such that $d_T \in [0, 1]$. In fact, d_T represents the control signal that regulates v_O during the line and load disturbances.

The non-ideal large-signal averaged model in (1) and (2) emulates the dynamics of the actual power converter. Hence, it can be used to investigate the tracking and regulation performance of the proposed state feedback controller in MATLAB/SIMULINK.

2.2. Linearized State-Space Averaged Model

The small-signal ac model of the dc-dc converter must be derived to design the state feedback with the integral controller. Therefore, the nonlinear model should be linearized around the equilibrium point. To simplify the control design process, the parasitic components in (1) and (2) are neglected. Thus, an ideal large-signal state-space averaged model is obtained

$$\begin{bmatrix} \frac{di_L}{dt} \\ \frac{dv_C}{dt} \end{bmatrix} = \begin{bmatrix} 0 & \frac{\bar{d}_T}{L} \\ \frac{-\bar{d}_T}{C} & \frac{-1}{rC} \end{bmatrix} \begin{bmatrix} i_L \\ v_C \end{bmatrix} + \begin{bmatrix} \frac{v_I}{L} \\ 0 \end{bmatrix} d_T, \tag{3}$$

where $v_C = v_O$.

The steady-state values of the inductor current I_L and output voltage V_C of the inverting dc-dc buck-boost converter can be written as

$$\begin{cases} I_L = \frac{V_C}{RD_T} \\ V_C = \frac{-D_T V_I}{\bar{D}_T}, \end{cases} \tag{4}$$

where D_T, V_I, and R are the steady-state values of the duty cycle, input voltage, and load resistance, respectively. Next, the linearized small-signal averaged model can be derived by linearizing (3) around the equilibrium point given in (4), which gives

$$\begin{bmatrix} \frac{d\tilde{i}_L}{dt} \\ \frac{d\tilde{v}_C}{dt} \end{bmatrix} = \begin{bmatrix} 0 & \frac{\bar{D}_T}{L} \\ \frac{-\bar{D}_T}{C} & \frac{-1}{RC} \end{bmatrix} \begin{bmatrix} \tilde{i}_L \\ \tilde{v}_C \end{bmatrix} + \begin{bmatrix} \frac{V_I - V_C}{L} \\ \frac{I_L}{C} \end{bmatrix} \tilde{d}, \tag{5}$$

and

$$\tilde{v}_O = \begin{bmatrix} 0 & 1 \end{bmatrix} \begin{bmatrix} \tilde{i}_L \\ \tilde{v}_C \end{bmatrix} + [0]\tilde{d}. \tag{6}$$

The small-signal ac quantities of the inductor current, capacitor voltage, and duty cycle are $\tilde{i}_L$, $\tilde{v}_C$, and $\tilde{d}$, respectively. The small-signal model can also be represented in compact form as

$$\begin{cases} \dot{x} = Ax + Bu \\ y = Cx + Du \end{cases} \tag{7}$$

The state variables vector x contains $\tilde{i}_L$ and $\tilde{v}_C$, while the input u and output y represent $\tilde{d}$ and $\tilde{v}_O$, respectively. The matrices A, B, C, and D are defined in (5) and (6). The parameters of the dc-dc buck-boost converter are given in Table 2.

Table 2. Parameters of dc-dc buck-boost converter [24].

Description	Parameter	Value
Inductor	L	30 μH
Output capacitor	C	2.2 mF
Load resistance	R	(1.2–12) Ω
Inductor ESR	r_L	0.050 Ω
Output capacitor ESR	r_C	0.006 Ω
MOSFET on-resistance	r_{DS}	0.110 Ω
Diode forward resistance	r_F	0.020 Ω
Diode threshold voltage	V_F	0.700 V
Input voltage	V_I	28 ± 4 V
Output voltage	V_O	12 V
Switching frequency	f_s	100 kHz

3. State-Feedback with Integral Control Design

3.1. Control Law Design

The block diagram of the state feedback with integral control system is shown in Figure 2. The control objective is to find the controller gains that place the closed-loop poles arbitrarily at the desired location on the s-plane and obtain the desired system response. If the state variables are available for measurements, the pole placement can be achieved if the system is controllable [25], which means that the controllability matrix

$$Co = \begin{bmatrix} B & AB & A^2B & \ldots & A^{n-1}B \end{bmatrix} \qquad (8)$$

has a full rank.

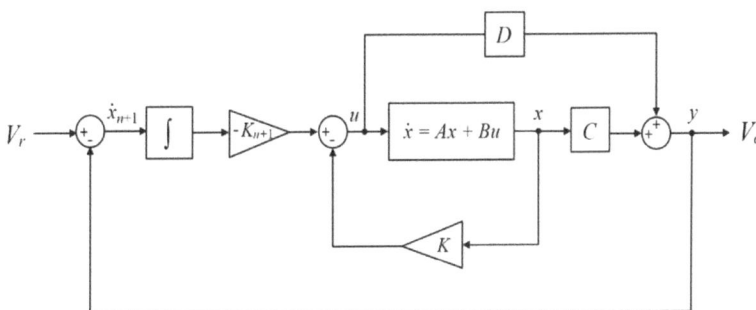

Figure 2. The block diagram of the state feedback with integral control system.

Furthermore, the power converter output voltage should track the desired reference voltage V_r. Hence, an integral part is added to the control scheme, which adds a new state x_{n+1} to the system with an integral gain K_{n+1}. From Figure 2, we have

$$\dot{x}_{n+1} = V_r - (Cx + Du). \qquad (9)$$

A control law u can be selected as

$$u = -Kx - K_{n+1}x_{n+1}, \qquad (10)$$

where K is a $1 \times n$ vector of constant gains. Substituting (10) back into (9) gives

$$\dot{x}_{n+1} = V_r - (C - DK)x + DK_{n+1}x_{n+1} \qquad (11)$$

On the other hand, if (10) is substituted into the open-loop state Equation (7), one obtains

$$\dot{x} = Ax - B(Kx + K_{n+1}x_{n+1}) \qquad (12)$$

Rearranging (12) results in

$$\dot{x} = (A - BK)x - BK_{n+1}x_{n+1} \qquad (13)$$

Now, based on (11) and (13), the augmented state-space model of the power converter can be written as

$$\begin{bmatrix} \dot{x} \\ \dot{x}_{n+1} \end{bmatrix} = \begin{bmatrix} A - BK & -BK_{n+1} \\ -C + DK & DK_{n+1} \end{bmatrix} \begin{bmatrix} x \\ x_{n+1} \end{bmatrix} + \begin{bmatrix} \Theta \\ 1 \end{bmatrix} V_r \qquad (14)$$

where Θ is an $n \times 1$ vector of zeros. Hence, the closed-loop dc-dc buck-boost converter dynamics become

$$\begin{cases} \dot{\bar{x}} = \left(\bar{A} - \bar{B}\bar{K}\right)\bar{x} + \begin{bmatrix} \Theta \\ 1 \end{bmatrix} V_r. \\ \bar{y} = \bar{C}\bar{x} \end{cases} \qquad (15)$$

The matrices $\bar{A}$, $\bar{B}$, $\bar{C}$, and $\bar{K}$ are given by

$$\bar{A} = \begin{bmatrix} A & \Theta \\ -C & 0 \end{bmatrix} \tag{16}$$

$$\bar{B} = \begin{bmatrix} B \\ -D \end{bmatrix} \tag{17}$$

$$\bar{C} = \begin{bmatrix} C - DK & -DK_{n+1} \end{bmatrix} \tag{18}$$

$$\bar{K} = \begin{bmatrix} K & K_{n+1} \end{bmatrix}. \tag{19}$$

It should be noted that the pair $[\bar{A}, \bar{B}]$ must be completely controllable in order to place the eigenvalues of the matrix $(\bar{A} - \bar{B}\bar{K})$ arbitrarily [25]. Thus, the controller gains vector $\bar{K}$ and can place the closed-loop poles of the system dynamics in (15) at the desired location on the s-plane.

The vector $\bar{K}$ can be computed manually via comparing the characteristic polynomial of the matrix $(\bar{A} - \bar{B}\bar{K})$ with the desired characteristic polynomial CP

$$CP = s^{n+1} + \alpha_n s^n + \ldots + \alpha_1 s + \alpha_0. \tag{20}$$

The parameters $\alpha_0, \alpha_1, \ldots \alpha_n$ are real constants, which are determined based on the desired closed-loop poles as illustrated in the following subsection.

3.2. Controller Gains Selection

In this research, the control objective is to obtain a transient response with a percentage overshoot $PO \leq 5\%$ and settling time $t_s \leq 5$ ms. The desired specifications are selected based on the buck-boost simulation results reported in [24]. It is also required to track a time-varying reference voltage V_r, regulate the output voltage, and reject the line and load variations.

To simplify the design process, the linearized ideal small-signal model in (5) is considered. The dominant closed-loop poles can be obtained using the characteristic equation of the second order system

$$s^2 + 2\zeta\omega_n s + \omega_n^2 = 0. \tag{21}$$

In [25], the relationship between the settling time, damping ratio, and natural frequency is defined by

$$t_s \cong \frac{4.6}{\zeta\omega_n}. \tag{22}$$

Based on (22), if the desired settling time t_s and damping ratio ζ are set to 1.5 ms and 0.688, respectively, the natural frequency ω_n is 4489.5 rad/s. It should be noted that the choice of t_s and ζ is not unique. The designer can choose different values for t_s and ζ that give excellent results. However, the values of the controller gains must be maintained to avoid any issues with the practical implementation of the electronic control circuit.

Using (21), ζ, and ω_n, the dominant closed-loop poles are $s_{1,2} = -3089 \pm j3258$. However, since the closed-loop control system in (15) comprises three state variables (inductor current, capacitor voltage, and output voltage error), a third pole should be placed far to the left at $s_3 = -12000$ on the s-plane, so that the desired transient response is not affected. Thus, the desired closed-loop poles of the state feedback with integral control system yield

$$P = \begin{bmatrix} -3089 + j3258 & -3089 - j3258 & -12000 \end{bmatrix}. \tag{23}$$

Next, (16) and (17) can be used to evaluate the matrices $\bar{A}$ and $\bar{B}$ based on the parameters of the buck-boost converter given in Table 2. In MATLAB, it can be verified that the pair $[\bar{A}, \bar{B}]$ has a full rank and the system is controllable. Thus, the feedback gain vector $\bar{K}$ can easily be computed using (acker) command in MATLAB, which gives

$$\bar{K} = \begin{bmatrix} 0.011 & -0.170 & 600 \end{bmatrix}. \tag{24}$$

The unit step response of the compensated small-signal linearized model of the inverting dc-dc buck-boost converter in CCM is shown in Figure 3. It can be seen that the output voltage v_O tracks the desired trajectory, while the percentage peak overshoot PO and settling time t_s are about 4.7% and 1.7 ms, respectively.

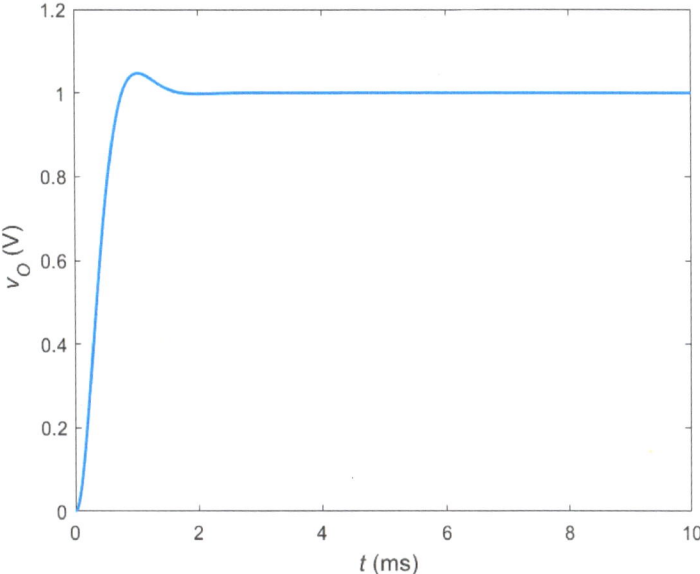

Figure 3. The unit step response of the compensated dc-dc buck-boost converter.

It is worth noting that the gains of the state feedback with integral control law in (24) are designed based on the linearized ideal dc-dc buck-boost converter model. Hence, when the simulation is conducted with a nonlinear power converter model with all the parasitic components included, the transient response characteristics will be different from the response shown in Figure 3. It will exhibit a longer settling time and larger PO. This is true because the linearized model does not include all the information on the actual dc-dc power converter dynamics. However, the state feedback controller gains can be tuned to compensate for the parasitic components effects and obtain the desired transient response characteristics.

3.3. Structure of Proposed Control System

The MATLAB/SIMULINK model of the state feedback with integral control of the PWM dc-dc buck-boost converter in CCM is shown in Figure 4.

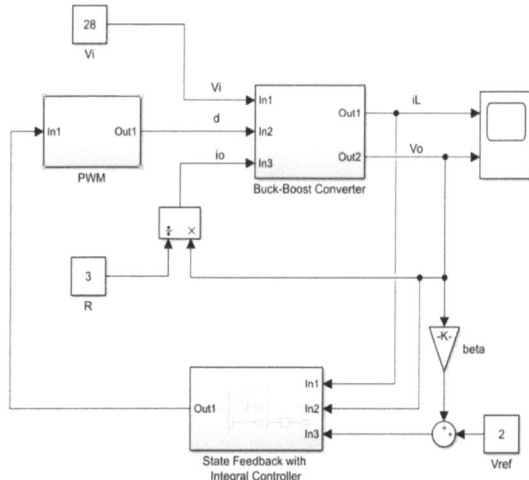

Figure 4. MATLAB/SIMULINK model of state feedback with integral control system of inverting dc-dc buck-boost converter.

The closed-loop control system of the dc-dc buck-boost converter is made up of the following parts:

- Pulse-Width Modulator: The PWM subsystem contains a comparator that compares the state feedback with integral control law with the ramp voltage V_T to generate the duty cycle d_T that drives the nonlinear power converter model;
- Power Converter: The large-signal non-ideal dc-dc buck-boost converter model is built in MATLAB/SIMULINK using s-function based on the state-space equations given in (1) and (2). The nonlinear model emulates the dc-dc buck-boost converter dynamics;
- State Feedback with Integral Controller: The controller subsystem comprises the state feedback with integral control law given in (10) along with the state feedback controller gains defined in (24).

4. Realization of Analog Control Circuit

The control scheme given in Figure 4 should be converted to an analog control circuit that can easily be built using electronic components. The schematic of the proposed control circuit is given in Figure 5. The control circuit is made up of op-amps, resistors, and a capacitor. Despite the nonidealities and tolerances of the electronic elements, the overall control circuit must reflect the mathematical expression of the given control law, which is designed via the pole placement technique.

The design steps of the state feedback with an integral control circuit are summarized as follows:

- Voltage sensor gain β: The buck-boost converter is designed to convert 28 V to 12 V. If the reference voltage $V_r = 2$ V, then the feedback network gain β is $\frac{V_r}{V_o} = \frac{2}{12} = \frac{1}{6}$;
- Summing, inverting, and differential op-amps: The gain of the summing, inverting, and differential op-maps in the control circuit is unity. Thus, the resistors of the summing op-amps R_{S1}, R_{S2}, and R_{S3}, inverting op-amp R_{I1} and R_{I2}, and differential op-amp R_{F1} and R_{F2} are set to 5.1 kΩ;
- Pulse-Width Modulator: The peak ramp voltage V_T is set to 2 V, whereas the switching frequency f_s is 100 kHz.
- Inductor current gain K_1: In the control design section, the gain of the inductor current K_1 has been computed as 0.011. Since the gain $K_1 = \frac{R_{L2}}{R_{L1}}$, the resistor R_{L1} and R_{L2} can be set to 100 kΩ and 1.1 kΩ, respectively;

- Output voltage gain K_2: In the control design section, the gain of the output voltage K_2 has been computed as 0.17. Since the gain $K_2 = \frac{R_{V2}}{R_{V1}}$, the resistor R_{V2} and R_{V1} can be set to 100 kΩ and 17 kΩ, respectively;
- Integral gain K_3: As reported in [26], the integral gain is defined as $K_3 = \frac{1}{R_1 C_1}$. In the control design section, the gain K_3 has been computed as 600. If the resistor R_1 is assumed to be 33 kΩ, then the capacitor C_1 is 56 nF;

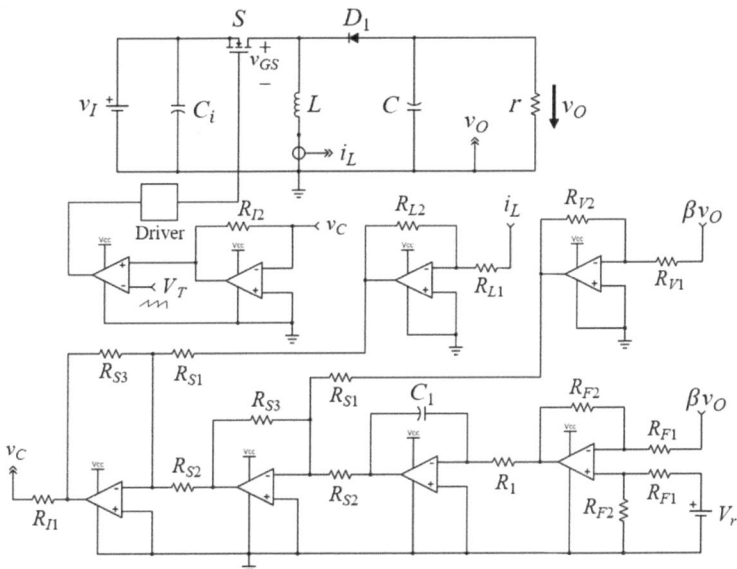

Figure 5. Schematic of state-feedback with integral controlled PWM dc-dc buck-boost converter circuit.

It should be noted that accurate output voltage and inductor current sensors are required to measure the control state variables. The inductor current measurement is important in the state feedback with an integral control system to improve the transient response characteristics and handle the non-minimum phase power converter [27]. General-purpose op-amps such as LF357 can be utilized to build the state feedback with the integral control circuit.

Additionally, the design procedure of the control circuit given above does not include the selection of the pulse-width modulator and the high-side gate driver of the MOSFET. The mitigations for over-voltage protection, over-current protection, EMC/EMI compatibility, and other practical engineering aspects should also be considered to develop an experimental prototype for testing and evaluation.

5. Flowchart of State-Feedback with Integral Control Design

The step-by-step design procedure of the state-feedback with integral control of the dc-dc buck-boost converter is summarized in a flowchart as shown in Figure 6.

First, the linearized small-signal averaged model of the power converter is derived in state-space form as defined in (5) and (6). The next step is to construct the closed-loop power converter dynamics as shown in (15), from which the matrices $\bar{A}$, $\bar{B}$, and $\bar{C}$ are obtained. Subsequently, the rank of the controllability matrix is computed to confirm that the pair $[\bar{A}, \bar{B}]$ is controllable.

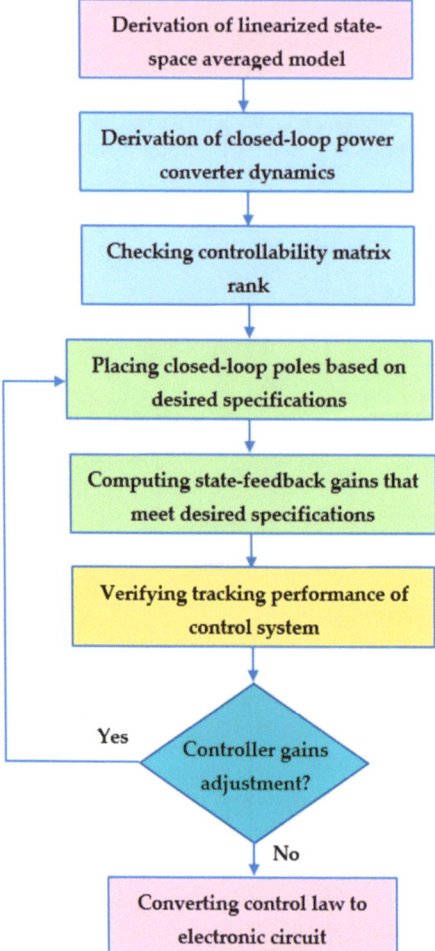

Figure 6. Flowchart of state-feedback with integral control design of dc-dc buck-boost converter.

The dominant closed-loop poles are obtained using the characteristic equation of the second-order system given in (21). Next, based on (22), the desired percentage overshoot and settling time yield the required damping ratio and natural frequency, which give the desired dominant poles. Since the augmented model contains three state variables, the third pole should be placed far to the left on the s-plane in order to maintain the desired transient response. Then, the desired closed-loop poles are lumped together as shown in (23), and the state-feedback control gains given in (24) are computed using the acker command in MATLAB.

Finally, the SIMULINK model of the state-feedback with an integral-controlled PWM dc-dc buck-boost converter is simulated to verify the tracking performance of the control system. If the desired response is achieved, the control equation is converted to an electronic circuit as explained in Section 4. However, if the system response requires further enhancement, the closed-loop poles' location can be adjusted and the controller gains are re-calculated for verification.

6. Results and Discussion

6.1. Validation of Control Design Approach

The schematic of state-feedback with an integral control circuit in Figure 5 has been constructed using Simscape Electrical in MATLAB. In order to validate the control design methodology, the electronic control circuit has been compared with the MATLAB/SIMULINK nonlinear model of the closed-loop control system given in Figure 4. The power converter parameters are defined in Table 2. The proposed state feedback controller gains are given in (24), whereas the corresponding electronic control circuit elements are defined in Section 4.

The MATLAB/SIMULINK model and the closed-loop power converter circuit in Simscape Electrical are simulated and compared under nominal operating conditions (load resistance $R = 3\ \Omega$ and input voltage $V_I = 28$ V). The simulation of the two closed-loop control schemes is conducted in MATLAB using (Automatic) solver and 0.1 µs step-size. The waveforms of the ramp voltage V_T, control voltage u, gate-to-source voltage v_{GS}, the inductor current i_L, and output voltage v_O during steady-state are shown in Figure 7. The simulation results of the mathematical closed-loop power converter model in SIMULINK and the corresponding closed-loop power converter circuit in Simscape Electrical are depicted in Figures 7a and 7b, respectively.

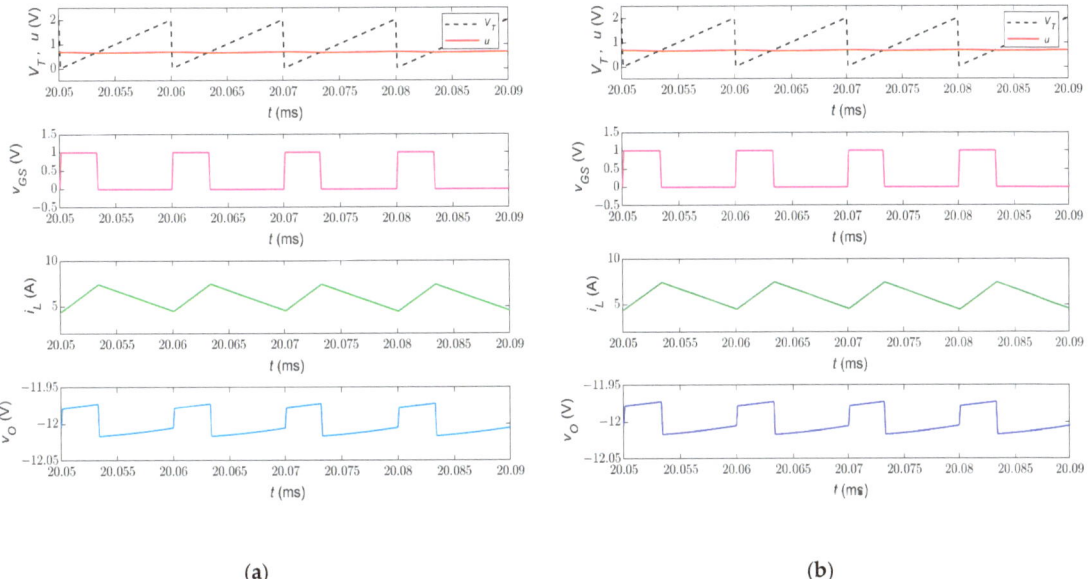

Figure 7. Steady-state waveforms of (**a**) MATLAB/SIMULINK model and (**b**) Simscape Electrical circuit of the state feedback with integral control of PWM dc-dc buck-boost converter in CCM. The figures show the control input u, ramp voltage V_T, gate-to-source voltage v_{GS}, inductor current i_L, and output voltage v_O.

It can be seen that the waveforms obtained from the mathematical model in Figure 7a and those obtained from the corresponding electronic circuit in Figure 7b are identical. That means the mathematical model of the power converter mathematical model emulates the power converter circuit dynamics successfully. Additionally, the state feedback with integral control law has been represented by the analog control circuit properly, which validates the control circuit design approach.

Notably, the dc output voltage is regulated at −12 V with a duty cycle of 0.336, whereas the switching frequency of the ramp voltage waveform V_T is 100 kHz. The negative output voltage is due to the topology of the inverting dc-dc buck-boost converter. It can also be seen that the power converter operates in CCM because the inductor current waveform is maintained above zero. The average value of the inductor current is around 5.99 A.

6.2. Rejection of Line and Load Variations

The performance of the state feedback with an integral control system has been investigated considering step change in input voltage v_I and load current i_O. The output voltage response during line variation is shown in Figure 8. In Figure 8a, as v_I changes from 28 V to 33 V, the percentage overshoot PO and settling time t_s are about 2.6% and 5.50 ms, respectively. Moreover, when the input voltage v_I changes from 28 V to 23 V as shown in Figure 8b, the maximum PO and t_s are around 3.5% and 5.5 ms, respectively. In both cases, it can be noticed that v_O is regulated at the desired value while maintaining consistent dynamics during the line variations.

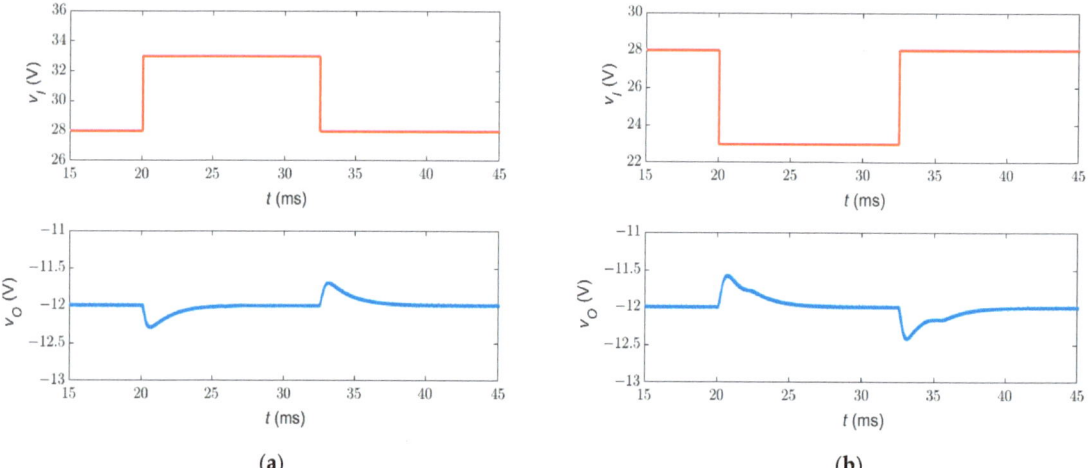

Figure 8. The tracking performance of the state feedback with integral control of inverting dc-dc buck-boost converter under line disturbance. (**a**) The output voltage response v_O when the input voltage v_I changes from 28 V to 33 V during the time interval $20 \leq t \leq 32.5$ ms. (**b**) The output voltage response v_O when the input voltage v_I changes from 28 V to 23 V during the time interval $20 \leq t \leq 32.5$ ms.

On the other hand, the output voltage responses to a step change in load current i_O are depicted in Figure 9. As shown in Figure 9a, when the load current i_O increases from 4 A to 6 A, the output voltage v_O exhibits a maximum percentage overshoot PO of 2% with settling time t_s of 4 ms. However, when the load current i_O decreases from 4 A to 2.5 A, Figure 9b shows that the output voltage v_O has a maximum percentage undershoot PO of 1% and reaches the steady-state value after 3.5 ms.

The simulation results show the disturbance rejection capability of the proposed control system. Although the control design is conducted based on the linearized ideal state-space model, the control circuit can still handle the nonlinear dynamics of the dc-dc buck-boost converter. In addition, the percentage overshoot and settling time of the output voltage response remain within the desired limits (maximum percentage overshoot $PO \leq 5\%$ and settling time $t_s \leq 5$ ms).

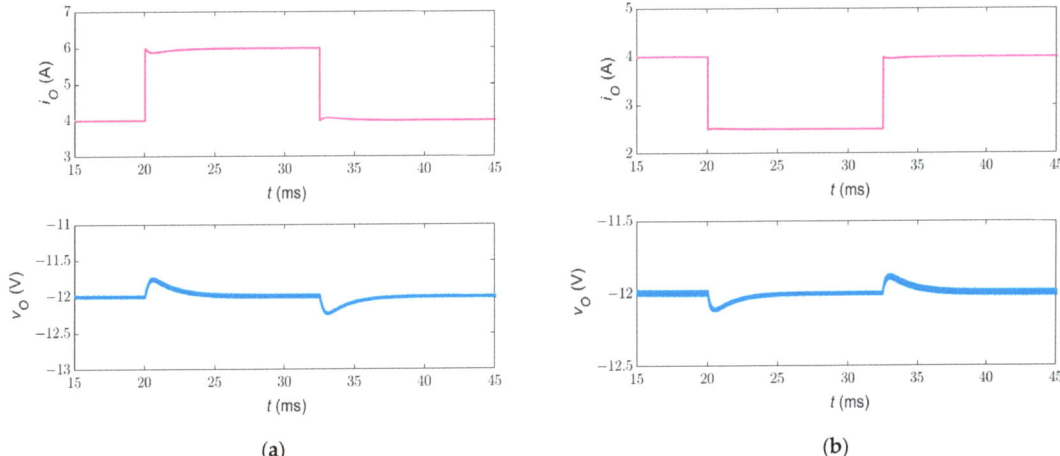

(a)   (b)

Figure 9. The tracking performance of the state feedback with integral control of inverting dc-dc buck-boost converter under load disturbance. (**a**) The output voltage response v_O when the load current i_O changes from 4 A to 6 A during the time interval $20 \leq t \leq 32.5$ ms. (**b**) The output voltage response v_O when the load current i_O changes from 4 A to 2.5 A during the time interval $20 \leq t \leq 32.5$ ms.

6.3. Tracking of Time-Varying Reference Voltage

The output voltage response v_O during step changes in the reference voltage V_r is shown in Figure 10. The power converter operates at nominal operating conditions (load resistance $R = 3\ \Omega$ and input voltage $V_I = 28$ V). It can be noticed that when the reference voltage V_r steps down from 2 V to 1.5 V, the output voltage v_O follows the desired trajectory v_d and shifts down from -12 V to -9 V. Likewise, when the reference voltage V_r steps up from 2 V to 2.5 V, then the output voltage v_O tracks the desired trajectory v_d and shifts down from -12 V to -15 V. In both cases, the output voltage v_O takes about 5.5 ms with no percentage overshoot to reach the steady-state value. Thus, the simulation results show that the proposed control circuit tracks the desired trajectory effectively.

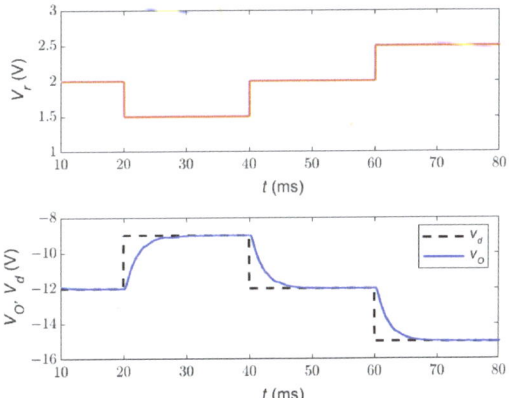

Figure 10. The output voltage response v_O of the state feedback with integral control of PWM dc-dc buck-boost converter in CCM during a time-varying reference voltage V_r. The upper sub-figure shows the step changes in reference voltage V_r. The lower sub-figure shows the tracking performance of the output voltage response v_O with respect to the desired trajectory v_d.

However, the output voltage response of the closed-loop nonlinear power converter circuit in Figure 10 exhibits a longer settling time as compared to that of the closed-loop ideal linearized power converter model shown in Figure 3. The discrepancy between the characteristics of the two responses is due to the inclusion of the nonlinearity and parasitic components of the dc-dc converter and the control circuit, which are not considered in the linearized closed-loop power converter model. Thus, the nonlinearities and modeling uncertainty of the power converter increase the settling time of the closed-loop system response.

Table 3 summarizes the characteristics of the state-feedback with an integral controlled dc-dc buck-boost converter under step changes in input voltage, load current, and the reference voltage. It can be noticed that the output voltage is maintained at -12 V during line and load variations. However, when the reference voltage changes, the output voltage follows the new desired trajectory as shown in Figure 10.

Table 3. Characteristics of proposed control circuit response during step changes in load current, input voltage, and reference voltage.

Disturbance Type (Δi_O, Δv_I, ΔV_r)	Overshoot/Undershoot (%)	Settling Time (ms)	Output Voltage (V)
$\Delta i_O \rightarrow$ 4 A to 6.0 A	2	4	-12
$\Delta i_O \rightarrow$ 4 A to 2.5 A	1	3.5	-12
$\Delta v_I \rightarrow$ 28 V to 33 V	2.6	5.5	-12
$\Delta v_I \rightarrow$ 28 V to 23 V	3.5	5.5	-12
$\Delta V_r \rightarrow$ 2 V to 2.5 V	0	5.5	-15
$\Delta V_r \rightarrow$ 2 V to 1.5 V	0	5.5	-9

7. Conclusions

The state feedback with integral control circuit using the pole placement technique has been developed for the inverting PWM dc-dc buck-boost converter in CCM. The control design methodology and the realization of the proposed control circuit have been introduced. The SIMULINK model and the corresponding Simscape Electrical circuit of the closed-loop power converter have been simulated in MATLAB to validate the design approach. It has been observed that the simulation results of the nonlinear closed-loop power converter model and the corresponding closed-loop power converter circuit are in good agreement. The pole placement technique results in a control law that places the closed-loop poles at the desired location on the left-half plane (LHP) and achieves the desired transient response. Furthermore, the state feedback with integral control eliminates the steady-state error at the output voltage and provides precise tracking performance. It has been shown that the line variation of 5 V and load variation of 2 A around the nominal operating point have been rejected with a percentage overshoot of 3.5% and settling time of 5.5 ms.

The state feedback with an integral control scheme is simple and implementable using op-amps and analog components, which is attractive for commercial and low-cost industrial applications. The proposed control design approach is flexible, which allows the designer to freely choose the closed-loop poles' location, compute the controller gains that meet the requirements, and convert the control equation to an electronic control circuit. The controller gains of the control circuit can further be tuned to compensate for actual power converter dynamics and improve the transient response characteristics. On the contrary, if a digital signal processor is chosen to implement the state feedback control algorithm, then the control law must be discretized, and further analysis is required in the z-domain to maintain the stability of the digital control system. Hence, the proposed design technique introduces a competitive alternative for embedded system-based control implementation.

Author Contributions: Conceptualization, H.A.-B.; methodology, H.A.-B.; software, H.A.-B.; validation, M.K.K.; formal analysis, H.A.-B.; resources, M.K.K.; writing—original draft preparation, A.S.; writing—review and editing, M.K.K. visualization, A.S.; All authors have read and agreed to the published version of the manuscript.

Funding: This research received no external funding.

Institutional Review Board Statement: Not applicable.

Informed Consent Statement: Not applicable.

Data Availability Statement: Not applicable.

Conflicts of Interest: The authors declare no conflict of interest.

Abbreviations

List of Acronyms

PWM	pulse-width modulated
EV	electric vehicle
NIOC	neural inverse optimal control
EKF	extended Kalman filter
MPC	model predictive control
LHP	left-half-plane
CPL	constant power load
HIL	hardware-in-the-loop
CCM	continuous conduction mode
MOSFET	metal-oxide-semiconductor field-effect transistor
ESR	equivalent series resistance
PO	percentage overshoot
CP	characteristic polynomial
EMC	electromagnetic compatibility
EMI	electromagnetic interference

List of Symbols

S	MOSFET
D_1	Diode
L	Inductor
C	Output capacitor
r_L	Inductor ESR
r_C	Capacitor ESR
r_F	Diode forward resistance
V_F	Diode threshold voltage
r_{DS}	MOSFET on-resistance
v_I	Large-signal input voltage
v_O	Large-signal output voltage
r	Large-signal load resistance
i_O	Large-signal load current
d_T	Large-signal time interval when S is ON
$\bar{d}_T$	Large-signal time interval when S is OFF
i_L	Large-signal inductor current
v_C	Large-signal capacitor voltage
V_I	Steady-state input voltage
V_O	Steady-state output voltage
R	Steady-state load resistance
I_L	Steady-state inductor current
D_T	Steady-state time interval when S is ON
$\bar{D}_T$	Steady-state time interval when S is OFF
$\tilde{i}_L$	Small-signal ac inductor current

$\tilde{v}_C$	Small-signal ac capacitor voltage
$\tilde{d}$	Small-signal ac duty cycle
x	State variables vector
A	State matrix
B	Input matrix
C	Output matrix
D	Direct transmission matrix
Co	Controllability matrix
u	System input
y	System output
V_r	Desired reference voltage
K	Constant gains vector
Θ	Zeros vector
t_s	Settling time
ζ	Damping ratio
ω_n	Natural frequency
P	Desired closed-loop poles vector
β	Voltage sensor gain
V_T	Peak ramp voltage
f_s	Switching frequency
K_1	Inductor current gain
K_2	Output voltage gain
K_3	Integral gain

References

1. Ruz-Hernandez, J.A.; Djilali, L.; Canul, M.A.R.; Boukhnifer, M.; Sanchez, E.N. Neural Inverse Optimal Control of a Regenerative Braking System for Electric Vehicles. *Energies* **2022**, *15*, 8975. [CrossRef]
2. Sankar, R.S.R.; Deepika, K.K.; Alsharef, M.; Alamri, B. A Smart ANN-Based Converter for Efficient Bidirectional Power Flow in Hybrid Electric Vehicles. *Electronics* **2022**, *11*, 3564. [CrossRef]
3. Danyali, S.; Aghaei, O.; Shirkhani, M.; Aazami, R.; Tavoosi, J.; Mohammadzadeh, A.; Mosavi, A. A New Model Predictive Control Method for Buck-Boost Inverter-Based Photovoltaic Systems. *Sustainability* **2022**, *14*, 11731. [CrossRef]
4. Murillo-Yarce, D.; Riffo, S.; Restrepo, C.; González-Castaño, C.; Garcés, A. Model Predictive Control for Stabilization of DC Microgrids in Island Mode Operation. *Mathematics* **2022**, *10*, 3384. [CrossRef]
5. Kahani, R.; Jamil, M.; Iqbal, M.T. Direct Model Reference Adaptive Control of a Boost Converter for Voltage Regulation in Microgrids. *Energies* **2022**, *15*, 5080. [CrossRef]
6. Jiang, Y.; Jin, X.; Wang, H.; Fu, Y.; Ge, W.; Yang, B.; Yu, T. Optimal Nonlinear Adaptive Control for Voltage Source Converters via Memetic Salp Swarm Algorithm: Design and Hardware Implementation. *Processes* **2019**, *7*, 490. [CrossRef]
7. Hamed, S.B.; Hamed, M.B.; Sbita, L.; Bajaj, M.; Blazek, V.; Prokop, L.; Misak, S.; Ghoneim, S.S.M. Robust Optimization and Power Management of a Triple Junction Photovoltaic Electric Vehicle with Battery Storage. *Sensors* **2022**, *22*, 6123. [CrossRef]
8. Lu, Y.; Zhu, H.; Huang, X.; Lorenz, R.D. Inverse-System Decoupling Control of DC/DC Converters. *Energies* **2019**, *12*, 179. [CrossRef]
9. Solsona, J.A.; Jorge, S.G.; Busada, C.A. Modeling and Nonlinear Control of dc–dc Converters for Microgrid Applications. *Sustainability* **2022**, *14*, 16889. [CrossRef]
10. Broday, G.R.; Lopes, L.A.C.; Damm, G. Exact Feedback Linearization of a Multi-Variable Controller for a Bi-Directional DC-DC Converter as Interface of an Energy Storage System. *Energies* **2022**, *15*, 7923. [CrossRef]
11. Broday, G.R.; Damm, G.; Pasillas-Lépine, W.; Lopes, L.A.C. A Unified Controller for Multi-State Operation of the Bi-Directional Buck–Boost DC-DC Converter. *Energies* **2021**, *14*, 7921. [CrossRef]
12. Csizmadia, M.; Kuczmann, M.; Orosz, T. A Novel Control Scheme Based on Exact Feedback Linearization Achieving Robust Constant Voltage for Boost Converter. *Electronics* **2023**, *12*, 57. [CrossRef]
13. Chen, P.; Liu, J.; Xiao, F.; Zhu, Z.; Huang, Z. Lyapunov-Function-Based Feedback Linearization Control Strategy of Modular Multilevel Converter–Bidirectional DC–DC Converter for Vessel Integrated Power Systems. *Energies* **2021**, *14*, 4691. [CrossRef]
14. Sira-Ramirez, H.; Silva-Ortigoza, R. *Control Design Techniques in Power Electronics Devices*; Springer: London, UK, 2006.
15. Bajoria, N.; Sahu, P.; Nema, R.K.; Nema, S. Overview of different control schemes used for controlling of DC-DC converters. In Proceedings of the 2016 International Conference on Electrical Power and Energy Systems (ICEPES), Bhopal, India, 14–16 December 2016; pp. 75–82.
16. Gkizas, G.; Yfoulis, C.; Amanatidis, C.; Stergiopoulos, F.; Giaouris, D.; Ziogou, C.; Voutetakis, S.; Papadopoulou, S. Digital state-feedback control of an interleaved DC-DC boost converter with bifurcation analysis. *Cont. Eng. Pract.* **2018**, *73*, 100–111. [CrossRef]

17. Pegueroles-Queralt, J.; Bianchi, F.D.; Gomis-Bellmunt, O. A Power Smoothing System Based on Supercapacitors for Renewable Distributed Generation. *IEEE Trans. Ind. Electron.* **2015**, *62*, 343–350. [CrossRef]
18. Hajizadeh, A.; Shahirinia, A.H.; Namjoo, N.; Yu, D.C. Self-tuning indirect adaptive control of non-inverting buck-boost converter. *IET Power Electron.* **2015**, *8*, 2299–2306. [CrossRef]
19. Czarkowski, D.; Kazimierczuk, M.K. Application of state feedback with integral control to pulse-width modulated push-pull DC-DC convertor. *IEE Proc.-Control Theory Appl.* **1994**, *141*, 99–103. [CrossRef]
20. Abdurraqeeb, A.M.; Al-Shamma'a, A.A.; Alkuhayli, A.; Noman, A.M.; Addoweesh, K.E. RST Digital Robust Control for DC/DC Buck Converter Feeding Constant Power Load. *Mathematics* **2022**, *10*, 1782. [CrossRef]
21. Koundi, M.; El Idrissi, Z.; El Fadil, H.; Belhaj, F.Z.; Lassioui, A.; Gaouzi, K.; Rachid, A.; Giri, F. State-Feedback Control of Interleaved Buck–Boost DC–DC Power Converter with Continuous Input Current for Fuel Cell Energy Sources: Theoretical Design and Experimental Validation. *World Electr. Veh. J.* **2022**, *13*, 124. [CrossRef]
22. Al-Baidhani, H.; Salvatierra, T.; Ordonez, R.; Kazimierczuk, M.K. Simplified nonlinear voltage-mode control of PWM DC-DC buck converter. *IEEE Trans. Energy Conv.* **2021**, *36*, 431–440. [CrossRef]
23. Al-Baidhani, H.; Kazimierczuk, M.K. Simplified Double-Integral Sliding-Mode Control of PWM DC-AC Converter with Constant Switching Frequency. *Appl. Sci.* **2022**, *12*, 10312. [CrossRef]
24. Al-Baidhani, H.; Kazimierczuk, M.K.; Ordóñez, R. Nonlinear Modelling and Control of PWM DC-DC Buck-Boost Converter for CCM. In Proceedings of the IECON 2018—44th Annual Conference of the IEEE Industrial Electronics Society, Washington, DC, USA, 21–23 October 2018; pp. 1374–1379.
25. Golnaraghi, F.; Kuo, B. *Automatic Control Systems*, 9th ed.; John Wiley & Sons: Hoboken, NJ, USA, 2010.
26. Kazimierczuk, M.K. *Pulse-Width Modulated DC-DC Power Converters*, 2nd ed.; John Wiley & Sons: Chichester, UK, 2016.
27. Al-Baidhani, H.; Kazimierczuk, M.K. Simplified Nonlinear Current-Mode Control of DC-DC Cuk Converter for Low-Cost Industrial Applications. *Sensors* **2023**, *23*, 1462. [CrossRef]

Disclaimer/Publisher's Note: The statements, opinions and data contained in all publications are solely those of the individual author(s) and contributor(s) and not of MDPI and/or the editor(s). MDPI and/or the editor(s) disclaim responsibility for any injury to people or property resulting from any ideas, methods, instructions or products referred to in the content.

Article

Adaptive Super-Twisting Sliding Mode Control of Active Power Filter Using Interval Type-2-Fuzzy Neural Networks

Jiacheng Wang [1], Yunmei Fang [2] and Juntao Fei [1,*]

[1] College of Information Science and Engineering, Jiangsu Key Laboratory of Power Transmission and Distribution Equipment Technology, Hohai University, Changzhou 213022, China
[2] College of Mechanical and Electrical Engineering, Hohai University, Changzhou 213022, China
* Correspondence: jtfei@hhu.edu.cn

Abstract: Aiming at the unknown uncertainty of an active power filter system in practical operation, combining the advantages of self-feedback structure, interval type-2 fuzzy neural network, and super-twisting sliding mode, an adaptive super-twisting sliding mode control method of interval type-2 fuzzy neural network with self-feedback recursive structure (IT2FNN-SFR STSMC) is proposed in this paper. IT2FNN has an uncertain membership function, which can enhance the nonlinear ability and robustness of the network. The historical information will be stored and utilized by the self-feedback recursive structure (SFR) at runtime. Therefore, the novel IT2FNN-SFR is designed to improve the dynamic approximation effect of the neural network and reduce the dependence of the controller on the actual mathematical model. The adaptive rate of each weight of the neural network is designed by the Lyapunov method and gradient descent (GD) algorithm to ensure the convergence and stability of the system. Super-twisting sliding mode control (STSMC) has strong robustness, which can effectively reduce system chattering, and improve control accuracy and system performance. The gain of the integral term in the STSMC is set as a constant, and the other gain is changed adaptively whose adaptive rate is deduced through the stability proof of the neural network, which greatly reduces the difficulty of parameter adjustment. The harmonic suppression ability of the designed control strategy is verified by simulation experiments.

Keywords: active power filter (APF); interval type-2 fuzzy neural network (IT2FNN); STSMC; self-feedback recursive structure

MSC: 68T07; 93C40

Citation: Wang, J.; Fang, Y.; Fei, J. Adaptive Super-Twisting Sliding Mode Control of Active Power Filter Using Interval Type-2-Fuzzy Neural Networks. *Mathematics* 2023, 11, 2785. https://doi.org/10.3390/math11122785

Academic Editors: Adrian Olaru, Gabriel Frumusanu and Catalin Alexandru

Received: 13 May 2023
Revised: 16 June 2023
Accepted: 19 June 2023
Published: 20 June 2023

Copyright: © 2023 by the authors. Licensee MDPI, Basel, Switzerland. This article is an open access article distributed under the terms and conditions of the Creative Commons Attribution (CC BY) license (https://creativecommons.org/licenses/by/4.0/).

1. Introduction

Nowadays, as countries around the world increasingly favor new energy power generation to improve the energy structure, the power grid is facing increasingly complex nonlinear loads, which will generate a large number of harmonics, presenting significant challenges to maintain the safe and stable operation of the grid [1–3]. Therefore, addressing harmonics in the grid is of utmost importance. One common solution is harmonic compensation, which achieves the goal by simply installing a parallel or series compensation device on the load side. Active power filters (APF) have become the primary equipment for harmonic compensation due to the flexibility, excellent compensation capabilities, and superior controllability when compared to passive power filters (PPF) [4,5]. The traditional harmonic current compensation tracking control (HCCTC) cannot fully utilize the superior characteristics of APF. It is the research direction of many researchers to design the novel HCCTC with better performance by incorporating various intelligent control theories.

Sliding mode control (SMC) is a highly robust, discontinuous nonlinear control method that is commonly used to address uncertainties and disturbances within a system [6–8]. In order to ameliorate chattering in SMC, researchers have come up with different methods.

In some studies, scholars improved the reaching law of sliding mode to reduce chattering and improve the convergence of the controller [9,10]. In one study, a novel nonsingular terminal sliding mode control (NTSMC) is designed to weaken the chattering in [11]. Due to the convergence property of the fractional-order algorithm, it is applied in the design of the SMC to reduce the chattering around the sliding surface and to enhance the robustness of the controller [12,13]. In [13], a controller is designed to combine the advantages of the nonsingular terminal sliding mode control and fractional-order sliding mode control, which has fast convergence, flexible control, and is chatter-free. Adaptive sliding mode control (ASMC) is also an approach to handle the chattering problem [14]. Roy et al. [15,16] developed two novel ASMC strategies successively, one of which removes the assumption of prior bounded uncertainty, and the other can solve the problem of overestimation or underestimation, further improving the robustness of the controller and reducing the chattering. In [17], a high-order sliding mode control (HOSMC) is proposed, to minimize chattering by removing the limitations of input–output relative order. A novel adaptive HOSMC is developed to enhance the chattering suppression ability [18]. As one of the HOSMC, second-order sliding mode control can effectively solve the chattering problem in the control of underactuated mechanical systems [19]. The super-twisting sliding mode (STSMC) applies to systems (typically of any order) where the control is present in the first derivative of the sliding variable. At the same time, it can produce continuous control signals to weaken the chattering [20]. In the field of speed and altitude tracking control of air-breathing hypersonic vehicles, a new composite controller combining STSMC, high-order disturbance observer, and backstepping control is designed [21]. In one study, a model-free controller was designed, which is combining STSMC and iterative learning laws, to be applied in the unknown dynamics tray indexing systems [22]. However, it may be difficult to adjust the gain of STSMC manually, and the performance of STSMC cannot be fully utilized. In this paper, the gain parameter of the integral term is selected manually, and the other gain is adjusted adaptively to reduce the difficulty of parameter selection.

The performance of STSMC can be further improved if the dynamic equations of the system can be predicted in advance and feed-forward compensation can be performed. Neural networks are commonly utilized due to their strong approximation ability [23–26]. In [23], a novel PID controller based on the NN is developed to adjust the minimum bandwidth for different situations automatically. For the design of the virtual unmodeled dynamic compensator, three-layer NNs were used to estimate unknowns in [24]. A neural network controller is proposed to adjust the sliding mode gain adaptively to reduce chattering [25]. Li et al. utilized the RBFNNs to estimate the unknown function in nonlinear systems [26]. The combination of fuzzy logic rules and neural networks can reduce the number of nodes of neural networks and improve the ability to deal with nonlinearity and robustness [27–30]. However, due to the fact that the Type-1 fuzzy method uses accurate and clear membership functions, the overall performance of the system could not reduce or eliminate the uncertainties effectively caused by changes in the environment and other factors. In view of the above problems, some scholars extend the interval type-1 to interval type-2, which makes the membership function uncertain, to further improve the nonlinear ability and robustness of the network [31–33]. However, FNNs could be regarded as static mapping, which has the defect of poor ability to deal with dynamics. To overcome the disadvantage of the poor dynamic ability of NNs, some scholars put forward a recursive neural network (RNN) by combining feedback structure with the network [34–40]. A fully connected recurrent NN is designed to approximate the exact mathematical dynamics in a real-time scene, which has a better ability for dynamic response, function approximation ability, and convergence speed [35]. A self-organizing RNN with novel adaptive algorithms is adopted to improve the accuracy of model prediction.

Motivated by the above research, an adaptive super-twisting sliding mode control method of interval type-2 fuzzy neural network with self-feedback recursive structure (IT2FNN-SFR STSMC) is proposed to ensure that the compensation current in the APF system can track the desired current quickly and accurately in this paper. The advantages of the IT2FNN and RNN are concentrated in the novel IT2FNN-SFR, resulting in its better dynamic approximation capacity. The adaptive laws of parameters in the proposed IT2FNN-SFR STSMC are derived by the Lyapunov method and GD algorithm, and the stability and convergence of the designed controller are also proved by the Lyapunov method. Moreover, the simulation experiments have been carried out to verify that the harmonic suppression capability is in line with international standards and has good steady-state response and dynamic performance. In summary, the major contributions of this article are described as:

(1) A new structure of IT2FNN, namely IT2FNN-SFR, has been proposed, which has the ability of strong robustness as IT2FNN and great dynamic response as RNN. The new NN is error-driven and online optimization, which means it is less dependent on accurate and detailed information about the system. The recursive structure in the NN will store and take advantage of the historical information to improve the accuracy of estimation and dynamic approximation effect;

(2) STSMC not only has the advantages of strong robustness and simple control principle of traditional SMC but also overcomes the chattering problem. In order to reduce the inaccuracy and complexity of manual parameter setting, a sliding mode gain adaptive law is deduced to realize a set of gain optimal solutions.

The remainder of this article is structured as follows. In Section 2, the principle and ideal dynamic equation of a single-phase APF are presented. Section 3 details a novel control strategy and provides proof of its stability. Section 4 shows the simulation experiment to verify the effectiveness of the proposed method. Finally, Section 5 concludes this article.

2. Principle of Active Power Filter

Due to the high reliability of compensation performance and flexible use, APF is widely used in power systems for harmonic suppression. The main circuit of a single APF consists of a PWM inverter circuit and a large capacitor on the DC side for energy storage. The current and voltage signals of the power grid circuit are collected for harmonic analysis and compensation through the sensors. The pulse width modulation (PWM) signal is generated by processing these signals by the controller. Finally, the PWM inverter works out to generate the corresponding current for the system current.

Figure 1 depicts the typical block diagram of a single-phase APF system control, where u_s, U_{dc}, i_s, i_c, i_L are the supply voltage, the DC-link voltage, the grid current, the compensation current, and the load current of the APF circuit, respectively. L and R are the equivalent inductance and resistance of the APF circuit.

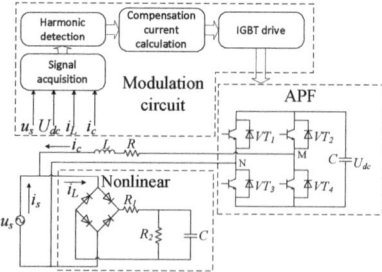

Figure 1. The block diagram of a single-phase active power filter.

According to Kirchhoff's voltage and current laws, the following equation can be obtained:

$$u_s = -L\frac{di_c}{dt} - Ri_c + U_{MN} \tag{1}$$

where U_{MN} is the output voltage of the DC side capacitor voltage modulated by the PWM inverter circuit.

Assuming that the control law of each cycle is $u \in [0,1]$, and the IGBTs are ideal, the working state is expressed as:

$$\begin{cases} VT_1, VT_4 \text{ is } on \text{ and } VT_2, VT_3 \text{ is } off. \, t \in [0, u] \\ VT_1, VT_4 \text{ is } off \text{ and } VT_2, VT_3 \text{ is } on. \, t \in [u, 1] \end{cases} \quad (2)$$

Due to $U_{MN} \in [-U_{dc}, U_{dc}]$, the following relation can be obtained:

$$U_{MN} = (2u - 1)U_{dc} \quad (3)$$

From (1) and (3), the ideal equation of compensation current is rewritten as:

$$\dot{i}_c = -\frac{R}{L}i_c - \frac{u_s}{L} - \frac{U_{dc}}{L} + \frac{2U_{dc}}{L}u \quad (4)$$

Hence, it can be abbreviated as:

$$\dot{x} = f_1(x) + bu \quad (5)$$

where x is the compensation current i_c.

$$f_1(x) = -\frac{R}{L}i_c - \frac{u_s}{L} - \frac{U_{dc}}{L} \quad (6)$$

$$b = \frac{2U_{dc}}{L} \quad (7)$$

Considering the internal parameters of the system often change uncertainly and the unknown external disturbances are existed in the system, the total disturbance containing parameter perturbation and external disturbances is defined as $g(t)$, assuming that it is bounded. The actual dynamic equation is given as:

$$\dot{x} = f + bu \quad (8)$$

where $f = f_1(x) + g(t) = -\frac{R}{L}i_c - \frac{u_s}{L} - \frac{U_{dc}}{L} + g(t)$.

3. Controller Design and Analysis

In order to improve the robustness of the system and reduce the chattering, the interval type-2-fuzzy neural networks with self-feedback recursive structure adaptive super-twisting sliding mode control (IT2FNN-SFR STSMC) strategy is designed. The block diagram is depicted in Figure 2. First, the structure of the novel networks is shown in Figure 3 and the basic functions and signal transmission of each layer are introduced as follows:

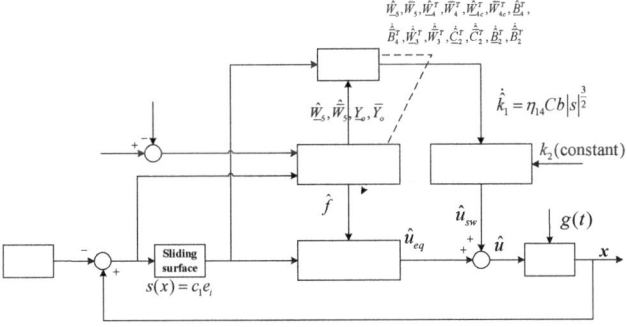

Figure 2. The block diagram of IT2FNN-SFR STSMC.

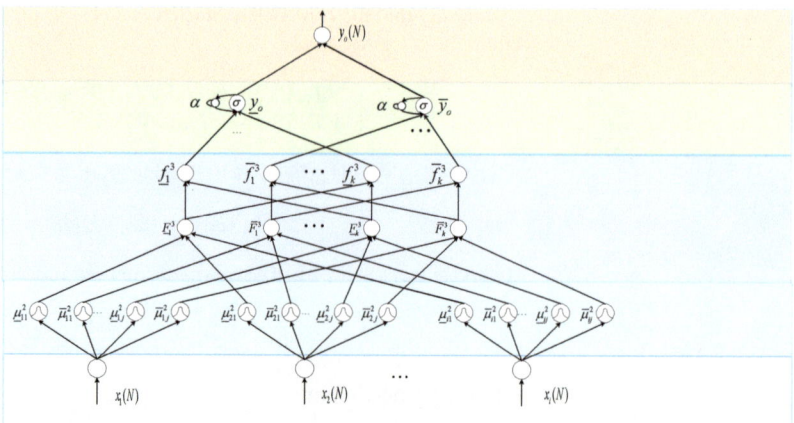

Figure 3. The structure of IT2FNN-SFR.

(1) Layer I (input layer): The input signals $X = \begin{bmatrix} x_1 & x_2 & \cdots & x_i \end{bmatrix}^T$ are transmitted in the layer, where the number of neurons in this layer is determined by the actual needs of the specific problem. In this paper, the current error and DC-side voltage error are used as the input of neural network, i.e., $i = 2$. The output of this layer is expressed as:

$$net_i^1(N) = x_i^1(N) \tag{9}$$

$$y_i^1(N) = f_i^1\left(net_i^1(N)\right) = net_i^1(N) \tag{10}$$

where $x_i^1(N)$ and $y_i^1(N)$ represent the input and output of the ith node in the first layer, respectively; N represents the number of sampling times.

(2) Layer II (membership function layer/fuzzy layer): In this layer, the Gaussian function with center vector and uncertain base width is selected to enhance the ability of neural network to deal with nonlinearity. Assume that the j-group type-2 fuzzy output set is $[\underline{\mu}_{ij}^2, \overline{\mu}_{ij}^2]$. In this paper, the number of output set is selected as 3, i.e., $j = 3$. The input–output transfer relation is expressed as:

$$x_{ij}^2(N) = y_i^1(N) \tag{11}$$

$$\underline{\mu}_{ij}^2(N) = \exp\left[-\frac{1}{2}\left(\frac{x_{ij}^2(N) - \underline{c}_{ij}^2}{\underline{\sigma}_{ij}^2}\right)^2\right] \tag{12}$$

$$\overline{\mu}_{ij}^2(N) = \exp\left[-\frac{1}{2}\left(\frac{x_{ij}^2(N) - \overline{c}_{ij}^2}{\overline{\sigma}_{ij}^2}\right)^2\right] \tag{13}$$

where μ_{ij}^2, c_{ij}^2, σ_{ij}^2 represent type-2 fuzzy output, center vector and base width of type 2 Gaussian fuzzy membership function, respectively; "$\underline{*}$", "$\overline{*}$" represent the lower and upper bounds of each variable, respectively.

(3) Layer III (Rule Layer): This layer integrates input signals from the Layer II. The number of neurons in the layer is the equal to the number of neurons in each group of the Layer II, i.e., $j = k = 3$. The signal transmission functions are expressed as:

$$\underline{F}_k^3 = \prod_{i=1}^I \underline{\mu}_{ij}^2 \tag{14}$$

$$\overline{F}_k^3 = \prod_{i=1}^{I} \overline{\mu}_{ij}^2 \tag{15}$$

$$\underline{f}_k^3 = \frac{\underline{F}_k^3}{\sum_{k=1}^{K} \underline{F}_k^3} \tag{16}$$

$$\overline{f}_k^3 = \frac{\overline{F}_k^3}{\sum_{k=1}^{K} \overline{F}_k^3} \tag{17}$$

where f_k^3 is the output of the rule layer after normalization.

(4) Layer IV (self-feedback recursive layer): This layer integrates the current and historical output information from the rule layer, and the temporary recursive layer is applied to reserve and process the history output information. In this layer, the group number of the layer is 1, i.e., $o = 1$. The output can be expressed as:

$$\underline{x}^4(N) = \sum \underline{\omega}_k^3 \underline{f}_k^3 \tag{18}$$

$$\underline{x}^{4c}(N) = \alpha \underline{x}^{4c}(N-1) + \underline{y}^4(N-1) \tag{19}$$

$$\underline{net}^4(N) = \underline{\omega}^4 \underline{x}^4(N) + \underline{\omega}^{4c} \underline{x}^{4c}(N) + \underline{b}^4 \tag{20}$$

$$\underline{y}^4(N) = \sigma\left(\underline{net}^4(N)\right) \tag{21}$$

$$\overline{x}^4(N) = \sum \overline{\omega}_k^3 \overline{f}_k^3 \tag{22}$$

$$\overline{x}^{4c}(N) = \alpha \overline{x}^{4c}(N-1) + \overline{y}^4(N-1) \tag{23}$$

$$\overline{net}^4(N) = \overline{\omega}^4 \overline{x}^4(N) + \overline{\omega}^{4c} \overline{x}^{4c}(N) + \overline{b}^4 \tag{24}$$

$$\overline{y}^4(N) = \sigma\left(\overline{net}^4(N)\right) \tag{25}$$

$$\sigma(z) = \frac{1}{1 + e^{-z}} \tag{26}$$

where $x^4(N)$ represents the signal from the rule layer at the current sampling time; $x^{4c}(N)$ represents the historical information in the temporary recursive layer; ω^{4c} represents a self-feedback weight; ω^4 represents the weight of the SFR layer; α is the self-feedback parameter, which determines the proportion of historical information; b_l^4 is a bias term; $\sigma(\cdot)$ is a sigmod activation function.

(5) Layer V (output layer): the network is defuzzification through weighting. The expression is given as:

$$y_o(N) = \underline{\omega}^5 \underline{y}^4(N) + \overline{\omega}^5 \overline{y}^4(N) \tag{27}$$

Next, a new neural network is designed to estimate the actual model and reduce the dependence of the controller on the model. Meanwhile, super twisted sliding mode control (STSMC) is introduced to reduce the chattering. However, the traditional STSMC is complicated in adjusting parameters. In this paper, the adaptive rate of gain k_1 is also designed to reduce the difficulty of parameter adjustment when designing the adaptive rate of network parameters as shown in Figure 2.

Define the tracking error of APF system as:

$$e = x - r \tag{28}$$

where x is a tracking current, r is the reference current.

Design the sliding surface as:

$$s = Ce \tag{29}$$

where C is the gain of the sliding surface, which is the positive constant to be designed.

The derivative of Equation (29) is obtained as:

$$\dot{s} = C(\dot{x} - \dot{r}) = C(f + bu - \dot{r}) \tag{30}$$

The equivalent control law can be derived when $\dot{s} = 0$:

$$u_{eq} = \frac{1}{b}(-f + \dot{r}) \tag{31}$$

Adaptive STSMC is used to design the switching control law:

$$u_{sw} = -\hat{k}_1 |s|^{\frac{1}{2}} \mathrm{sgn}(s) - \int k_2 \mathrm{sgn}(s) dt \tag{32}$$

where $k_1 > 0, k_2 > 0$.

Substituting $\hat{f}$ into the control law, the control strategy can be established:

$$u = \frac{1}{b}(-\hat{f} + \dot{r}) - \hat{k}_1 |s|^{\frac{1}{2}} \mathrm{sgn}(s) - \int k_2 \mathrm{sgn}(s) dt \tag{33}$$

From the proposed neural network, the following results can be obtained:

$$\begin{aligned} f &= \underline{W}_5^{*T} \underline{\Phi}^*(\underline{W}_4^*, \underline{W}_{4c}^*, \underline{B}_4^*, \underline{W}_3^*, \underline{B}_2^*, \underline{C}_2^*) \\ &+ \overline{W}_5^{*T} \overline{\Phi}^*(\overline{W}_4^*, \overline{W}_{4c}^*, \overline{B}_4^*, \overline{W}_3^*, \overline{B}_2^*, \overline{C}_2^*) + \varepsilon \end{aligned} \tag{34}$$

where ε is the error between the optimal value and the actual value, and $\underline{C}_2^*, \overline{C}_2^*, \underline{B}_2^*, \overline{B}_2^*, \underline{W}_3^*, \overline{W}_3^*, \underline{W}_4^*, \overline{W}_4^*, \underline{W}_{4c}^*, \overline{W}_{4c}^*, \underline{W}_5^*, \overline{W}_5^*, \underline{B}_4^*, \overline{B}_4^*$ are the best parameters.

For the sake of further proof of brevity, define that:

$$\underline{\Phi} = \Phi(\underline{W}_4, \underline{W}_{4c}, \underline{B}_4, \underline{W}_3, \underline{B}_2, \underline{C}_2) \tag{35}$$

$$\overline{\Phi} = \Phi(\overline{W}_4, \overline{W}_{4c}, \overline{B}_4, \overline{W}_3, \overline{B}_2, \overline{C}_2) \tag{36}$$

The output of IT2FNN-SFR is used to replace the f, denoted as:

$$\hat{f} = y_o = \hat{\underline{W}}_5^T \hat{\underline{\Phi}} + \hat{\overline{W}}_5^T \hat{\overline{\Phi}} \tag{37}$$

where the superscript "∧" represent the estimated values of the corresponding parameters.

The estimation error of the designed neural network is:

$$\begin{aligned} \tilde{f} &= f - \hat{f} = \underline{W}_5^{*T} \underline{\Phi}^* + \overline{W}_5^{*T} \overline{\Phi}^* + \varepsilon - (\hat{\underline{W}}_5^T \hat{\underline{\Phi}} + \hat{\overline{W}}_5^T \hat{\overline{\Phi}}) \\ &= \hat{\underline{W}}_5^T \tilde{\underline{\Phi}} + \tilde{\underline{W}}_5^T \hat{\underline{\Phi}} + \hat{\overline{W}}_5^T \tilde{\overline{\Phi}} + \tilde{\overline{W}}_5^T \hat{\overline{\Phi}} + \varepsilon_0 \end{aligned} \tag{38}$$

where the superscript "∼" represent the approximation errors; $\varepsilon_0 = \tilde{\underline{W}}_5^T \tilde{\underline{\Phi}} + \tilde{\overline{W}}_5^T \tilde{\overline{\Phi}} + \varepsilon$ is the total approximation error.

By expanding Taylor $\underline{\Phi}^*$ and $\overline{\Phi}^*$, respectively, we can get the following results:

$$\begin{aligned}\widetilde{\underline{\Phi}} &= \tfrac{\partial \underline{\Phi}}{\partial \underline{W}_4}\Big|_{\underline{W}_4=\hat{\underline{W}}_4}(\underline{W}_4^*-\hat{\underline{W}}_4) + \tfrac{\partial \underline{\Phi}}{\partial \underline{W}_{4c}}\Big|_{\underline{W}_{4c}=\hat{\underline{W}}_{4c}}(\underline{W}_{4c}^*-\hat{\underline{W}}_{4c}) \\ &+ \tfrac{\partial \underline{\Phi}}{\partial \underline{B}_4}\Big|_{\underline{B}_4=\hat{\underline{B}}_4}(\underline{B}_4^*-\hat{\underline{B}}_4) + \tfrac{\partial \underline{\Phi}}{\partial \underline{W}_3}\Big|_{\underline{W}_3=\hat{\underline{W}}_3}(\underline{W}_3^*-\hat{\underline{W}}_3) \\ &+ \tfrac{\partial \underline{\Phi}}{\partial \underline{C}_2}\Big|_{\underline{C}_2=\hat{\underline{C}}_2}(\underline{C}_2^*-\hat{\underline{C}}_2) + \tfrac{\partial \underline{\Phi}}{\partial \underline{B}_2}\Big|_{\underline{B}_2=\hat{\underline{B}}_2}(\underline{B}_2^*-\hat{\underline{B}}_2) + \underline{O}_h \\ &= \underline{\Phi}_{\underline{W}_4}\widetilde{\underline{W}}_4 + \underline{\Phi}_{\underline{W}_{4c}}\widetilde{\underline{W}}_{4c} + \underline{\Phi}_{\underline{B}_4}\widetilde{\underline{B}}_4 + \underline{\Phi}_{\underline{W}_3}\widetilde{\underline{W}}_3 + \underline{\Phi}_{\underline{C}_2}\widetilde{\underline{C}}_2 \\ &+ \underline{\Phi}_{\underline{B}_2}\widetilde{\underline{B}}_2 + \underline{O}_h \end{aligned} \quad (39)$$

$$\begin{aligned}\widetilde{\overline{\Phi}} &= \overline{\Phi}_{\overline{W}_4}\widetilde{\overline{W}}_4 + \overline{\Phi}_{\overline{W}_{4c}}\widetilde{\overline{W}}_{4c} + \overline{\Phi}_{\overline{B}_4}\widetilde{\overline{B}}_4 + \overline{\Phi}_{\overline{W}_3}\widetilde{\overline{W}}_3 + \overline{\Phi}_{\overline{C}_2}\widetilde{\overline{C}}_2 \\ &+ \overline{\Phi}_{\overline{B}_2}\widetilde{\overline{B}}_2 + \overline{O}_h\end{aligned} \quad (40)$$

where $\underline{O}_h$ and $\overline{O}_h$ are the vectors of higher order terms, and the above partial derivatives are consistent with the Jacobian matrix arrangement, and the result is:

$$\underline{\Phi}_{\underline{W}_4} = \left[\frac{\partial \underline{\phi}}{\partial \underline{\omega}^4}\right]_{1\times 1} \quad (41)$$

$$\underline{\Phi}_{\underline{W}_{4c}} = \left[\frac{\partial \underline{\phi}}{\partial \underline{\omega}^{4c}}\right]_{1\times 1} \quad (42)$$

$$\underline{\Phi}_{\underline{B}_4} = \left[\frac{\partial \underline{\phi}}{\partial \underline{b}^4}\right]_{1\times 1} \quad (43)$$

$$\underline{\Phi}_{\underline{W}_3} = \left[\frac{\partial \underline{\phi}}{\partial \underline{\omega}_1^3}, \frac{\partial \underline{\phi}}{\partial \underline{\omega}_2^3}, \cdots, \frac{\partial \underline{\phi}}{\partial \underline{\omega}_k^3}\right]_{1\times k} \quad (44)$$

$$\underline{\Phi}_{\underline{C}_2} = \left[\frac{\partial \underline{\phi}}{\partial \underline{c}_{11}^2}, \cdots, \frac{\partial \underline{\phi}}{\partial \underline{c}_{1j}^2}, \frac{\partial \underline{\phi}}{\partial \underline{c}_{21}^2}, \cdots, \frac{\partial \underline{\phi}}{\partial \underline{c}_{ij}^2}\right]_{1\times (i\times j)} \quad (45)$$

$$\underline{\Phi}_{\underline{B}_2} = \left[\frac{\partial \underline{\phi}}{\partial \underline{b}_{11}^2}, \cdots, \frac{\partial \underline{\phi}}{\partial \underline{b}_{1j}^2}, \frac{\partial \underline{\phi}}{\partial \underline{b}_{21}^2}, \cdots, \frac{\partial \underline{\phi}}{\partial \underline{b}_{ij}^2}\right]_{1\times (i\times j)} \quad (46)$$

$$\overline{\Phi}_{\overline{W}_4} = \left[\frac{\partial \overline{\phi}}{\partial \overline{\omega}^4}\right]_{1\times 1} \quad (47)$$

$$\overline{\Phi}_{\overline{W}_{4c}} = \left[\frac{\partial \overline{\phi}}{\partial \overline{\omega}^{4c}}\right]_{1\times 1} \quad (48)$$

$$\overline{\Phi}_{\overline{B}_4} = \left[\frac{\partial \overline{\phi}}{\partial \overline{b}^4}\right]_{1\times 1} \quad (49)$$

$$\overline{\Phi}_{\overline{W}_3} = \left[\frac{\partial \overline{\phi}}{\partial \overline{\omega}_1^3}, \frac{\partial \overline{\phi}}{\partial \overline{\omega}_2^3}, \cdots, \frac{\partial \overline{\phi}}{\partial \overline{\omega}_k^3}\right]_{1\times k} \quad (50)$$

$$\overline{\Phi}_{\overline{C}_2} = \left[\frac{\partial \overline{\phi}}{\partial \overline{c}_{11}^2}, \cdots, \frac{\partial \overline{\phi}}{\partial \overline{c}_{1j}^2}, \frac{\partial \overline{\phi}}{\partial \overline{c}_{21}^2}, \cdots, \frac{\partial \overline{\phi}}{\partial \overline{c}_{ij}^2}\right]_{1\times (i\times j)} \quad (51)$$

$$\overline{\Phi}_{\overline{B}_2} = \left[\frac{\partial \overline{\phi}}{\partial \overline{b}_{11}^2}, \cdots, \frac{\partial \overline{\phi}}{\partial \overline{b}_{1j}^2}, \frac{\partial \overline{\phi}}{\partial \overline{b}_{21}^2}, \cdots, \frac{\partial \overline{\phi}}{\partial \overline{b}_{ij}^2}\right]_{1\times (i\times j)} \quad (52)$$

By substituting (39) and (40) into (38), it can be obtained:

$$\begin{aligned}\tilde{f} =& \hat{\underline{W}}_5^T \Phi_{\underline{W}_4} \tilde{\underline{W}}_4 + \hat{\underline{W}}_5^T \Phi_{\underline{W}_{4c}} \tilde{\underline{W}}_{4c} + \hat{\underline{W}}_5^T \Phi_{\underline{B}_4} \tilde{\underline{B}}_4 \\ &+ \hat{\underline{W}}_5^T \Phi_{\underline{W}_3} \tilde{\underline{W}}_3 + \hat{\underline{W}}_5^T \Phi_{\underline{C}_2} \tilde{\underline{C}}_2 + \hat{\underline{W}}_5^T \Phi_{\underline{B}_2} \tilde{\underline{B}}_2 + \hat{\underline{W}}_5^T \hat{\underline{\Phi}} \\ &+ \hat{\overline{W}}_5^T \Phi_{\overline{W}_4} \tilde{\overline{W}}_4 + \hat{\overline{W}}_5^T \Phi_{\overline{W}_{4c}} \tilde{\overline{W}}_{4c} + \hat{\overline{W}}_5^T \Phi_{\overline{B}_4} \tilde{\overline{B}}_4 + \hat{\overline{W}}_5^T \Phi_{\overline{W}_3} \tilde{\overline{W}}_3 \\ &+ \hat{\overline{W}}_5^T \Phi_{\overline{C}_2} \tilde{\overline{C}}_2 + \hat{\overline{W}}_5^T \Phi_{\overline{B}_2} \tilde{\overline{B}}_2 + \hat{\overline{W}}_5^T \hat{\overline{\Phi}} + \Delta_0\end{aligned} \tag{53}$$

where the total higher order approximation error is:

$$\Delta_0 = \hat{\underline{W}}_5 O_h + \hat{\overline{W}}_5 \overline{O}_h + \varepsilon_0 \tag{54}$$

Theorem 1. *Considering the system of APF in (8), the proposed IT2FNN-SFR STSMC strategy is guaranteed to be stable if the controller is designed as (33) and the parameter adaptive laws are designed properly as (55)–(69):*

$$\dot{\hat{\underline{W}}}_4^T = -\dot{\tilde{\underline{W}}}_4^T = \eta_1 \text{Csgn}(s) \hat{\underline{W}}_5^T \Phi_{\underline{W}_4} \tag{55}$$

$$\dot{\hat{\overline{W}}}_4^T = -\dot{\tilde{\overline{W}}}_4^T = \eta_2 \text{Csgn}(s) \hat{\overline{W}}_5^T \Phi_{\overline{W}_4} \tag{56}$$

$$\dot{\hat{\underline{W}}}_{4c}^T = -\dot{\tilde{\underline{W}}}_{4c}^T = \eta_3 \text{Csgn}(s) \hat{\underline{W}}_5^T \Phi_{\underline{W}_{4c}} \tag{57}$$

$$\dot{\hat{\overline{W}}}_{4c}^T = -\dot{\tilde{\overline{W}}}_{4c}^T = \eta_4 \text{Csgn}(s) \hat{\overline{W}}_5^T \Phi_{\overline{W}_{4c}} \tag{58}$$

$$\dot{\hat{\underline{B}}}_4^T = -\dot{\tilde{\underline{B}}}_4^T = \eta_5 \text{Csgn}(s) \hat{\underline{W}}_5^T \Phi_{\underline{B}_4} \tag{59}$$

$$\dot{\hat{\overline{B}}}_4^T = -\dot{\tilde{\overline{B}}}_4^T = \eta_6 \text{Csgn}(s) \hat{\overline{W}}_5^T \Phi_{\overline{B}_4} \tag{60}$$

$$\dot{\hat{\underline{W}}}_3^T = -\dot{\tilde{\underline{W}}}_3^T = \eta_7 \text{Csgn}(s) \hat{\underline{W}}_5^T \Phi_{\underline{W}_3} \tag{61}$$

$$\dot{\hat{\overline{W}}}_3^T = -\dot{\tilde{\overline{W}}}_3^T = \eta_8 \text{Csgn}(s) \hat{\overline{W}}_5^T \Phi_{\overline{W}_3} \tag{62}$$

$$\dot{\hat{\underline{C}}}_2^T = -\dot{\tilde{\underline{C}}}_2^T = \eta_9 \text{Csgn}(s) \hat{\underline{W}}_5^T \Phi_{\underline{C}_2} \tag{63}$$

$$\dot{\hat{\overline{C}}}_2^T = -\dot{\tilde{\overline{C}}}_2^T = \eta_{10} \text{Csgn}(s) \hat{\overline{W}}_5^T \Phi_{\overline{C}_2} \tag{64}$$

$$\dot{\hat{\underline{B}}}_2^T = -\dot{\tilde{\underline{B}}}_2^T = \eta_{11} \text{Csgn}(s) \hat{\underline{W}}_5^T \Phi_{\underline{B}_2} \tag{65}$$

$$\dot{\hat{\overline{B}}}_2^T = -\dot{\tilde{\overline{B}}}_2^T = \eta_{12} \text{Csgn}(s) \hat{\overline{W}}_5^T \Phi_{\overline{B}_2} \tag{66}$$

$$\dot{\hat{\underline{W}}}_5 = -\dot{\tilde{\underline{W}}}_5 = \eta_{13} \text{Csgn}(s) \hat{\underline{\Phi}} \tag{67}$$

$$\dot{\hat{\overline{W}}}_5 = -\dot{\tilde{\overline{W}}}_5 = \eta_{14} \text{Csgn}(s) \hat{\overline{\Phi}} \tag{68}$$

$$\dot{\hat{k}}_1 = -\dot{\tilde{k}}_1 = \eta_{15} Cb|s|^{\frac{1}{2}} \tag{69}$$

where $\eta_1 \sim \eta_{15}$ are the learning rates of the corresponding adaptive gains, which are positive constants.

Proof. When $s \neq 0$, a Lyapunov function candidate is given as:

$$\begin{aligned} V &= |s| + \frac{1}{2\eta_1} tr\left(\tilde{\underline{W}}_4^T \tilde{\underline{W}}_4\right) + \frac{1}{2\eta_2} tr\left(\tilde{\overline{W}}_4^T \tilde{\overline{W}}_4\right) \\ &+ \frac{1}{2\eta_3} tr\left(\tilde{\underline{W}}_{4c}^T \tilde{\underline{W}}_{4c}\right) + \frac{1}{2\eta_4} tr\left(\tilde{\overline{W}}_{4c}^T \tilde{\overline{W}}_{4c}\right) \\ &+ \frac{1}{2\eta_5} tr\left(\tilde{\underline{B}}_4^T \tilde{\underline{B}}_4\right) + \frac{1}{2\eta_6} tr\left(\tilde{\overline{B}}_4^T \tilde{\overline{B}}_4\right) \\ &+ \frac{1}{2\eta_7} tr\left(\tilde{\underline{W}}_3^T \tilde{\underline{W}}_3\right) + \frac{1}{2\eta_8} tr\left(\tilde{\overline{W}}_3^T \tilde{\overline{W}}_3\right) \\ &+ \frac{1}{2\eta_9} tr\left(\tilde{\underline{C}}_2^T \tilde{\underline{C}}_2\right) + \frac{1}{2\eta_{10}} tr\left(\tilde{\overline{C}}_2^T \tilde{\overline{C}}_2\right) \\ &+ \frac{1}{2\eta_{11}} tr\left(\tilde{\underline{B}}_2^T \tilde{\underline{B}}_2\right) + \frac{1}{2\eta_{12}} tr\left(\tilde{\overline{B}}_2^T \tilde{\overline{B}}_2\right) \\ &+ \frac{1}{2\eta_{13}} tr\left(\tilde{\underline{W}}_5^T \tilde{\underline{W}}_5\right) + \frac{1}{2\eta_{14}} tr\left(\tilde{\overline{W}}_5^T \tilde{\overline{W}}_5\right) \\ &+ \frac{1}{2\eta_{15}} \tilde{k}_1^2 + \frac{Cbk_2}{2}\left(\int sgn(s)dt\right)^2 \end{aligned} \tag{70}$$

The derivative of Equation (70) is obtained:

$$\begin{aligned} \dot{V} &= sgn(s) \cdot \dot{s} + \frac{1}{\eta_1} \dot{\tilde{\underline{W}}}_4^T \tilde{\underline{W}}_4 + \frac{1}{\eta_2} \dot{\tilde{\overline{W}}}_4^T \tilde{\overline{W}}_4 + \frac{1}{\eta_3} \dot{\tilde{\underline{W}}}_{4c}^T \tilde{\underline{W}}_{4c} \\ &+ \frac{1}{\eta_4} \dot{\tilde{\overline{W}}}_{4c}^T \tilde{\overline{W}}_{4c} + \frac{1}{\eta_5} \dot{\tilde{\underline{B}}}_4^T \tilde{\underline{B}}_4 + \frac{1}{\eta_6} \dot{\tilde{\overline{B}}}_4^T \tilde{\overline{B}}_4 + \frac{1}{\eta_7} \dot{\tilde{\underline{W}}}_3^T \tilde{\underline{W}}_3 \\ &+ \frac{1}{\eta_8} \dot{\tilde{\overline{W}}}_3^T \tilde{\overline{W}}_3 + \frac{1}{\eta_9} \dot{\tilde{\underline{C}}}_2^T \tilde{\underline{C}}_2 + \frac{1}{\eta_{10}} \dot{\tilde{\overline{C}}}_2^T \tilde{\overline{C}}_2 + \frac{1}{\eta_{11}} \dot{\tilde{\underline{B}}}_2^T \tilde{\underline{B}}_2 \\ &+ \frac{1}{\eta_{12}} \dot{\tilde{\overline{B}}}_2^T \tilde{\overline{B}}_2 + \frac{1}{\eta_{13}} \dot{\tilde{\underline{W}}}_5^T \tilde{\underline{W}}_5 + \frac{1}{\eta_{14}} \tilde{\overline{W}}_5^T \dot{\tilde{\overline{W}}}_5 + \frac{1}{\eta_{15}} \tilde{k}_1 \cdot \dot{\tilde{k}}_1 \\ &+ Cbk_2 sgn(s) \int sgn(s) dt \end{aligned} \tag{71}$$

To simplify the proof process, define that:

$$\begin{aligned} \dot{H} &= \frac{1}{\eta_1} \dot{\tilde{\underline{W}}}_4^T \tilde{\underline{W}}_4 + \frac{1}{\eta_2} \dot{\tilde{\overline{W}}}_4^T \tilde{\overline{W}}_4 + \frac{1}{\eta_3} \dot{\tilde{\underline{W}}}_{4c}^T \tilde{\underline{W}}_{4c} + \frac{1}{\eta_4} \dot{\tilde{\overline{W}}}_{4c}^T \tilde{\overline{W}}_{4c} \\ &+ \frac{1}{\eta_5} \dot{\tilde{\underline{B}}}_4^T \tilde{\underline{B}}_4 + \frac{1}{\eta_6} \dot{\tilde{\overline{B}}}_4^T \tilde{\overline{B}}_4 + \frac{1}{\eta_7} \dot{\tilde{\underline{W}}}_3^T \tilde{\underline{W}}_3 + \frac{1}{\eta_8} \dot{\tilde{\overline{W}}}_3^T \tilde{\overline{W}}_3 \\ &+ \frac{1}{\eta_9} \dot{\tilde{\underline{C}}}_2^T \tilde{\underline{C}}_2 + \frac{1}{\eta_{10}} \dot{\tilde{\overline{C}}}_2^T \tilde{\overline{C}}_2 + \frac{1}{\eta_{11}} \dot{\tilde{\underline{B}}}_2^T \tilde{\underline{B}}_2 + \frac{1}{\eta_{12}} \dot{\tilde{\overline{B}}}_2^T \tilde{\overline{B}}_2 \\ &+ \frac{1}{\eta_{13}} \tilde{\underline{W}}_5^T \dot{\tilde{\underline{W}}}_5 + \frac{1}{\eta_{14}} \tilde{\overline{W}}_5^T \dot{\tilde{\overline{W}}}_5 + \frac{1}{\eta_{15}} \tilde{k}_1 \cdot \dot{\tilde{k}}_1 \end{aligned} \tag{72}$$

Substituting (30) and (72) into (71) obtains:

$$\begin{aligned} \dot{V} &= Csgn(s) \cdot (f + bu - \dot{r}) + \dot{H} + Cbk_2 sgn(s) \int sgn(s)dt \\ &= Cs(f - \hat{f}) - Cb\hat{k}_1 |s|^{\frac{1}{2}} + \dot{H} \end{aligned} \tag{73}$$

If the conditions in Theorem 1 are satisfied, one can obtained:

$$\begin{aligned} \dot{V} &= Csgn(s) \cdot \Delta_0 - Cb\hat{k}_1 |s|^{\frac{1}{2}} + \frac{1}{\eta_{15}} \tilde{k}_1 \cdot \dot{\tilde{k}}_1 \\ &= Csgn(s) \cdot \Delta_0 - Cbk_1^* |s|^{\frac{1}{2}} \\ &\leq -Cbk_1^* |s|^{\frac{1}{2}} + C|\Delta_0| \end{aligned} \tag{74}$$

Suppose that the total higher order approximation error $|\Delta_0|$ has an upper bound, that is, $|\Delta_0| < \Delta_{max}$. It can be obtained that:

$$\begin{cases} \dot{V} \leq 0 & |s| \geq (\frac{\Delta_{max}}{bk_1^*})^2 \\ \dot{V} > 0 & |s| < (\frac{\Delta_{max}}{bk_1^*})^2 \end{cases} \quad (75)$$

When the system is stable, it is clear that s will tend to 0, and when $\dot{V} > 0$, the changes of s is unknown. However, it can be concluded that s will eventually converge to:

$$-(\frac{\Delta_{max}}{bk_1^*})^2 \leq s \leq (\frac{\Delta_{max}}{bk_1^*})^2 \quad (76)$$

When it comes to $s = 0$, it is easy to find that it meets the requirement in Equation (76).

$$-\frac{1}{C}(\frac{\Delta_{max}}{bk_1^*})^2 \leq e \leq \frac{1}{C}(\frac{\Delta_{max}}{bk_1^*})^2 \quad (77)$$

As for e, it can be easily to conclude that e is bounded and the system is stable. □

Remark 1. *When the neural network performs parameter self-learning, matrix operations and partial derivative calculations are required, which will inevitably increase the amount of calculation. Table 1 gives the calculation time of the four control methods. Although the proposed method increases the computational complexity, the computational burden of the design method is relatively small while meeting the APF performance requirements. However, further optimization remains to be done.*

Table 1. The computational time of four control method.

Strategy	ASMC	FNNASMC-SFR based on LESO [41]	IT2FNN-SFR STSMC	CTSMC-MLNN [42]
Time(s)	5.5619	15.795	28.234	32.829

4. Numerical Verification

The effectiveness and reliability of the novel control strategy for APF system is verified on MATLAB/Simulink. The simulation experiment and comparative results are introduced below.

In the simulation process, the CPU is i5-8300H(2.30 GHz), the system is 64-bit and the version of MATLAB is 2019b.

A traditional PI controller is used in the DC voltage control of APF. The parameters of PI controller are given as: $K_P = 0.15$, $K_I = 0.02$. The designed controller in this paper is applied in the current control of APF system. The parameters of the membership function of IT2FNN in this article are selected by experts as follows:

$$\underline{C_2} = \begin{bmatrix} 0.4 & 0.02 & -0.02 \\ 8 & 1 & -1 \end{bmatrix}; \underline{B_2} = \begin{bmatrix} 1 & 1 & 1 \\ 5 & 5 & 5 \end{bmatrix}$$

$$\overline{C_2} = \begin{bmatrix} 0.6 & 0.05 & -0.05 \\ 10 & 2 & -2 \end{bmatrix}; \overline{B_2} = \begin{bmatrix} 1.2 & 1.2 & 1.2 \\ 3 & 3 & 3 \end{bmatrix}$$

In addition, the parameters of the STSMC and the related parameters of the adaptive law are selected as follows: $C = 10$, $k_2 = 2.5$, $\eta_1 = 2250$, $\eta_2 = 2250$, $\eta_3 = 50$, $\eta_4 = 50$, $\eta_5 = 375$, $\eta_6 = 375$, $\eta_7 = 75$, $\eta_8 = 75$, $\eta_9 = 15$, $\eta_{10} = 15$, $\eta_{11} = 40$, $\eta_{12} = 40$, $\eta_{13} = 20,000$, $\eta_{14} = 22,500$, $\eta_{15} = 0.0009$.

The other parameters used in the simulation process are shown in Table 2.

Table 2. Parameters for Simulation.

Parameters	Values
Supply voltage	24 V/50 Hz
APF main circuit	$L = 18$ mH, $R = 0.1$ Ω, $C_0 = 2200$ μF, $U_{dc} = 50$ V
Non-linear load at steady state	$R_1 = 5$ Ω, $R_2 = 15$ Ω, $C_1 = 1000$ uF
Additional non-linear load in parallel	$R_1 = 15$ Ω, $R_2 = 15$ Ω, $C_2 = 1000$ uF
Sampling time	$T_s = 1\,e^{-5}$ s

When the APF does not work out, the power supply current is the same as the load current, which is shown in Figure 4. It is not difficult to find that the power supply current is severely polluted by distortion, and the THD of the power supply current is 40.30% (as shown in Figure 5). At 0 s, APF is integrated into the power grid for harmonic compensation. The power supply current is shown in Figure 6, and it recovers to a sine wave in a very short time. The THD of supply current, as shown in Figure 7, is only 2.37%, which greatly eliminates harmonics and significantly improves current quality. Figures 8 and 9 are the compensation current tracking curve and tracking error curve under the designed controller. In the Figure 8, the red line represents the reference current and the blue line represents the compensation current. It can be seen that the compensation current i_c can track the expected reference current i_r perfectly, the tracking error can approach to zero in a very short time, which further verifies that the designed method has a great ability of harmonic compensation.

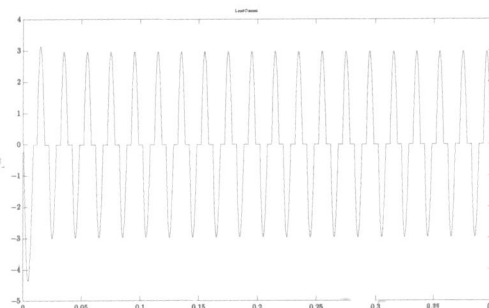

Figure 4. The power supply current before compensation.

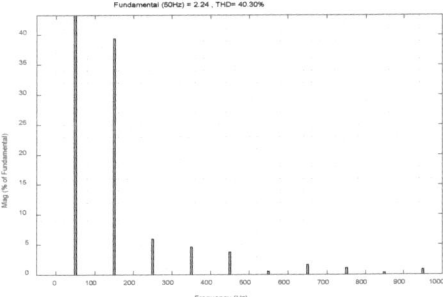

Figure 5. The spectrum analysis of power supply current before compensation.

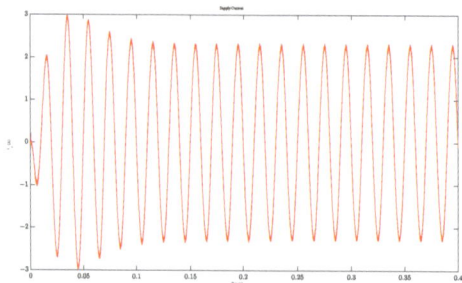

Figure 6. The power supply current after compensation.

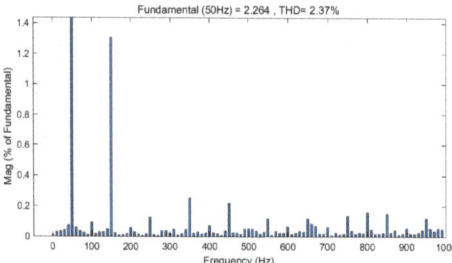

Figure 7. The spectrum analysis of power supply current after compensation.

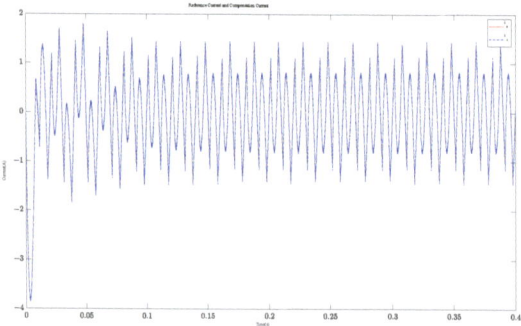

Figure 8. The compensation current tracking curve.

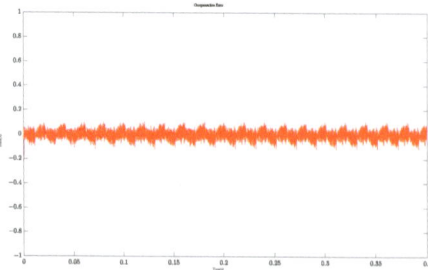

Figure 9. The compensation current tracking error curve.

At the same time, to verify the dynamic performance of the designed method, an additional nonlinear load was connected to the load at 0.5 s to make the total load of the circuit smaller, and the parallel load was disconnected at 0.7 s to make the total load of the circuit larger. Figure 10 describes the power supply current curves of 0~0.9 s, 0.4~0.6 s, and 0.6~0.8 s from top to bottom, respectively. It can be seen from both the overall power supply current diagram and the amplification diagram of 0.1 s before and after the load change that the power supply current can maintain the sinusoidal waveform. According to Figures 11 and 12, the compensation current tracking curve and error curve still have good tracking performance when the load changes. The frequency spectrum analysis of the power supply current at 0.6 s and 0.8 s is shown in Figures 13 and 14. The THD is 1.64% after load connection and 2.58% after load disconnection, both of which have good performance.

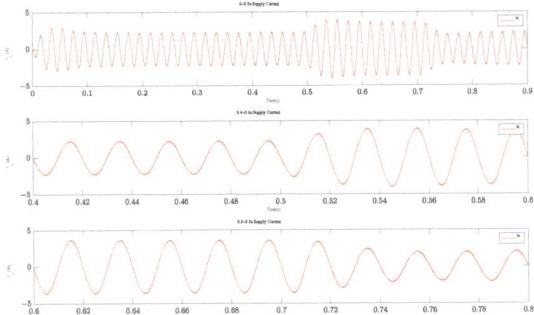

Figure 10. The power supply current of simulation experiment.

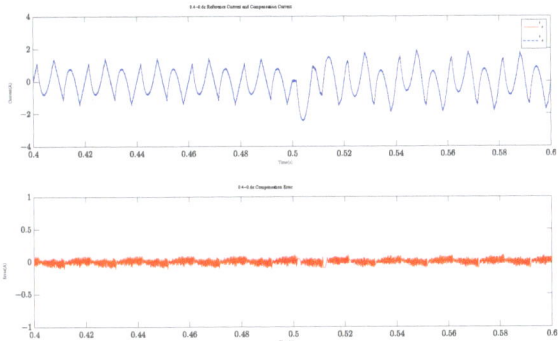

Figure 11. Current tracking and error curve when load is connected.

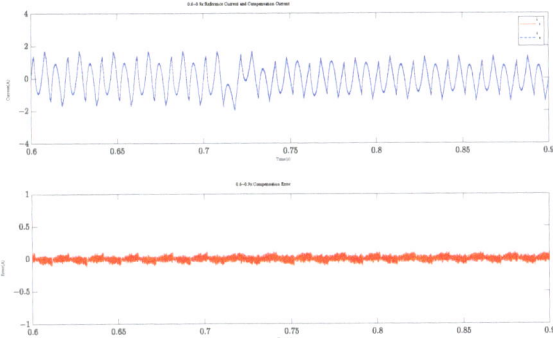

Figure 12. Current tracking and error curve when load is disconnected.

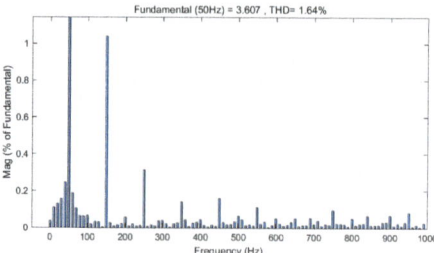

Figure 13. Power supply current spectrum analysis after load is connected.

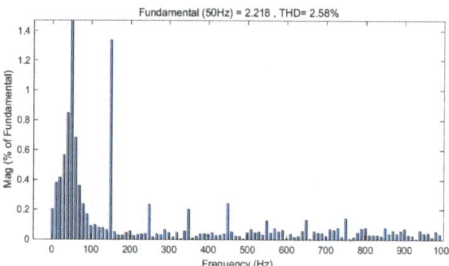

Figure 14. Power supply current spectrum analysis after load is disconnected.

In this paper, the output of the IT2FNN-SFR is taken as the practical mathematical model of APF, and the estimated curve is shown in Figure 15. The estimated curve has almost the same phase and similar amplitude compared with the ideal curve. Although there are many burrs and mutations, the estimated curve is more consistent with the actual situation for the system with actual parameter perturbation and external disturbance. Meanwhile, the adaptive gain of STSMC is shown in Figure 16, which estimates the stable value in a very short time, and its stable value is about 0.3.

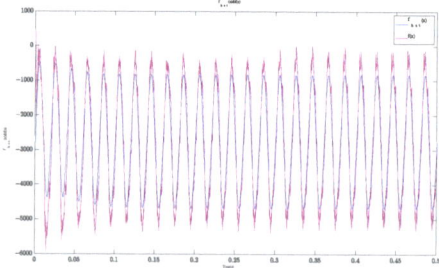

Figure 15. The output of the IT2FNN-SFR.

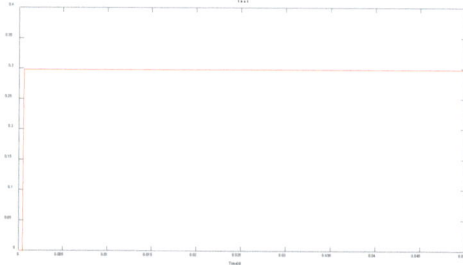

Figure 16. The gain of the adaptive STSMC.

At the same time, the adaptive curves of internal parameters in the designed neural network are depicted in Figure 17a–g. As shown in the figure, internal parameters will be adjusted adaptively and stabilized to the optimal value in a short time, which signifies that the NN has strong robustness and self-adaptive performance, and further verifies the superiority of the designed network.

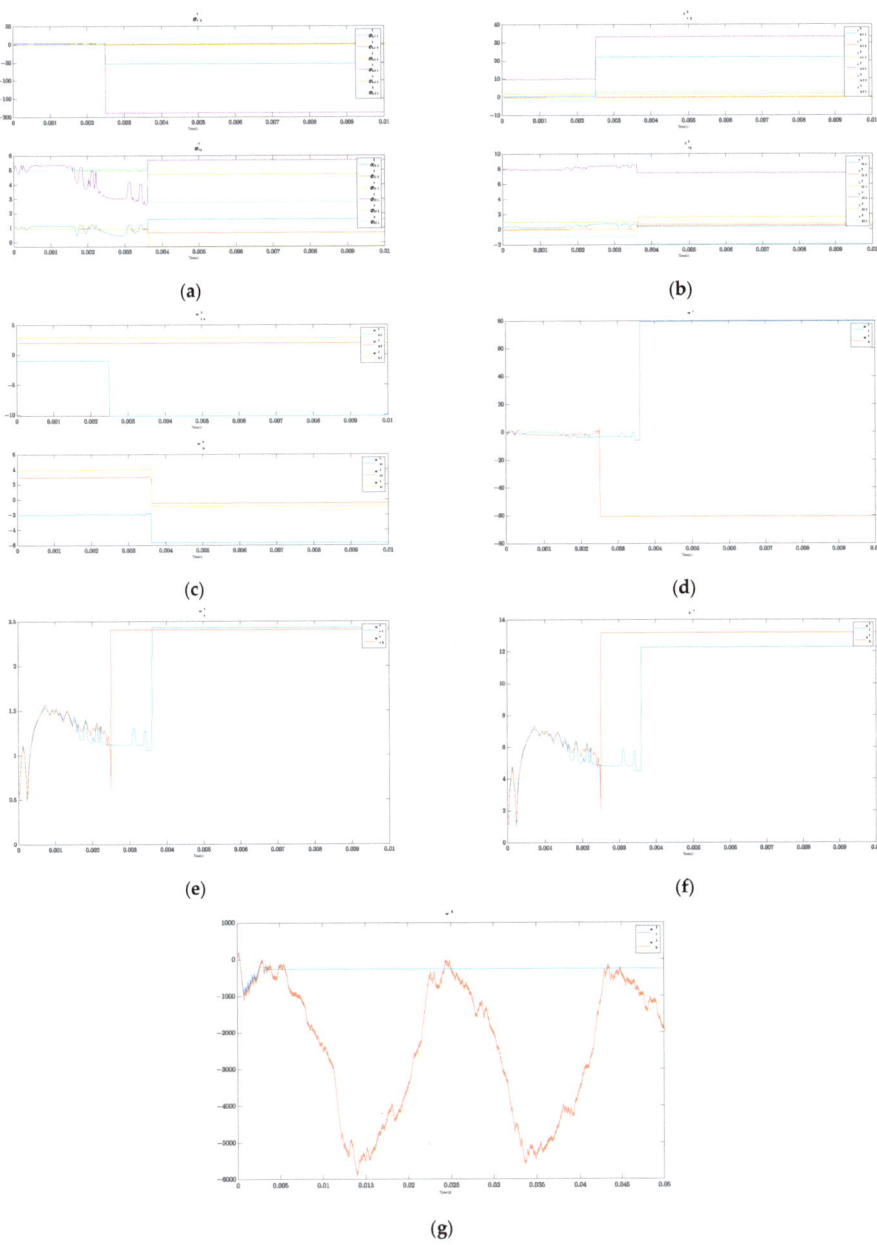

Figure 17. The adaptive curves of internal parameters in the IT2FNN-SFR. (**a**) The base width curves of membership function. (**b**) The center vector curves of membership function. (**c**) The input weight curves of SFR layer. (**d**) The weight curves of the SFR layer. (**e**) The self-feedback weight curves of SFR layer. (**f**) The bias term curves of SFR layer. (**g**) The output weight curves.

To verify that the proposed method in this paper can significantly weaken the inherent chattering in SMC, the control output comparison curve is given in Figure 18. Compared with the ASMC method (red curve), IT2FNN-SFR STSMC (blue curve) has a smaller control output amplitude range in a certain time period (take 0.2~0.3 s as an example). In addition, Table 3 gives the variance index of the control output, which can also verify the above conclusions.

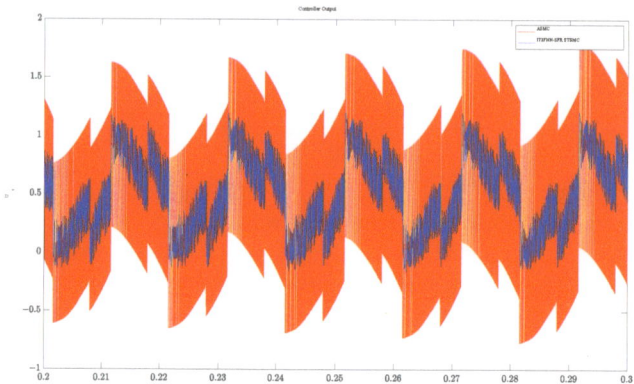

Figure 18. The control output comparison curve.

Table 3. The output variance of the two controllers.

Strategy Index	IT2FNN-SFR STSMC	ASMC
Output variance	0.0650	0.1362

Remark 2. *APF is a high-frequency switching power electronics, and its system output chattering is an important indicator to measure system performance. Variance is a statistic that measures the degree of dispersion of data, but large outliers may cause data skew, so the output is standardized as follows:*

$$u^*(t) = \frac{u(t) - u(t)_{max}}{u(t)_{max} - u(t)_{min}} \tag{78}$$

The variance of the standardized control $u(t)^*$ *can be easily calculated, and the results are shown in Table 3.*

Meanwhile, the superiority of IT2FNN-SFR STSMC controller is illustrated by comparing with the adaptive sliding mode controller (ASMC). Figure 19 shows the error comparison curves of the two controllers. Compared with ASMC, the designed controller can track the desired signal more quickly and has a faster response speed at the initial moment. The overall tracking error is slightly less than that of ASMC controller, which indicates the compensation effect of the IT2FNN-SFR STSMC controller is better. The corresponding THD of the two controllers is also given in Table 4. It can be seen that the THD of the designed IT2FNN-SFR STSMC controller is slightly better than that of the ASMC controller in startup, steady state, and load changes.

Table 4. The corresponding THD of the two controllers.

Control Strategy	0 s	0.2 s	0.6 s	0.8 s
IT2FNN-SFR STSMC	16.44%	2.37%	1.64%	2.58%
ASMC	20.03%	2.66%	1.68%	2.92%

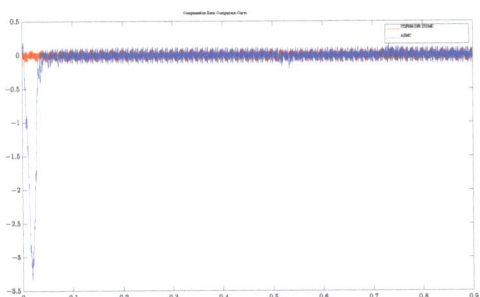

Figure 19. The compensation current error contrast curve.

5. Conclusions

In this paper, an adaptive super-twisting sliding mode control strategy for single-phase active power filter is proposed by combining interval type-2 fuzzy neural network with self-feedback recursion structure. The advantages of the IT2FNN and RNN are combined in the novel IT2FNN-SFR to obtain better dynamic approximation capacity. The new IT2FNN-SFR estimator is designed to approximate the unknown model of the system, improving the nonlinear processing ability and accuracy by using interval tpye-2 fuzzy membership function. At the same time, the self-feedback recursive structure can use the historical information to adjust the output, which improves the approximation ability of the network. The recursive structure in the NN will store the historical information to improve the accuracy of estimation and dynamic approximation character. The STSMC has cannot only effectively reduce system chattering, improve control accuracy, robustness, and system performance., but also simplify the parameter setting process. Finally, the simulation results show that the controller has good steady-state performance and dynamic response, and can effectively suppress the harmonics of the APF system.

Author Contributions: Conceptualization, J.F.; methodology, J.W.; writing—original draft preparation, J.W.; writing—review and editing, J.F. and Y.F. All authors have read and agreed to the published version of the manuscript.

Funding: This work is partially supported by National Science Foundation of China under Grant No. 62273131.

Data Availability Statement: All relevant data are within the manuscript.

Conflicts of Interest: The authors declare no conflict of interest.

References

1. Wu, Z.; Yang, Z.; Ding, K.; He, G. Order-Domain-Based Harmonic Injection Method for Multiple Speed Harmonics Suppression of PMSM. *IEEE Trans. Power Electron.* **2021**, *36*, 4478–4487. [CrossRef]
2. RGregory, R.; Azevedo, C.; Santos, I. Study of Harmonic Distortion Propagation from a Wind Park. *IEEE Lat. Am. Trans.* **2020**, *18*, 1077–1084. [CrossRef]
3. Jiao, N.; Wang, S.; Ma, J.; Chen, X.; Liu, T.; Zhou, D.; Yang, Y. The Closed-Loop Sideband Harmonic Suppression for CHB Inverter with Unbalanced Operation. *IEEE Trans. Power Electron.* **2022**, *37*, 5333–5341. [CrossRef]
4. Fu, J.; Chen, L.; Zhao, H.; Zhang, P. High and Low Frequency Control Strategy for APF DC Side Ripple Voltage under Unbalanced Load. In Proceedings of the 2019 IEEE 2nd International Conference on Electronics and Communication Engineering (ICECE), Xi'an, China, 9–11 December 2019; pp. 326–330.
5. Vahedi, H.; Shojaei, A.A.; Dessaint, L.-A.; Al-Haddad, K. Reduced DC-Link Voltage Active Power Filter Using Modified PUC5 Converter. *IEEE Trans. Power Electron.* **2018**, *33*, 943–947. [CrossRef]
6. Incremona, G.P.; Rubagotti, M.; Ferrara, A. Sliding Mode Control of Constrained Nonlinear Systems. *IEEE Trans. Autom. Control* **2017**, *62*, 2965–2972. [CrossRef]
7. Hou, L.; Ma, J.; Wang, W. Sliding Mode Predictive Current Control of Permanent Magnet Synchronous Motor with Cascaded Variable Rate Sliding Mode Speed Controller. *IEEE Access* **2022**, *10*, 33992–34002. [CrossRef]
8. Qu, L.; Qiao, W.; Qu, L. An Extended-State-Observer-Based Sliding-Mode Speed Control for Permanent-Magnet Synchronous Motors. *IEEE J. Emerg. Sel. Top. Power Electron.* **2021**, *9*, 1605–1613. [CrossRef]

9. Chen, X.; Li, Y.; Ma, H.; Tang, H.; Xie, Y. A Novel Variable Exponential Discrete Time Sliding Mode Reaching Law. *IEEE Trans. Circuits Syst. II Express Briefs* **2021**, *68*, 2518–2522. [CrossRef]
10. Liu, H.; Wang, H.; Sun, J. Attitude control for QTR using exponential nonsingular terminal sliding mode control. *J. Syst. Eng. Electron.* **2019**, *30*, 191–200.
11. Ren, L.; Lin, G.; Zhao, Y.; Liao, Z.; Peng, F. Adaptive Nonsingular Finite-Time Terminal Sliding Mode Control for Synchronous Reluctance Motor. *IEEE Access* **2021**, *9*, 51283–51293. [CrossRef]
12. Fei, J.; Wang, H.; Fang, Y. Novel Neural Network Fractional-Order Sliding-Mode Control with Application to Active Power Filter. *IEEE Trans. Syst. Man Cybern. Syst.* **2022**, *52*, 3508–3518. [CrossRef]
13. Bo, L.; Peng, J.; Chong, K.T.; Rodriguez, J.; Guerrero, J.M. Robust Fuzzy-Fractional-Order Nonsingular Terminal Sliding-Mode Control of LCL-Type Grid-Connected Converters. *IEEE Trans. Ind. Electron.* **2022**, *69*, 5854–5866.
14. Liu, Y.-J.; Chen, H. Adaptive Sliding Mode Control for Uncertain Active Suspension Systems with Prescribed Performance. *IEEE Trans. Syst. Man Cybern. Syst.* **2021**, *51*, 6414–6422. [CrossRef]
15. Roy, S.; Roy, S.B.; Lee, J.; Baldi, S. Overcoming the Underestimation and Overestimation Problems in Adaptive Sliding Mode Control. *IEEE/ASME Trans. Mechatron.* **2019**, *24*, 2031–2039. [CrossRef]
16. Roy, S.; Baldi, S.; Fridman, L.M. On adaptive sliding mode control without a priori bounded uncertainty. *Automatica* **2020**, *111*, 108650. [CrossRef]
17. Levant, A. Higher-order sliding modes, differentiation and output-feedback control. *Int. J. Control.* **2003**, *76*, 924–941. [CrossRef]
18. Guo, J. The Load Frequency Control by Adaptive High Order Sliding Mode Control Strategy. *IEEE Access* **2022**, *10*, 25392–25399. [CrossRef]
19. Shah, I.; Rehman, F.U. Smooth Second Order Sliding Mode Control of a Class of Underactuated Mechanical Systems. *IEEE Access* **2018**, *6*, 7759–7771. [CrossRef]
20. Chalanga, A.; Kamal, S.; Fridman, L.M.; Bandyopadhyay, B.; Moreno, J.A. Implementation of Super-Twisting Control: Super-Twisting and Higher Order Sliding-Mode Observer-Based Approaches. *IEEE Trans. Ind. Electron.* **2016**, *63*, 3677–3685. [CrossRef]
21. Wu, Y.; Ma, F.; Liu, X.; Hua, Y.; Liu, X.; Li, G. Super Twisting Disturbance Observer-Based Fixed-Time Sliding Mode Backstepping Control for Air-Breathing Hypersonic Vehicle. *IEEE Access* **2021**, *8*, 17567–17583. [CrossRef]
22. Wang, W.; Ma, J.; Li, X.; Cheng, Z.; Zhu, H.; Teo, C.S.; Lee, T.H. Iterative Super-Twisting Sliding Mode Control for Tray Indexing System with Unknown Dynamics. *IEEE Trans. Ind. Electron.* **2020**, *68*, 9855–9865. [CrossRef]
23. Merayo, N.; Juárez, D.; Aguado, J.C.; de Miguel, I.; Durán, R.J.; Fernández, P.; Lorenzo, R.M.; Abril, E.J. PID Controller Based on a Self-Adaptive Neural Network to Ensure QoS Bandwidth Requirements in Passive Optical Networks. *J. Opt. Commun. Netw.* **2017**, *9*, 433–445. [CrossRef]
24. Dai, W.; Zhang, L.; Fu, J.; Chai, T.; Ma, X. Dual-Rate Adaptive Optimal Tracking Control for Dense Medium Separation Process Using Neural Networks. *IEEE Trans. Neural Netw. Learn. Syst.* **2021**, *32*, 4202–4216. [CrossRef]
25. Yang, Y.; Ye, Y. Backstepping sliding mode control for uncertain strict-feedback nonlinear systems using neural-network-based adaptive gain scheduling. *J. Syst. Eng. Electron.* **2018**, *29*, 580–586.
26. Li, K.; Li, Y. Adaptive Neural Network Finite-Time Dynamic Surface Control for Nonlinear Systems. *IEEE Trans. Neural Netw. Learn. Syst.* **2021**, *32*, 5688–5697. [CrossRef]
27. Liu, X.; Su, C.-Y.; Yang, F. FNN Approximation-Based Active Dynamic Surface Control for Suppressing Chatter in Micro-Milling with Piezo-Actuators. *IEEE Trans. Syst. Man Cybern. Syst.* **2017**, *47*, 2100–2113. [CrossRef]
28. Wai, R.; Yao, J.; Lee, J. Backstepping Fuzzy-Neural-Network Control Design for Hybrid Maglev Transportation System. *IEEE Trans. Neural Netw. Learn. Syst.* **2015**, *26*, 302–317. [PubMed]
29. Wai, R.-J.; Chen, M.-W.; Liu, Y.-K. Design of Adaptive Control and Fuzzy Neural Network Control for Single-Stage Boost Inverter. *IEEE Trans. Ind. Electron.* **2015**, *62*, 5434–5445. [CrossRef]
30. Liu, L.; Fei, J. Extended State Observer Based Interval Type-2 Fuzzy Neural Network Sliding Mode Control with Its Application in Active Power Filter. *IEEE Trans. Power Electron.* **2022**, *37*, 5138–5154. [CrossRef]
31. Wang, J.; Luo, W.; Liu, J.; Wu, L. Adaptive Type-2 FNN-Based Dynamic Sliding Mode Control of DC–DC Boost Converters. *IEEE Trans. Syst. Man, Cybern. Syst.* **2021**, *51*, 2246–2257. [CrossRef]
32. Gao, Y.; Liu, J.; Wang, Z.; Wu, L. Interval Type-2 FNN-Based Quantized Tracking Control for Hypersonic Flight Vehicles with Prescribed Performance. *IEEE Trans. Syst. Man Cybern. Syst.* **2021**, *51*, 1981–1993. [CrossRef]
33. Kim, C.-J.; Chwa, D. Obstacle Avoidance Method for Wheeled Mobile Robots Using Interval Type-2 Fuzzy Neural Network. *IEEE Trans. Fuzzy Syst.* **2015**, *23*, 677–687. [CrossRef]
34. Wai, R.-J.; Lin, C.-M.; Peng, Y.-F. Adaptive Hybrid Control for Linear Piezoelectric Ceramic Motor Drive Using Diagonal Recurrent CMAC Network. *IEEE Trans. Neural Netw.* **2004**, *15*, 1491–1506. [CrossRef] [PubMed]
35. Yogi, S.C.; Tripathi, V.K.; Behera, L. Adaptive Integral Sliding Mode Control Using Fully Connected Recurrent Neural Network for Position and Attitude Control of Quadrotor. *IEEE Trans. Neural Netw. Learn. Syst.* **2021**, *32*, 5595–5609. [CrossRef]
36. Fei, J.; Chen, Y.; Liu, L.; Fang, Y. Fuzzy Multiple Hidden Layer Recurrent Neural Control of Nonlinear System Using Terminal Sliding-Mode Controller. *IEEE Trans. Cybern.* **2022**, *52*, 9519–9534. [CrossRef]
37. Fei, J.; Liu, L. Real-Time Nonlinear Model Predictive Control of Active Power Filter Using Self-Feedback Recurrent Fuzzy Neural Network Estimator. *IEEE Trans. Ind. Electron.* **2022**, *69*, 8366–8376. [CrossRef]

38. Han, H.-G.; Zhang, L.; Hou, Y.; Qiao, J.-F. Nonlinear Model Predictive Control Based on a Self-Organizing Recurrent Neural Network. *IEEE Trans. Neural Networks Learn. Syst.* **2016**, *27*, 402–415. [CrossRef]
39. Fei, J.; Wang, Z.; Pan, Q. Self-Constructing Fuzzy Neural Fractional-Order Sliding Mode Control of Active Power Filter. *IEEE Trans. Neural Networks Learn. Syst.* **2022**. [CrossRef]
40. Fei, J.; Wang, Z.; Fang, Y. Self-Evolving Chebyshev Fuzzy Neural Fractional-Order Sliding Mode Control for Active Power Filter. *IEEE Trans. Ind. Inform.* **2022**, *19*, 2729–2739. [CrossRef]
41. Wang, J.; Fei, J. Self-Feedback Neural Network Sliding Mode Control with Extended State Observer for Active Power Filter. *IEEE Internet Things J.* **2023**. [CrossRef]
42. Zhang, L.; Fei, J. Intelligent Complementary Terminal Sliding Mode Using Multi-Loop Neural Network for Active Power Filter. *IEEE Trans. Power Electron.* **2023**. [CrossRef]

Disclaimer/Publisher's Note: The statements, opinions and data contained in all publications are solely those of the individual author(s) and contributor(s) and not of MDPI and/or the editor(s). MDPI and/or the editor(s) disclaim responsibility for any injury to people or property resulting from any ideas, methods, instructions or products referred to in the content.

Article

An Extended-State Observer Based on Smooth Super-Twisting Sliding-Mode Controller for DC-DC Buck Converters

Dian Jiang [1], Yunmei Fang [2] and Juntao Fei [1,*]

[1] Jiangsu Key Laboratory of Power Transmission and Distribution Equipment Technology, College of Information Science and Engineering, Hohai University, Changzhou 213022, China
[2] College of Mechanical and Electrical Engineering, Hohai University, Changzhou 213022, China
* Correspondence: jtfei@hhu.edu.cn

Abstract: This paper designs a novel smooth super-twisting extended-state observer (SSTESO)-based smooth super-twisting sliding-mode control (SSTSMC) scheme to promote the robust ability and voltage-tracking performance of DC-DC buck converters. First, an SSTESO is proposed to estimate the unknown lumped disturbance and compensate for the estimation of the voltage controller. The SSTESO is realized by constructing a novel smooth function to replace the nonlinear sign function in STESO, which can provide a faster convergence speed and higher estimation accuracy. The SSTSM controller is designed by adopting a similar smooth function to further suppress chattering and improve dynamic response. Comprehensive simulation results demonstrate that the proposed SSTESO-based SSTSMC scheme can improve the robustness and transient response of a DC-DC buck converter system in the presence of external disturbance and parameter uncertainties.

Keywords: smooth super-twisting sliding-mode control (SSTSMC); smooth super-twisting extended-state observer (SSTESO); DC-DC buck converter; unknown lumped disturbance

MSC: 93C40; 93C10

Citation: Jiang, D.; Fang, Y.; Fei, J. An Extended-State Observer Based on Smooth Super-Twisting Sliding-Mode Controller for DC-DC Buck Converters. *Mathematics* **2023**, *11*, 2835. https://doi.org/10.3390/math11132835

Academic Editors: Adrian Olaru, Gabriel Frumusanu and Catalin Alexandru

Received: 24 May 2023
Revised: 16 June 2023
Accepted: 21 June 2023
Published: 24 June 2023

Copyright: © 2023 by the authors. Licensee MDPI, Basel, Switzerland. This article is an open access article distributed under the terms and conditions of the Creative Commons Attribution (CC BY) license (https://creativecommons.org/licenses/by/4.0/).

1. Introduction

The sustainable development of humanity requires a wider use of renewable energy sources for electricity generation that have the advantages of high reliability, little need for maintenance, and independence from the supply of fossil fuels. Considering that renewable energy sources usually provide variable DC output voltage, DC-DC buck converters have played an important role in providing adequate power sources for electronic systems and have been extensively adopted in photovoltaic systems [1–3], fuel-cell hybrid systems [4], energy storage systems [5], etc. In addition, DC-DC buck converters are also used in many intelligent fields, such as modular drivers for LEDs [6] and off-chip components for Internet of Things applications [7–9]. The function of DC-DC buck converters is to convert DC input voltage into another fixed or adjustable DC output voltage, to realize the stable flow of energy.

The main control target of the DC-DC buck converter is to regulate the output voltage and track reference voltage accurately and quickly. However, as a typical nonlinear system, the DC-DC buck converter system contains both external disturbances and parameter uncertainties. It may be difficult to obtain excellent performance using a conventional linear control algorithm. In addition, some application scenarios produce higher voltage accuracy and more stable current for DC-DC buck converters. Therefore, maintaining high-precision voltage-tracking performance and superior robustness in the buck converter has become a research hotspot.

In the early literature, linear controllers such as PI and PID, which maintain acceptable performance around a specific operating point, were widely used in buck converters. However, these linear controllers were sensitive to time-varying external disturbances.

In recent decades, more nonlinear control strategies have been applied to DC-DC buck converters, such as model predictive control [10], neural network control [11], adaptive control [12], optimal control [13], sliding-mode control (SMC) [14–16], etc.

Among the above-mentioned control methods, SMC for buck converters has attracted significant attention due to its superior precision and robustness. However, chattering is a problem for SMC. To address the chattering problem, a saturation function is proposed to replace the sign function in the conventional sliding-mode algorithm [17]. Nevertheless, indefinite steady errors remain. Super-twisting SMC (STSMC) [18] is another way to suppress chattering by adding an integration element and hiding the discontinuous sign function in the integral term. However, STSMC is still essentially a nonsmooth control algorithm, and a control lag of the integration part also exists. Therefore, a chattering problem still exists as the system parameters change. Another problem with SMC is that its robustness is always limited. Better robustness often depends on increasing the value of the switching gain, which may sacrifice both dynamic and steady-state performance.

To break the constrained relationship between the switching gain and robustness in SMC, extended-state observer (ESO)-based composite sliding-mode controllers have been proposed in [19–21]. ESO is the core part of active disturbance rejection control (ADRC), which can estimate the lumped disturbance of a controlled plant using a special mechanism [22,23]. The fact that the estimation error can only be guaranteed to converge to zero asymptotically means that the disturbance will take a long time to be estimated accurately. To speed up the convergence process of a conventional ESO, a super-twisting algorithm was adopted to construct a super-twisting extended-state observer (STESO) by the authors in [24,25]. However, with the introduction of nonlinear functions, a chattering problem is also introduced to STESO. The chattering of disturbance estimation will eventually be superimposed on the control signal, which may make the performance of the composite controller worse. Some intelligent control methods have been developed to control the dynamics systems [26–34].

Based on the above-mentioned analysis, a smooth super-twisting sliding-mode controller (SSTSMC) combined with a smooth STESO (SSTESO) is proposed in this paper to enhance the robustness and dynamic performance of a DC-DC buck converter output-voltage regulation system. To improve the convergence speed and the smoothness of a conventional super-twisting algorithm (STA), a novel smooth switch function is constructed. Replacing the sign function in conventional STSMC and STESO with the proposed smooth function, a novel SSTESO-based SSTSMC scheme is obtained. A widely used Lyapunov function is employed to demonstrate the stability of the presented smooth STA (SSTA). Due to the characteristics of the proposed smooth function, the SSTSMC not only accelerates the convergence process but also improves steady-state and robustness performance compared with the conventional STSMC. With SSTESO, this combines the advantages of conventional ESO and STESO, which greatly accelerates the convergence of the estimation error without introducing the chattering problem into the extended-state observer. Then, the proposed control scheme, combining the SSTESO with SSTSMC, can effectively increase the dynamic response speed and improve steady-state performance and robustness. The main contributions and novelty of this paper are as follows:

(1) A pair of novel smooth functions is constructed to replace the sign function in conventional STA, and the stability of the optimized SSTA is demonstrated. Two sets of SSTESO are designed to estimate the matched and mismatched disturbance in a DC-DC buck converter system. Compared to the traditional ESO, the SSTESO not only accelerates the convergence of estimation error but also guarantees the accuracy of the disturbance estimate.

(2) A smooth STSMC is proposed by adopting the SSTA to increase the dynamic response speed and further reduce chattering. The proposed SSTESO-based composite SSTSMC scheme is successfully applied to the DC-DC buck converter. Performance comparison experiments among the STSMC, SSTSMC, ESO-based SSTSMC, STESO-based

SSTSMC, and SSTESO-based SSTSMC schemes are carried out in simulations that validate the superiority of the proposed control scheme.

2. Conventional STESO-Based STSM Controller Design

2.1. Modeling of a DC-DC Buck Converter

The basic topology of a DC-DC buck converter is shown in Figure 1, which comprises a DC voltage input v_{in}, a PWM gate drive-controlled switch device Q, a diode D, an output filter inductor L, an output filter capacitor C, and a load resistance R. The switch ON and OFF cases of the DC-DC buck converter are shown with dashed lines 1 and 2, respectively.

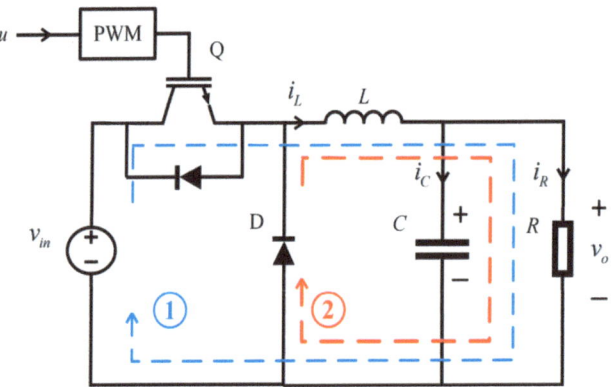

Figure 1. The topology of DC-DC buck converter.

The state-space method is used here to analyze the buck converter system. The dynamic model can be written as:

$$\begin{cases} \frac{di_L}{dt} = -\frac{v_o}{L} + \frac{v_{in}}{L}u \\ \frac{dv_o}{dt} = \frac{i_L}{C} - \frac{v_o}{RC} \end{cases} \quad (1)$$

where v_0 is an output voltage, i_L is an inductor current, and the duty ratio $u \in [0,1]$ denotes the control signal.

The desired output voltage is denoted as v_r. The tracking error can be expressed as $x_1 = v_o - v_r$. It should be noted that the load resistance in practice is usually unknown, and the value of input voltage, filter capacitor, and inductor are not exact. Considering the uncertainties and external disturbances in DC-DC buck converters, the time derivative of tracking error x_1 is as follows:

$$\dot{x}_1 = \frac{i_L}{C_0} - \frac{v_o}{R_0 C_0} + d_1(t) \quad (2)$$

where C_0 and R_0 are the nominal values of capacitor C and load resistance R, respectively, and the lumped disturbance is denoted as $d_1(t) = (\frac{1}{C} - \frac{1}{C_0})i_L + (\frac{1}{R_0 C_0} - \frac{1}{RC})v_0$.

Then, we define $x_2 = \frac{i_L}{C_0} - \frac{v_o}{R_0 C_0}$. Using (1), the derivative of x_2 is written as follows:

$$\dot{x}_2 = \frac{uv_{in0}}{L_0 C_0} - \frac{x_1}{L_0 C_0} - \frac{x_2}{R_0 C_0} - \frac{v_r}{L_0 C_0} + d_2(t) \quad (3)$$

where v_{in0} and L_0 are the nominal values of input voltage v_{in} and inductor L, respectively, and the matched disturbance is denoted as

$$d_2(t) = -\frac{d_1}{R_0 C_0} + (\frac{v_{in}}{LC} - \frac{v_{in0}}{L_0 C_0})u + (\frac{1}{L_0 C_0} - \frac{1}{LC})v_o$$

Therefore, the dynamic model of the DC-DC buck converter can be rewritten as:

$$\begin{cases} \dot{x}_1 = x_2 + d_1(t) \\ \dot{x}_2 = \frac{uv_{in0}}{L_0 C_0} - \frac{x_1}{L_0 C_0} - \frac{x_2}{R_0 C_0} - \frac{v_r}{L_0 C_0} + d_2(t) \end{cases} \quad (4)$$

It can be seen that the dynamic model of the buck converter contains both matched and mismatched disturbances.

The objective of the DC-DC buck converter control is to promptly regulate the output voltage to the desired value, i.e., $v_0 \to v_d$ or $x_1 \to 0$. The closed-loop system should still exhibit good control performance in the case of external disturbances.

2.2. Conventional STESO-Based STSM Controller

In this subsection, the conventional STESO is first employed to estimate the lumped disturbances $d_1(t)$ and $d_2(t)$, which performs as feedforward compensation in the following STSM control scheme design.

Regarding the mismatched disturbance $d_1(t)$ as an extended system state, then the first equation of (4) can be reconstructed as

$$\begin{cases} \dot{z}_1 = x_2 + z_2 \\ \dot{z}_2 = \dot{d}_1(t) \end{cases} \quad (5)$$

where $z_1 = x_1$ and $z_2 = d_1(t)$.

From (5), the estimation of the mismatched disturbance $d_1(t)$ can be easily transformed into the problem of estimating the extended system states z_2. According to [35], the conventional ESO constructed for (5) is given as follows:

$$\begin{cases} e_1 = \hat{z}_1 - z_1 \\ \dot{\hat{z}}_1 = \hat{z}_2 + x_2 - l_1 e_1 \\ \dot{\hat{z}}_2 = -l_2 e_1 \end{cases} \quad (6)$$

where l_1 and l_2 are the parameters of ESO, and $\hat{z}_1$ and $\hat{z}_2$ are the estimations of z_1 and z_2, respectively.

According to [25], STESO can be constructed by replacing the linear term e_1 in ESO with the nonlinear functions $f_1(e_1)$ and $f_2(e_1)$, as follows:

$$\begin{cases} \dot{\hat{z}}_1 = \hat{z}_2 + x_2 - l_1 f_1(e_1) \\ \dot{\hat{z}}_2 = -l_2 f_2(e_1) \end{cases} \quad (7)$$

where nonlinear functions $f_1(e_1)$ and $f_2(e_1)$ are designed based on a generalized super-twisting technique, as follows:

$$\begin{cases} f_1(e_1) = k_1 |e_1|^{1/2} \text{sign}(e_1) \\ f_2(e_1) = k_1^2 \text{sign}(e_1) \end{cases} \quad (8)$$

with $k_1 > 0$.

Similarly, regarding the matched disturbance $d_2(t)$ as an extended system state, a new set of STESO can be constructed to estimate, as follows:

$$\begin{cases} e_3 = \hat{z}_3 - z_3 \\ \dot{\hat{z}}_3 = \hat{z}_4 + \frac{uv_{in0}}{L_0 C_0} - \frac{x_1}{L_0 C_0} - \frac{x_2}{R_0 C_0} - \frac{v_r}{L_0 C_0} - l_3 f_3(e_3) \\ \dot{\hat{z}}_4 = -l_4 f_4(e_3) \end{cases} \quad (9)$$

with

$$\begin{cases} f_3(e_3) = k_2|e_3|^{1/2}\text{sign}(e_3) \\ f_4(e_3) = k_2^2\text{sign}(e_3) \end{cases} \quad (10)$$

where $k_2 > 0$, l_3 and l_4 are the parameters of STESO, $z_3 = x_2$, $z_4 = d_2(t)$, and $\hat{z}_3$ and $\hat{z}_4$ are the estimations of z_3 and z_4, respectively. Then the estimated values can be compensated for by the controller to improve the robustness and transient response of the system.

Since the output voltage v_o of a DC-DC buck converter system is a DC voltage signal, improving the accuracy and stability of the output signal has become the primary control goal. Therefore, an STSM control algorithm with strong chattering suppression and robustness is adopted in the paper.

The STSM control algorithm was first proposed by Levant [36], and its main feature is to smooth out the discontinuous signal in the conventional first-order sliding-mode (FOSM) controller. For System (4), the STSM control law can be designed as follows:

$$\begin{cases} s = cx_1 + x_2 \\ u_e = \frac{1}{v_{in0}}(x_1 + \frac{L_0}{R_0}x_2 + v_r - cL_0C_0x_2) \\ u_s = -\mu_1|s|^{1/2}\text{sign}(s) + u_I \\ \dot{u}_I = -\mu_2\text{sign}(s) \\ u^* = u_e + \frac{L_0C_0}{v_{in0}}u_s \end{cases} \quad (11)$$

where s is the sliding-mode state variable, $c > 0$ is the sliding-mode constant, μ_1 and μ_2 are the control gains, u_e is the equivalent control term, u_s is the switching control term, u_I is the integral term in u_s, and u^* is the control signal.

The stability and finite-time convergence of the STSM controller are proved in previous literature [18].

After estimating the lumped disturbance by STESO, the composite STSM control law is designed as follows:

$$u^* = u_e + \frac{L_0C_0}{v_{in0}}(u_s - c\hat{z}_2 - \hat{z}_4 - \dot{\hat{z}}_2) \quad (12)$$

Consequently, the STESO-based STSM controller for a DC-DC buck converter system is constructed. The estimation of the mismatched disturbance $d_1(t)$ and matched disturbance $d_2(t)$ can be estimated by the actual system state x_1, x_2, and control signal u^*.

This composite control scheme provides a method that depends on accurate feedforward compensation to eliminate the influence of lumped disturbance without sacrificing other control performance. However, with the application of the super-twisting algorithm to ESO, which effectively accelerates the convergence of the estimation error, the chattering problem is also introduced into STESO. The chattering of the disturbance estimation will eventually be superimposed on the control signal, which may affect the dynamic response and static performance of the system. In addition, the chattering suppression ability of STSMC can be further improved by adopting a smooth function to replace the sign function.

3. SSTESO-Based Smooth STSMC Design

3.1. Design of SSTESO

To ensure both convergence speed and smoothness of disturbance estimation, a pair of smooth functions are constructed to replace the sign function in STESO, as follows:

$$\begin{cases} g_1(x) = |x|^{1/2}\arctan(\frac{x}{\alpha_1}) \\ g_2(x) = \arctan(\left|\frac{x}{\alpha_1}\right|)[\frac{1}{2}\arctan(\frac{x}{\alpha_1}) + \frac{x}{\alpha_1 + x^2/\alpha_1}] \end{cases} \quad (13)$$

with $\alpha_1 > 0$.

Remark 1. *A new parameter α_1 is introduced to improve the applicability of the novel smooth functions in different application scenarios. By setting the appropriate parameter α_1, both the response speed and smoothness of the system can be guaranteed, even if the state variables are of different orders of magnitude in different systems.*

Remark 2. *It should be noted that the function gain of $\arctan(e_1/\alpha_1)$ is larger than that of $\text{sign}(e_1)$ when the estimation error is far away from the origin, which can make the estimation error converge to the neighborhood of the origin more rapidly. In addition, the smaller gain of $\arctan(e_1/\alpha_1)$ when the estimation error is near the origin can guarantee the smoothness of disturbance estimation in the zero domain. By this simple analysis, it is concluded that this inverse tangent function is superior to the sign function.*

From (13), the SSTESO to estimate disturbance $d_1(t)$ can be constructed as follows:

$$\begin{cases} \dot{\hat{z}}_1 = \hat{z}_2 + x_2 - l_1 k_1 g_1(e_1) \\ \dot{\hat{z}}_2 = -l_2 k_1^2 g_2(e_1) \end{cases} \tag{14}$$

Letting $e_2 = \hat{z}_2 - z_2$ and subtracting (14) from (5), one obtains:

$$\begin{cases} \dot{e}_1 = -\theta_1 g_1(e_1) + e_2 \\ \dot{e}_2 = -\theta_2 g_2(e_1) - \dot{d}_1(t) \end{cases} \tag{15}$$

where $\theta_1 = l_1 k_1$ and $\theta_2 = l_2 k_1^2$.

Similarly, a new set of SSTESO to estimate matched disturbance $d_2(t)$ can be constructed as follows:

$$\begin{cases} \dot{\hat{z}}_3 = \hat{z}_4 + \frac{uv_{in0}}{L_0 C_0} - \frac{x_1}{L_0 C_0} - \frac{x_2}{R_0 C_0} - \frac{v_r}{L_0 C_0} - l_3 k_2 g_1(e_3) \\ \dot{\hat{z}}_4 = -l_4 k_2^2 g_2(e_3) \end{cases} \tag{16}$$

Letting $e_4 = \hat{z}_4 - z_4$, the estimation error equation can be written as follows:

$$\begin{cases} \dot{e}_3 = -l_3 k_2 g_1(e_3) + e_4 \\ \dot{e}_4 = -l_4 k_2^2 g_2(e_3) - \dot{d}_2(t) \end{cases} \tag{17}$$

which is similar to (15). Therefore, only the convergence of (e_1, e_2) in (15) is analyzed.

Assumption 1. *Considering the mismatched disturbance $d_1(t)$ and matched disturbance $d_2(t)$ in System (4) are continuous and assumed to satisfy the following condition:*

$$\left| \frac{\dot{d}_1(t,e)}{g_2(e_1)} \right| \leq D, \quad \left| \frac{\dot{d}_2(t,e)}{g_2(e_3)} \right| \leq D \tag{18}$$

where D is a positive constant and assume $D > 1$.

Remark 3. *The essence of SSTESO is to estimate the unknown lumped disturbance and compensate for the estimated values to the voltage controller, $|d_i(t,e)|(i = 1,2)$ vanishes as $e_i \to 0 (i = 1,2,3,4)$. Therefore, Assumption 1 is reasonable.*

Theorem 1. *Let $\theta_1 > D > 1$ and θ_2 satisfy*

$$\theta_2 > \frac{D^2}{\theta_1^2 - 1} + D \tag{19}$$

If Assumption 1 is satisfied, the SSTESO designed as (14) for System (5) can drive the estimation errors $(e_1, e_2) \to (0, 0)$.

Proof. Selecting the Lyapunov function for System (15) is as follows:

$$V(e_1, e_2) = \zeta^T P \zeta \quad (20)$$

where $\zeta^T = [|e_1|^{1/2} \arctan(e_1/\alpha_1), e_2]$, and P is a symmetrical matrix constructed as

$$P = \begin{bmatrix} r & -q \\ -q & 2 \end{bmatrix} = \begin{bmatrix} \theta_1^2 + 2\theta_2 - \theta_1 & -(\theta_1 - 1) \\ -(\theta_1 - 1) & 2 \end{bmatrix} \quad (21)$$

The determinant of P can be computed as:

$$\det(P) = \theta_1^2 + 4\theta_2 - 1 \quad (22)$$

From (19), it follows straightforwardly that $4\theta_2 > D > 1$, and hence $\det(P) > 0$. Concurrently, since the bottom-right entry of P is positive, P is positive definite. Consider that the time derivative of vector ζ is as follows:

$$\dot{\zeta} = \dot{g}_1(|e_1|) A(\delta(t, e)) \zeta \quad (23)$$

where

$$A(\delta(t, e_1)) = \begin{bmatrix} -\theta_1 & 1 \\ -\theta_2 + \delta(t, e_1) & 0 \end{bmatrix}, \quad \delta(t, e_1) = -\frac{\dot{d}_1(t)}{g_2(|e_1|) \mathrm{sign}(e_1)} \quad (24)$$

From Assumption 1, it is clear that $|\delta(t, e_1)| \leq D$. Then, the time derivative of the Lyapunov function can be given as:

$$\dot{V}(e_1, e_2) = -\dot{g}_1(|e_1|) \zeta^T Q(\delta(t, e_1)) \zeta$$

where

$$\begin{cases} a(t, e_1) = -\theta_2 + \delta(t, e), \\ Q(\delta(t, e_1)) = \begin{bmatrix} 2\theta_1 r - 2q|a(t, e_1)| & 2|a(t, e_1)| - r - \theta_1 q \\ 2|a(t, e_1)| - r - \theta_1 q & 2q \end{bmatrix} \end{cases}$$

For all possible values of $a(t, e_1)$, since $|\delta(t, e_1)| \leq D$ and $\theta_2 > D$, then $a(t, e_1) \in [-\theta_2 - D, -\theta_2 + D] \subset (-\infty, 0)$. Therefore, all possible values of $a(t, e_1)$ are negative and

$$\theta_2 - D \leq |a(t, e_1)| \leq \theta_2 + D$$

Then, the determinant of $Q(\delta(t, e_1))$ can be computed as follows:

$$\det(Q(\delta(t, e_1))) = -r^2 + c_1(t, e_1) r - c_0(t, e_1) \quad (25)$$

where

$$\begin{cases} c_1(t, e_1) = 4|a(t, e_1)| + 2\theta_1(\theta_1 - 1) \\ c_0(t, e_1) = 4|a(t, e_1)|(\theta_1 - 1)^2 + [\theta_1(\theta_1 - 1) - 2|a(t, e_1)|]^2 \end{cases}$$

Both $c_0(t, e_1)$ and $c_1(t, e_1)$ are positive. In addition, from (25), it can be computed that

$$c_1^2(t, e_1) - 4c_0(t, e_1) = 16|a(t, e_1)|(\theta_1^2 - 1) > 0$$

Therefore, the roots of $\det(Q(t, e_1))$ as a polynomial in r are always real. These roots are:

$$r_1 = \frac{c_1(t, e_1) - \sqrt{c_1^2(t, e_1) - 4c_0(t, e_1)}}{2},$$

$$r_2 = \frac{c_1(t, e_1) + \sqrt{c_1^2(t, e_1) - 4c_0(t, e_1)}}{2}$$

From (19), the minimum value of r_2 over all possible values of $|a(t,e_1)|$ is given by:

$$\begin{aligned} r_{2\min} &= 2(\theta_2 - D) + \theta_1^2 - \theta_1 + 2\sqrt{(\theta_2 - D)(\theta_1^2 - 1)} \\ &> \theta_1^2 + 2\theta_2 - \theta_1 = r \end{aligned} \quad (26)$$

In addition, using (19) again obtains

$$\begin{aligned} r_1 &< 2(\theta_2 + D) + \theta_1^2 - \theta_1 - 2\sqrt{(\theta_2 - D)(\theta_1^2 - 1)} \\ &< \theta_1^2 + 2\theta_2 - \theta_1 = r. \end{aligned} \quad (27)$$

Therefore, the inequality $r_1 < r < r_2$ holds for all possible values of $|a(t,e_1)|$, and it follows from (25) that $\det(Q(t,e_1)) > 0$. Concurrently, with the bottom-right entry of $Q(t,e_1)$ is $2q = 2(\theta_1 - 1) > 0$, and $Q(t,e_1)$ is positive definite.

It is clear that $\dot{g}_1(|e_1|) > 0$ for all $e_1 > 0$. This latter fact, jointly with (24), implies that

$$\dot{V}(e_1, e_2) \leq -\dot{g}_1(|e_1|) \lambda_{\min}(Q(\delta(t,e_1))) \|\xi\|^2 \leq 0 \quad (28)$$

Thus, the proposed SSTESO constructed as (14) will drive the estimation errors $(e_1, e_2) \to (0,0)$. Then, the unknown disturbance $d_1(t)$ can be estimated by the SSTESO. □

Similarly, the matched disturbance $d_2(t)$ can be estimated accurately when the estimation error of SSTESO converges to the origin. Moreover, the block diagram of the two sets of SSTESO is shown in Figure 2.

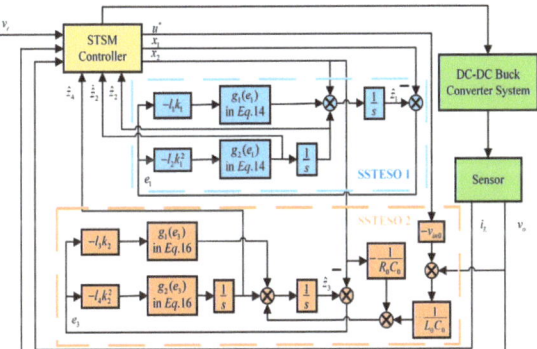

Figure 2. The block diagram of two sets of SSTESO.

Through the two sets of SSTESO designed above, the lumped disturbance $d_1(t)$, $d_2(t)$ and the time derivative of mismatched disturbance $\dot{d}_1(t)$ can be estimated accurately as $\hat{z}_2$, $\hat{z}_4$, and $\dot{\hat{z}}_2$, respectively. Then, these estimated values can be compensated for by the voltage controller to improve the robustness of the system.

3.2. SSTESO-Based SSTSMC Design

Super-twisting SMC, which is one of the high-order SMCs, can achieve chattering suppression and eliminate the steady error by adding the integration element. However, STSMC is still essentially a nonsmooth control algorithm, and the control lag of the integration part also exists.

Therefore, in this paper, an algorithm is proposed to realize the smoothness in the zero domain of the nonsmooth control algorithm. This new control algorithm, called SSTSMC, is realized by constructing a pair of novel simple smooth functions to replace the switch function in conventional STSMC and is constructed as:

$$\begin{cases} \dot{s} = -\mu_1 |s|^{1/2}\arctan(\frac{s}{\beta}) + u_I \\ \dot{u}_I = -\mu_2 \arctan(\left|\frac{s}{\beta}\right|)[\frac{1}{2}\arctan(\frac{s}{\beta}) + \frac{s}{\beta+s^2/\beta}] \end{cases} \quad (29)$$

where $\mu_1 > 0$ and $\mu_2 > 0$ are the control gains, $\beta > 0$ is an adjustable parameter. Then, the control scheme based on SSTSMC for the DC-DC buck converter can be constructed.

First, the sliding-mode surface for System (4) can be formulated as follows:

$$s = cx_1 + \dot{x}_1 = cx_1 + x_2 + d_1(t) \quad (30)$$

where $c > 0$ is the constant to be designed.

Subject to the nominal plant model (4), the time derivative of s is given by

$$\dot{s} = cx_2 + \frac{uv_{in0}}{L_0 C_0} - \frac{x_1}{L_0 C_0} - \frac{x_2}{R_0 C_0} - \frac{v_r}{L_0 C_0} + dis(t) \quad (31)$$

where $dis(t) = cd_1(t) + d_2(t) + \dot{d}_1(t)$ is the lumped disturbance.

Letting $\dot{s} = 0$ in (29), we obtain the equivalent controller as follows:

$$u_{eq} = \frac{1}{v_{in0}}(x_1 + \frac{L_0}{R_0}x_2 + v_r - cL_0 C_0 x_2) - \frac{L_0 C_0}{v_{in0}} dis(t) \quad (32)$$

However, the lumped disturbance is usually unknown in practice. Thus, the lumped disturbance $dis(t)$ in the equivalent controller can be replaced by the estimated values $\hat{dis}(t) = c\hat{z}_2 + \hat{z}_4 + \hat{z}_2$ from SSTESO proposed above. The new equivalent controller can be obtained as follows:

$$u_{eq} = \frac{1}{v_{in0}}(x_1 + \frac{L_0}{R_0}x_2 + v_r - cL_0 C_0 x_2) - \frac{L_0 C_0}{v_{in0}} \hat{dis}(t) \quad (33)$$

From (29), the reaching law can be obtained as follows:

$$\begin{cases} u_{sw} = -\mu_1 |s|^{1/2}\arctan(\frac{s}{\beta}) + u_I \\ \dot{u}_I = -\mu_2 \arctan(\left|\frac{s}{\beta}\right|)[\frac{1}{2}\arctan(\frac{s}{\beta}) + \frac{s}{\beta+s^2/\beta}] \end{cases} \quad (34)$$

Combining Equations (31) and (32), the control law u^* for System (4) can be obtained as follows:

$$u^* = u_{eq} + \frac{L_0 C_0}{v_{in0}} u_{sw} \quad (35)$$

Taking the control law (33) to (29), the time derivative of the sliding-mode variable s can be written as follows:

$$\begin{aligned} \dot{s} &= u_{sw} + c(d_1(t) - \hat{z}_2) + (d_2(t) - \hat{z}_4) + (\dot{d}_1(t) - \dot{\hat{z}}_2) \\ &= u_{sw} + ce_2 + e_4 - \dot{e}_2 \end{aligned} \quad (36)$$

By proof of SSTESO, the estimation error $e_i (i = 1, 2, 3, 4)$ will converge to zero. Thus, we assume that the estimated values of SSTESO are accurate and compensated for by the controller in time to make $e_i = 0 (i = 1, 2, 3, 4)$. Then, $\dot{s}$ can be obtained as follows:

$$\begin{cases} \dot{s} = -\mu_1 |s|^{1/2}\arctan(\frac{s}{\beta}) + u_I \\ \dot{u}_I = -\mu_2 \arctan(\left|\frac{s}{\beta}\right|)[\frac{1}{2}\arctan(\frac{s}{\beta}) + \frac{s}{\beta+s^2/\beta}] \end{cases} \quad (37)$$

which is similar to Equation (15). Therefore, the stability analysis of SSTSMC is similar to SSTESO, and is omitted here.

Consequently, the proposed SSTESO-based SSTSM control scheme is constructed completely, and the structure diagram of this controller is shown in Figure 3.

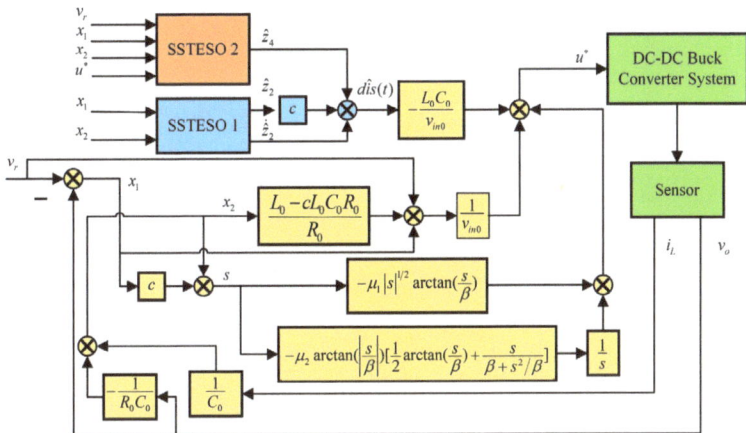

Figure 3. Control structure of SSTESO-based SSTSMC for a DC-DC buck converter.

4. Simulation Study

In this section, the effectiveness and reliability of the proposed SSTESO-based SSTSMC scheme for the DC-DC buck converter are verified using MATLAB/Simulink. In Simulink, the simulation step size is set to 1.0×10^{-5}s, and the switching frequency of the system is set to 50 kHz.

To illustrate the superiority of the proposed control strategy, two sets of comparative analysis are conducted: (1) comparing SSTSMC with STSMC; (2) comparing SSTESO+SSTSMC with conventional ESO+SSTSMC and STESO+SSTSMC schemes. The specific parameters of the DC-DC buck converter are listed in Table 1. Moreover, the reaching laws of conventional STSMC are shown in (36).

$$\begin{cases} u_{sw-ST} = -\mu_1 |s|^{1/2} \text{sign}(s) + u_1 \\ \dot{u}_1 = -\mu_2 \text{sign}(s) \end{cases} \tag{38}$$

Table 1. Nominal parameter values.

Description	Parameter	Value	Units
Inductor	L_0	6.0×10^{-3}	H
Capacitor	C_0	2.2×10^{-3}	F
Load resistance	R_0	$30 \rightarrow 20$	Ω
Input voltage	v_{in0}	25	V
Reference voltage	v_r	$12 \rightarrow 15$	V

For fair comparison, the parameters of STSMC, SSTSMC, ESO+SSTSMC, STESO+SSTSMC, and SSTESO+SSTESO are the same. All the parameters used in these controllers are obtained through a trial-and-error method to achieve better tracking and robustness performance, and the relevant values are shown in Table 2.

To compare the responses of these controllers under disturbance rejection and parameter uncertainty, the following simulation tests are performed in the buck converter system: (1) Startup-phase analysis; (2) Reference-voltage variation; (3) Linear load-resistance variation; (4) Input-voltage variation.

Table 2. Controller parameter values.

Controllers	Parameters and Values
STSMC	$c = 5.70 \times 10^6, \mu_1 = 4.05 \times 10^5, \mu_2 = 5.25 \times 10^9$
SSTSMC	$c = 5.70 \times 10^6, \mu_1 = 4.05 \times 10^5, \mu_2 = 5.25 \times 10^9, \beta = 400$
ESO+SSTSMC	$c = 5.70 \times 10^6, \mu_1 = 4.05 \times 10^5, \mu_2 = 5.25 \times 10^9, \beta = 400, l_1 = 126,$ $l_2 = 3969, l_3 = 1.68 \times 10^4, l_4 = 7.06 \times 10^7$
STESO+SSTSMC	$c = 5.70 \times 10^6, \mu_1 = 4.05 \times 10^5, \mu_2 = 5.25 \times 10^9, \beta = 400, l_1 = 126,$ $l_2 = 3969, l_3 = 1.68 \times 10^4, l_4 = 7.06 \times 10^7, k_1 = 48, k_2 = 89$
SSTESO+SSTSMC	$c = 5.70 \times 10^6, \mu_1 = 4.05 \times 10^5, \mu_2 = 5.25 \times 10^9, \beta = 400, l_1 = 126,$ $l_2 = 3969, l_3 = 1.68 \times 10^4, l_4 = 7.06 \times 10^7, k_1 = 48, k_2 = 89,$ $\alpha_1 = 5 \times 10^{-4}, \alpha_2 = 8 \times 10^3$

4.1. Controller Comparative Analysis

(1) Startup-Phase Analysis

In this simulation, the reference voltage V_{ref} is set to 12 V, and the load resistance remains unchanged at 30 Ω. The response curves of the output voltage during the startup phase are shown in Figure 4. In the voltage-rise phase, because the sliding-mode state variable s is far away from the origin, the function gain of $\arctan(s/\beta)$ is larger than $\text{sign}(s)$. Therefore, SSTSMC makes the output voltage reach the desired voltage faster than STSMC. Then, in the voltage-adjustment phase, s reaches the neighborhood of the origin and the function gain of $\arctan(s/\beta)$ is smaller than that of $\text{sign}(s)$. For this reason, SSTSMC makes the system have less overshoot and a shorter startup time. Compared to STSMC, SSTSMC takes less time to reach a steady state and reduces voltage overshoot, which proves that SSTSMC can accelerate the convergence and provide better transient characteristics.

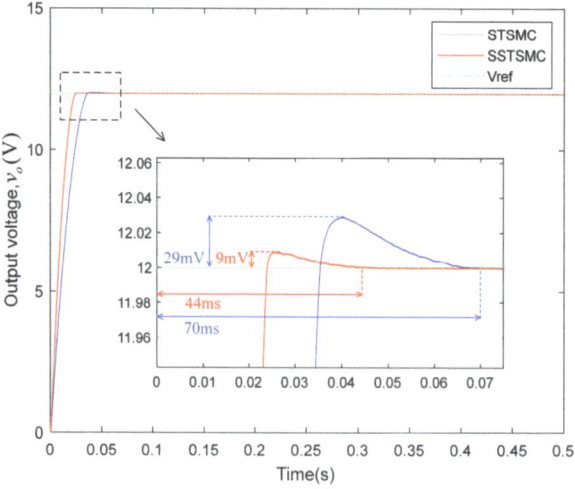

Figure 4. Output voltages of the two sliding-mode controllers at startup.

(2) Reference-Voltage Variations

The reference voltage is changed from 12 V to 15 V at 1 s, and the load resistance remains at 30 Ω. The dynamic processes of the output voltage using both strategies during the reference changes are shown in Figure 5. It can be seen from Figure 5 that because the rise of the reference voltage is small, there is no overshoot in the output voltage during the voltage-adjustment phase. Furthermore, the output voltage of the SSTSMC strategy tracks the new reference voltage successfully to within 12 ms, which is shorter than that of the STSMC strategy by 37%. This result proves that SSTSMC has superior tracking performance for reference trajectory tracking.

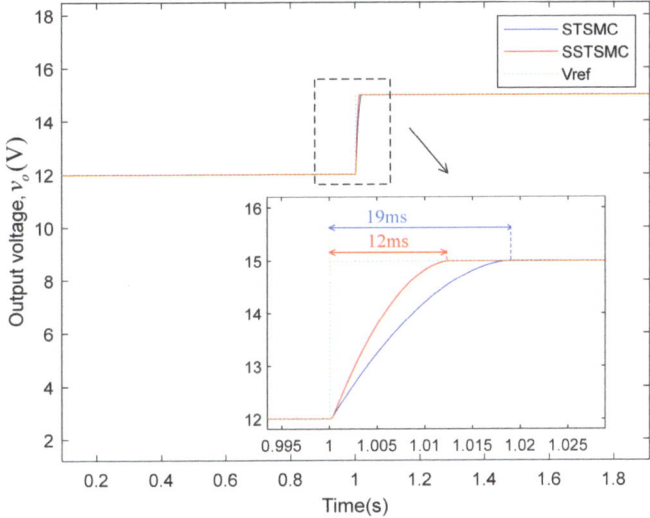

Figure 5. Output voltages of the two sliding-mode controllers when the reference changes.

(3) Linear Load-Resistance Variations

The voltage fluctuation owing to external disturbances is evaluated. The simulation conditions for this time are that the output-voltage reference value remains unchanged at 12 V, and the load resistance is changed from 30 Ω to 20 Ω at 1 s. The response curves of the output voltage during the load changes are shown in Figure 6. Since the sliding-mode gain of SSTSMC is larger than that of STSMC at the moment of the introduction of disturbance, SSTSMC responds more quickly to external disturbance. From Figure 6, the recovery time and drop voltage of SSTSMC is measured as 78 ms and 0.10 V, respectively, which are both smaller than that of STSMC.

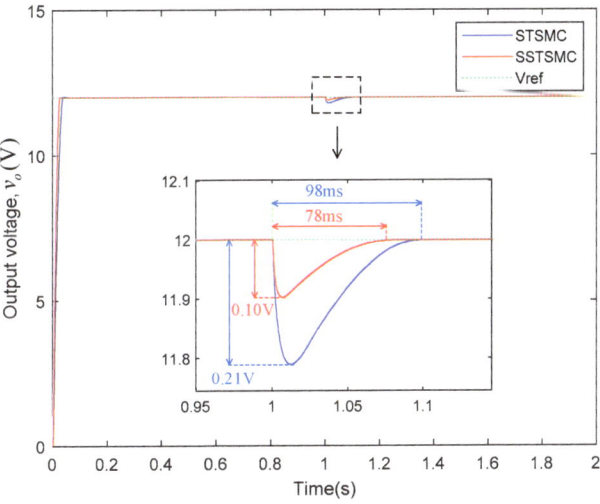

Figure 6. Output voltages of the two sliding-mode controllers when the load steps down.

According to the simulation results, the proposed SSTSMC control strategy has shown better robustness and resistance to disturbance ability compared to the STSMC. As with the load change, the performance of the proposed SSTSMC strategy is optimal.

(4) Input-Voltage Variations

Considering the input voltage cannot be kept at the nominal value all the time in practical engineering applications, the input voltage v_{in0} in the actual test will fluctuate boundedly around the nominal value. Therefore, to further investigate the robustness of the proposed SSTSMC strategy, a sinusoidal disturbance signal ($10\sin 1000\pi t$) is added on the nominal value of the input voltage v_{in0} with the output-voltage reference value remaining unchanged at 12 V and the load resistance remaining unchanged at 30 Ω. The simulation results are shown in Figure 7.

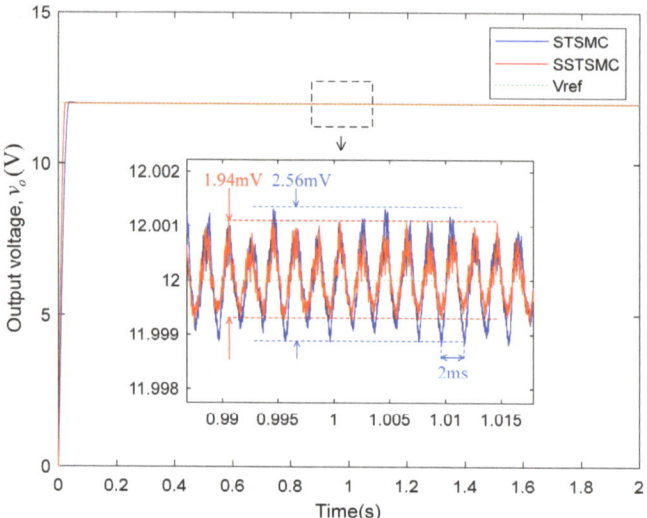

Figure 7. Output voltages of the two sliding-mode controllers when input voltage varies.

From Figure 7, the voltage fluctuations under the STSMC strategy are 2.56 mV, while the SSTSMC strategy reduces voltage fluctuations by 24% (1.94 mV), which means, from another perspective, that the SSTSMC strategy is more robust and has a better dynamic adjustment ability in face of input-voltage time-varying disturbance.

The simulation results mentioned above show that the SSTSM control strategy can greatly reduce the impact of disturbance, increase the dynamic response speed, and improve the robustness of the closed-loop control system. Therefore, the SSTSMC is chosen as the basic controller to compare the SSTESO with ESO and STESO to prove the superiority of the proposed SSTESO+SSTSMC scheme.

4.2. Extended-State Observer Comparative Analysis

In this section, the SSTSMC is selected as the controller and combined with different extended-state observers for simulation. The simulation conditions are similar to the previous section, and the controller parameters are shown in Table 2.

(1) Startup-Phase Analysis

The simulation results during the startup phase of the system using three control schemes are shown in Figure 8 and Table 3. It can be seen from Figure 8c,d that the SSTESO spends the shortest time among the three observers to make the disturbance estimation $\hat{z}_2$ and $\hat{z}_2$ converge to the origin when there is no external disturbance in the system. Before the disturbance estimation converges to the origin, the disturbance estimates will also be compensated for by the controller. Therefore, from Figure 8a, the output-voltage response speed of the controllers with ESO is faster than that of the controller without ESO in the voltage-rise phase. Furthermore, because the convergence speed of conventional ESO and STESO is not as fast as SSTESO, a long compensation will lead to a larger overshoot of the system output voltage and a longer startup time in the startup phase. Compared to the data in Table 3, the overshoot of output voltage of the SSTESO+SSTSMC scheme is even smaller than SSTSMC without the extended-state observer. Moreover, Figure 8c shows that there is a static error in the disturbance estimation of STESO but not in SSTESO, and the estimated disturbance value of STESO exhibits larger chatter than that of SSTESO.

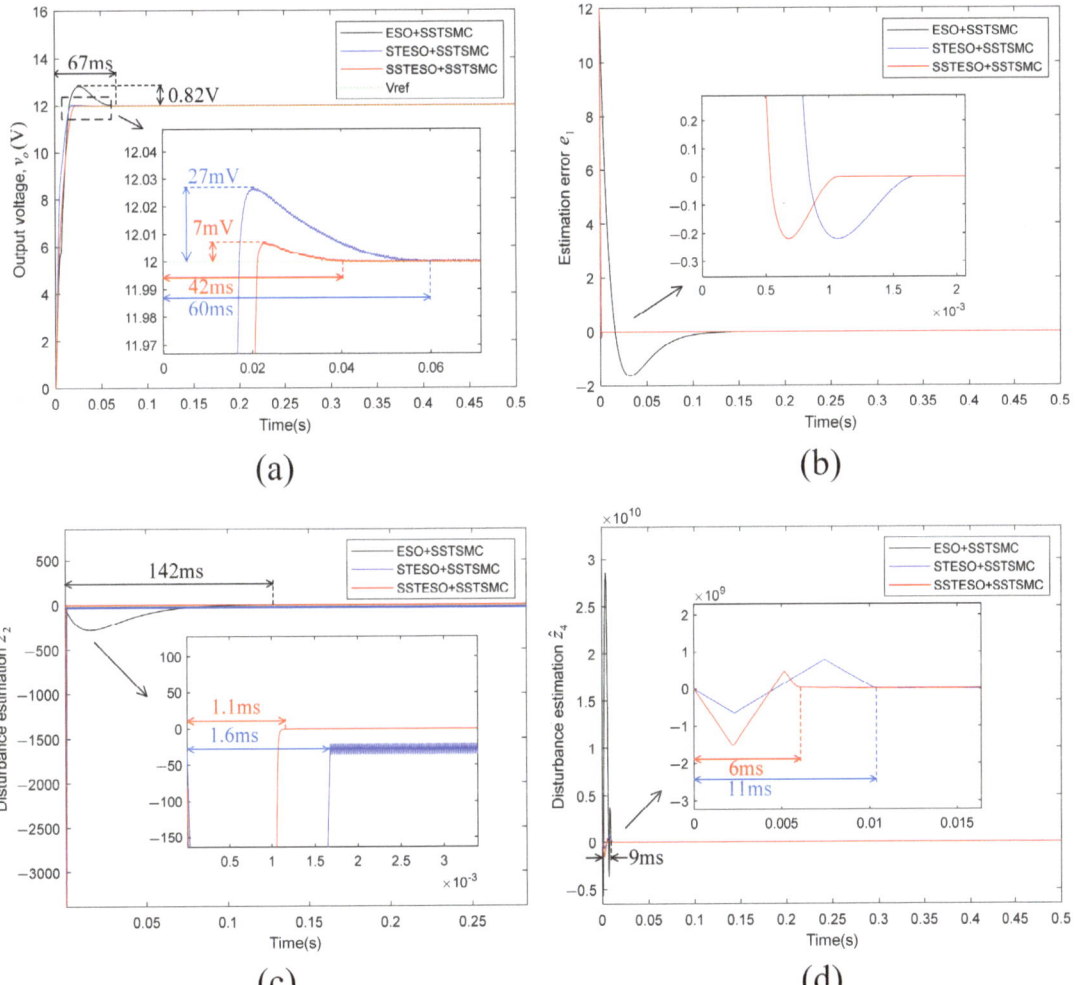

Figure 8. Response curves of the three control schemes at startup. (**a**) Output voltage v_o, V. (**b**) Convergence curve of estimation error e_1. (**c**) Disturbance estimation of $d_1(t)$. (**d**) Disturbance estimation of $d_2(t)$.

To sum up, SSTESO has a faster convergence speed and better disturbance estimation accuracy than STESO and conventional ESO. In addition, the SSTESO+SSTSMC scheme can accelerate the convergence rate of the system and provide better transient characteristics.

Table 3. Comparative study of control schemes under three simulations.

Simulation	Controller	V_r (mV)	t_s (ms)
(1)	STSMC	29	70
	SSTSMC	9	44
	ESO+SSTSMC	820	67
	STESO+SSTSMC	27	60
	SSTESO+SSTSMC	7	42
(2)	STSMC	-	19
	SSTSMC	-	12
	ESO+SSTSMC	75	47
	STESO+SSTSMC	-	11
	SSTESO+SSTSMC	-	11
(3)	STSMC	210	98
	SSTSMC	100	78
	ESO+SSTSMC	91	35
	STESO+SSTSMC	36	48
	SSTESO+SSTSMC	9	1

(2) Reference-Voltage Variations

The simulation curves of output voltage, estimation error, and disturbance estimate during the reference change are shown in Figure 9. As can be seen from Figure 9a, the curves of output voltage during the reference-voltage step-up are similar to those of the startup phase. However, because the rise in reference voltage is small, the output voltage of the STESO+SSTSMC and SSTESO+SSTSMC schemes tracks smoothly from 12 V to 15 V without overshoot within 11 ms. Furthermore, because of the slow convergence rate of ESO, the overshoot (75 mV) and adjustment time (47 ms) of ESO+SSTSMC in the adjustment phase are both large. It also can be seen from Figure 6b, c that the convergence rate of estimation error e_1 and disturbance estimation $\hat{z}_2$ of STESO and SSTESO is much faster than that of ESO. However, compared with SSTESO, there is still a static error and a large chattering in the disturbance estimation $\hat{z}_2$ of STESO. It can be seen in Figure 9d that, because the value of disturbance estimation $\hat{z}_4$ is very large and there is a linear term in ESO, the convergence time of ESO is slightly shorter than that of STESO and SSTESO.

It can be concluded that SSTESO combines the advantages of ESO and STESO, which have faster convergence rates and more accurate disturbance estimations.

(3) Linear Load-Resistance Variations

The dynamic processes of three strategies under linear load-resistance change conditions are shown in Figure 10. It can be seen from Figure 10c that, when the disturbance is introduced into the system at 1 s, the disturbance estimation $\hat{z}_2$ of STESO and SSTESO converge to a value quickly within 1 ms. Furthermore, it takes 158 ms for the $\hat{z}_2$ of conventional ESO to converge to the same value as SSTESO. There is a static error and a large chattering in the disturbance estimation of STESO when it is stable, while the disturbance estimation of SSTESO has better accuracy and smoothness. A similar conclusion can be drawn from Figure 10b. Figure 10d shows the curves of disturbance estimation $\hat{z}_4$, which outlines that the convergence rates of all three schemes are similar. Furthermore, the SSTESO has a smaller disturbance estimation to compensate for the controller when the disturbance becomes larger, making the controller respond faster to the disturbance. Moreover, for disturbance estimation $\hat{z}_4$, SSTESO has better smoothness than STESO. Then, in Figure 10a, because of the faster convergence rate and more accurate disturbance estimation of SSTESO, the output voltage of the SSTESO+SSTSMC scheme has the smallest drop in voltage and the shortest recovery time among all three schemes when the load resistance steps down. Simulation results show the proposed SSTESO+SSTSMC scheme has better robustness and resistance in the presence of disturbance.

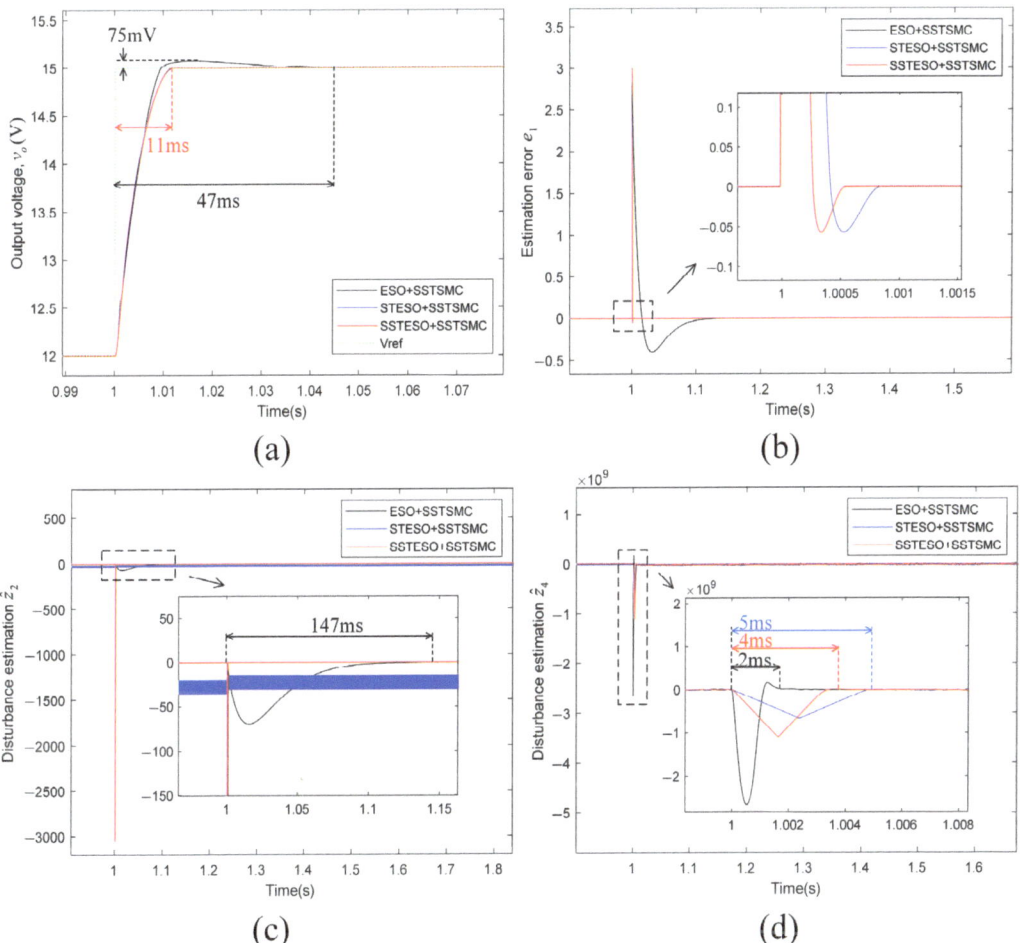

Figure 9. Response curves of the three control schemes when the reference changes. (**a**) Output voltage v_o, V. (**b**) Convergence curve of estimation error e_1. (**c**) Disturbance estimation of $d_1(t)$. (**d**) Disturbance estimation of $d_2(t)$.

(4) Input-Voltage Variations

A sinusoidal disturbance signal $(10\sin 1000\pi t)$ is added based on the nominal value of the input voltage v_{in0} to investigate the robustness of the proposed SSTESO+SSTSMC strategy, as shown in Figure 11. The output-voltage chattering of the STESO+SSTSMC scheme is larger than that of the other two schemes. That is because the external disturbance of this system caused by the input-voltage fluctuation is not large, and so the ESO and SSTESO with better smoothness perform better in this simulation. The fluctuation in the disturbance estimation in STESO is larger, and this chattering will be superimposed on the control signal eventually, which will aggravate the fluctuation in output voltage. The voltage fluctuation under the STESO+STSMC scheme is 2.48 mV, which is even larger than that of SSTSMC without an extended-state observer. Furthermore, the voltage fluctuation under the SSTESO+STSMC scheme is 1.89 mV, which is slightly smaller than the scheme with only the controller. The above analysis shows that, if the estimated value of the observer is not accurate enough, the robustness of the controller may become worse in the presence of input-voltage variations.

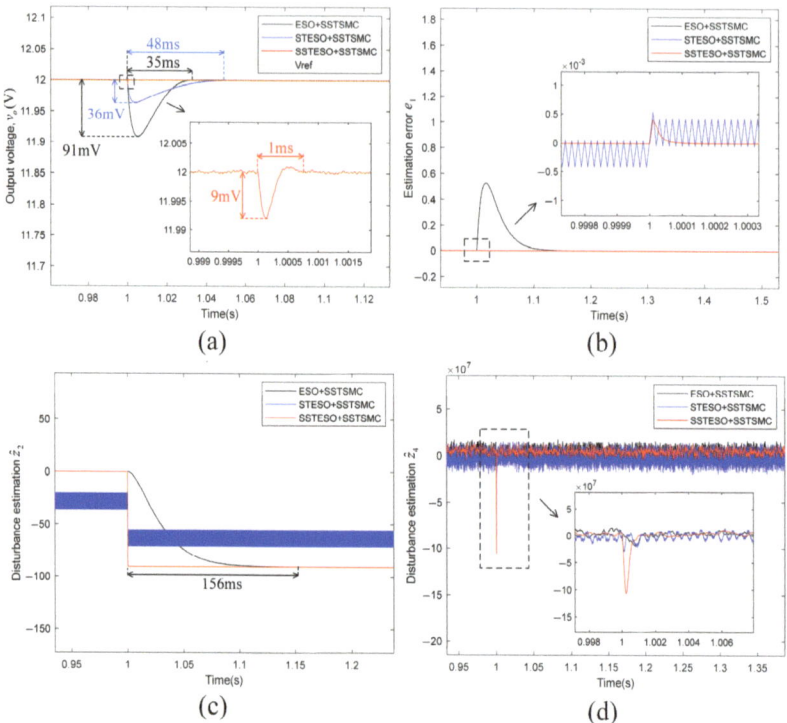

Figure 10. Response curves of the three control schemes when the load steps down. (**a**) Output voltage v_o, V. (**b**) Convergence curve of estimation error e_1. (**c**) Disturbance estimation of $d_1(t)$. (**d**) Disturbance estimation of $d_2(t)$.

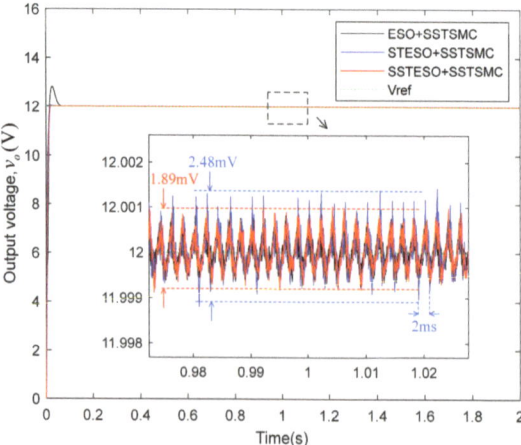

Figure 11. Output voltages of the three control schemes when input-voltage variations.

4.3. Detail Results Analysis and Summary

To compare the control schemes in a useful manner, performance criteria are very useful. In this article, two criteria have been favored, namely voltage maximum rise or fall v_r and adjustment time t_s. Table 3 presents detailed simulation results of different control schemes under the first three simulation tests. It can be concluded that, because of the characteristic of the proposed smooth function, the dynamic performance and robustness of SSTSMC are better than that of STSMC. In addition, for the extended-state observers, the proposed SSTESO can greatly improve the speed of convergence compared to conventional ESO. In addition, there is almost no fluctuation in the estimated value of SSTESO, which makes the compensation to the controller more accurate. The data in Table 3 shows that the proposed SSTESO+SSTSM control scheme can effectively improve the dynamic tracking performance and robustness of the system.

5. Conclusions

This paper proposes a smooth super-twisting extended-state observer-based smooth super-twisting sliding-mode control scheme for DC-DC buck converters with matched and mismatched disturbances. First, the improved smooth super-twisting algorithm not only accelerates the convergence speed but also ensures the smoothness of the system near the zero domain. Then, compared to the conventional ESO, the proposed SSTESO can make the estimation error converge to the origin faster, but it does not introduce chattering into disturbance estimation like STESO, which provides higher estimation accuracy. Simulation and experimental results demonstrate that the proposed SSTESO+SSTSMC scheme has a faster response time, better tracking performance, and stronger robustness against output-power variation and parameter uncertainties.

Author Contributions: Conceptualization, J.F.; methodology, D.J. and Y.F.; writing—original draft preparation, D.J.; writing—review and editing, J.F. All authors have read and agreed to the published version of the manuscript.

Funding: This work is partially supported by the National Science Foundation of China under Grant No. 62273131.

Conflicts of Interest: The authors declare no conflict of interest.

References

1. Sarikhani, A.; Allahverdinejad, B.; Hamzeh, M. A Nonisolated Buck–Boost DC–DC Converter with Continuous Input Current for Photovoltaic Applications. *IEEE J. Emerg. Sel. Top. Power Electron.* **2021**, *9*, 804–811. [CrossRef]
2. Vinnikov, D.; Chub, A.; Kosenko, R.; Zakis, J.; Liivik, E. Comparison of Performance of Phase-Shift and Asymmetrical Pulsewidth Modulation Techniques for the Novel Galvanically Isolated Buck–Boost DC–DC Converter for Photovoltaic Applications. *IEEE J. Emerg. Sel. Top. Power Electron.* **2017**, *5*, 624–637. [CrossRef]
3. Zapata, J.W.; Kouro, S.; Carrasco, G.; Renaudineau, H.; Meynard, T.A. Analysis of Partial Power DC–DC Converters for Two-Stage Photovoltaic Systems. *IEEE J. Emerg. Sel. Top. Power Electron.* **2019**, *7*, 591–603. [CrossRef]
4. Ramirez-Murillo, H.; Restrepo, C.; Konjedic, T.; Calvente, J.; Romero, A.; Baier, C.R.; Giral, R. An Efficiency Comparison of Fuel-Cell Hybrid Systems Based on the Versatile Buck–Boost Converter. *IEEE Trans. Power Electron.* **2018**, *33*, 1237–1246. [CrossRef]
5. Wu, H.; Sun, K.; Chen, L.; Zhu, L.; Xing, Y. High Step-Up/Step-Down Soft-Switching Bidirectional DC–DC Converter with Coupled-Inductor and Voltage Matching Control for Energy Storage Systems. *IEEE Trans. Ind. Electron.* **2016**, *63*, 2892–2903. [CrossRef]
6. Gobbato, C.; Kohler, S.; Souza, I.; Denardin, G.; Lopes, J. Integrated Topology of DC–DC Converter for LED Street Lighting System Based on Modular Drivers. *IEEE Trans. Ind. Appl.* **2018**, *54*, 3881–3889. [CrossRef]
7. Hong, W.; Lee, M. A 7.4-MHz Tri-Mode DC-DC Buck Converter with Load Current Prediction Scheme and Seamless Mode Transition for IoT Applications. *IEEE Trans. Circuits Syst. I Regul. Pap.* **2020**, *67*, 4544–4555. [CrossRef]
8. Kim, S.-Y.; Hwang, K.C.; Yang, Y.; Lee, K.-Y.; Park, Y.-J.; Ali, I.; Nga, T.T.K.; Ryu, H.-C.; Khan, Z.H.N.; Park, S.-M.; et al. Design of a High Efficiency DC–DC Buck Converter with Two-Step Digital PWM and Low Power Self-Tracking Zero Current Detector for IoT Applications. *IEEE Trans. Power Electron.* **2018**, *33*, 1428–1439. [CrossRef]
9. Zhao, M.; Li, M.; Song, S.; Hu, Y.; Yao, Y.; Bai, X.; Hu, R.; Wu, X.; Tan, Z. An Ultra-Low Quiescent Current Tri-Mode DC-DC Buck Converter with 92.1% Peak Efficiency for IoT Applications. *IEEE Trans. Circuits Syst. I Regul. Pap.* **2022**, *69*, 428–439. [CrossRef]

10. Bi, K.; Lv, H.; Chen, L.; Li, J.; Zhu, Y.; Huang, W.; Fan, Q. A Model Predictive Controlled Bidirectional Four Quadrant Flying Capacitor DC/DC Converter Applied in Energy Storage System. *IEEE Trans. Power Electron.* **2022**, *37*, 7705–7717. [CrossRef]
11. Chen, J.; Chen, Y.; Tong, L.; Peng, L.; Kang, Y. A Backpropagation Neural Network-Based Explicit Model Predictive Control for DC–DC Converters with High Switching Frequency. *IEEE J. Emerg. Sel. Top. Power Electron.* **2020**, *8*, 2124–2142. [CrossRef]
12. Liu, P.-J.; Chien, L.-H. A High-Efficiency Integrated Multimode Battery Charger with an Adaptive Supply Voltage Control Scheme. *IEEE Trans. Power Electron.* **2018**, *33*, 6869–6876. [CrossRef]
13. Wu, J.; Lu, Y. Decoupling and Optimal Control of Multilevel Buck DC–DC Converters with Inverse System Theory. *IEEE Trans. Ind. Electron.* **2020**, *67*, 7861–7870. [CrossRef]
14. Ding, S.; Zheng, W.X.; Sun, J.; Wang, J. Second-Order Sliding-Mode Controller Design and Its Implementation for Buck Converters. *IEEE Trans. Ind. Inform.* **2018**, *14*, 1990–2000. [CrossRef]
15. Komurcugil, H.; Biricik, S.; Guler, N. Indirect Sliding Mode Control for DC–DC SEPIC Converters. *IEEE Trans. Ind. Inform.* **2020**, *16*, 4099–4108. [CrossRef]
16. Lin, X.; Liu, J.; Liu, F.; Liu, Z.; Gao, Y.; Sun, G. Fractional-Order Sliding Mode Approach of Buck Converters with Mismatched Disturbances. *IEEE Trans. Circuits Syst. I Regul. Pap.* **2021**, *68*, 3890–3900. [CrossRef]
17. Jin, H.; Zhao, X. Approach Angle-Based Saturation Function of Modified Complementary Sliding Mode Control for PMLSM. *IEEE Access* **2019**, *7*, 126014–126024. [CrossRef]
18. Rakhtala, S.M.; Casavola, A. Real-Time Voltage Control Based on a Cascaded Super Twisting Algorithm Structure for DC–DC Converters. *IEEE Trans. Ind. Electron.* **2022**, *69*, 633–641. [CrossRef]
19. Linares-Flores, J.; Juarez-Abad, J.A.J.; Hernandez-Mendez, A.; Castro-Heredia, O.; Guerrero-Castellanos, J.F.; Heredia-Barba, R.R.; Curiel-Olivares, G. Sliding Mode Control Based on Linear Extended State Observer for DC-to-DC Buck–Boost Power Converter System with Mismatched Disturbances. *IEEE Trans. Ind. Appl.* **2022**, *58*, 940–950. [CrossRef]
20. Zhuo, S.; Gaillard, A.; Xu, L.; Bai, H.; Paire, D.; Gao, F. Enhanced Robust Control of a DC–DC Converter for Fuel Cell Application Based on High-Order Extended State Observer. *IEEE Trans. Transp. Electrif.* **2020**, *6*, 278–287. [CrossRef]
21. Zhuo, S.; Gaillard, A.; Xu, L.; Paire, D.; Gao, F. Extended State Observer-Based Control of DC–DC Converters for Fuel Cell Application. *IEEE Trans. Power Electron.* **2020**, *35*, 9923–9932. [CrossRef]
22. Lakomy, K.; Madonski, R.; Dai, B.; Yang, J.; Kicki, P.; Ansari, M.; Li, S. Active Disturbance Rejection Control Design with Suppression of Sensor Noise Effects in Application to DC–DC Buck Power Converter. *IEEE Trans. Ind. Electron.* **2022**, *69*, 816–824. [CrossRef]
23. Yang, J.; Cui, H.; Li, S.; Zolotas, A. Optimized Active Disturbance Rejection Control for DC-DC Buck Converters with Uncertainties Using a Reduced-Order GPI Observer. *IEEE Trans. Circuits Syst. I Regul. Pap.* **2018**, *65*, 832–841. [CrossRef]
24. Hou, Q.; Ding, S. Finite-Time Extended State Observer-Based Super-Twisting Sliding Mode Controller for PMSM Drives with Inertia Identification. *IEEE Trans. Transp. Electrif.* **2022**, *8*, 1918–1929. [CrossRef]
25. Zhao, L.; Zheng, C.; Wang, Y.; Liu, B. A Finite-Time Control for a Pneumatic Cylinder Servo System Based on a Super-Twisting Extended State Observer. *IEEE Trans. Syst. Man Cybern. Syst.* **2021**, *51*, 1164–1173. [CrossRef]
26. Fei, J.; Zhang, L.; Zhuo, J.; Fang, Y. Wavelet Fuzzy Neural Super-Twisting Sliding Mode Harmonic Control of Active Power Filter. *IEEE Trans. Fuzzy Syst.* **2023**, 1–12. [CrossRef]
27. Fei, J.; Wang, Z.; Pan, Q. Self-Constructing Fuzzy Neural Fractional-Order Sliding Mode Control of Active Power Filter. *IEEE Trans. Neural Netw. Learn. Syst.* **2022**, 1–12. [CrossRef]
28. Fei, J.; Wang, Z.; Fang, Y. Self-Evolving Chebyshev Fuzzy Neural Fractional-Order Sliding Mode Control for Active Power Filter. *IEEE Trans. Ind. Inform.* **2022**, *19*, 2729–2739. [CrossRef]
29. Tong, S.; Sun, K.; Sui, S. Observer-Based Adaptive Fuzzy Decentralized Optimal Control Design for Strict-Feedback Nonlinear Large-Scale Systems. *IEEE Trans. Fuzzy Syst.* **2018**, *26*, 569–584. [CrossRef]
30. Fei, J.; Chen, Y.; Liu, L.; Fang, Y. Fuzzy Multiple Hidden Layer Recurrent Neural Control of Nonlinear System Using Terminal Sliding-Mode Controller. *IEEE Trans. Cybern.* **2022**, *52*, 9519–9534. [CrossRef]
31. Fei, J.; Wang, H.; Fang, Y. Novel Neural Network Fractional-order Sliding Mode Control with Application to Active Power Filter. *IEEE Trans. Syst. Man Cybern. Syst.* **2022**, *52*, 3508–3518. [CrossRef]
32. Li, Y.; Zhang, J.; Tong, S. Fuzzy Adaptive Optimized Leader-Following Formation Control for Second-Order Stochastic Multiagent Systems. *IEEE Trans. Ind. Inform.* **2022**, *18*, 6026–6037. [CrossRef]
33. Fei, J.; Liu, L. Real-Time Nonlinear Model Predictive Control of Active Power Filter Using Self-Feedback Recurrent Fuzzy Neural Network Estimator. *IEEE Trans. Ind. Electron.* **2022**, *69*, 8366–8376. [CrossRef]
34. Fei, J.; Wang, Z.; Liang, X.; Feng, Z.; Xue, Y. Adaptive Fractional Sliding Mode Control of Micro gyroscope System Using Double Loop Recurrent Fuzzy Neural Network Structure. *IEEE Trans. Fuzzy Syst.* **2022**, *30*, 1712–1721. [CrossRef]
35. Zhou, R.; Fu, C.; Tan, W. Implementation of Linear Controllers via Active Disturbance Rejection Control Structure. *IEEE Trans. Ind. Electron.* **2021**, *68*, 6217–6226. [CrossRef]
36. Levant, A. Robust exact differentiation via sliding mode technique. *Automatica* **1998**, *34*, 379–384. [CrossRef]

Disclaimer/Publisher's Note: The statements, opinions and data contained in all publications are solely those of the individual author(s) and contributor(s) and not of MDPI and/or the editor(s). MDPI and/or the editor(s) disclaim responsibility for any injury to people or property resulting from any ideas, methods, instructions or products referred to in the content.

Article

Multi-Objective Optimization for Controlling the Dynamics of the Diabetic Population

Karim El Moutaouakil [1,*], Abdellatif El Ouissari [1], Vasile Palade [2,*], Anas Charroud [1], Adrian Olaru [3], Hicham Baïzri [4], Saliha Chellak [5] and Mouna Cheggour [4]

1. Engineering Science Laboratory, FPT, Sidi Mohamed Ben Abdellah University, Fez 30000, Morocco; abdellatif.elouissari@usmba.ac.ma (A.E.O.); anas.charroud@usmba.ac.ma (A.C.)
2. Centre for Computational Science and Mathematical Modelling, Coventry University, Priory Road, Coventry CV1 5FB, UK
3. Department of Robotics and Production System, University Politehnica of Bucharest, 020771 Bucharest, Romania; adrian.olaru2301@upb.ro
4. MorphoSciences Research Laboratory, Faculty of Medicine and Pharmacy, Cadi Ayyad University, Marrakech 40001, Morocco; hi.baizri@uca.ma (H.B.); mouna.cheggour@uca.ac.ma (M.C.)
5. Biosciences and Health Laboratory, Faculty of Medicine and Pharmacy, Cadi Ayyad University, Marrakech 40001, Morocco; sa.chellak@uca.ma
* Correspondence: karim.elmoutaouakil@usmba.ac.ma (K.E.M.); vasile.palade@coventry.ac.uk (V.P.)

Abstract: To limit the adverse effects of diabetes, a personalized and long-term management strategy that includes appropriate medication, exercise and diet has become of paramount importance and necessity. Compartment-based mathematical control models for diabetes usually result in objective functions whose terms are conflicting, preventing the use of single-objective-based models for obtaining appropriate personalized strategies. Taking into account the conflicting aspects when controlling the diabetic population dynamics, this paper introduces a multi-objective approach consisting of four steps: (a) modeling the problem of controlling the diabetic population dynamics using a multi-objective mathematical model, (b) discretizing the model using the trapezoidal rule and the Euler–Cauchy method, (c) using swarm-intelligence-based optimizers to solve the model and (d) structuring the set of controls using soft clustering methods, known for their flexibility. In contrast to single-objective approaches, experimental results show that the multi-objective approach obtains appropriate personalized controls, where the control associated with the compartment of diabetics without complications is totally different from that associated with the compartment of diabetics with complications. Moreover, these controls enable a significant reduction in the number of diabetics with and without complications, and the multi-objective strategy saves up to 4% of the resources needed for the control of diabetes without complications and up to 18% of resources for the control of diabetes with complications.

Keywords: diabetes mellitus (DM); dynamic control of diabetic population (DCDP); non-dominated sorting genetic algorithm II (NSGA-II); multi-objective firefly algorithm (MOFA); Fuzzy-CMeans (FCM); Gaussian mixture model (GMM); kernel convolution; fast Fourier transform (FFT)

MSC: 90C20; 90C29; 90C90; 93E20

1. Introduction

Diabetes is a permanent disease resulting from the pancreas' incapacity to generate insulin, or the body being incapable of utilizing the insulin properly. According to forecasts by the International Diabetes Federation (IDF), by 2045, one person in eight, or approximately 783 million individuals, will be suffering from diabetes, representing an average increase of 46% [1]. More than 90% of people living with diabetes are type 2 diabetics. Urbanization, an aging population, reduced physical activity and the increased prevalence of overweight and obesity are the major contributors to the increase in type 2 diabetes. According to the IDF's 2021 report, diabetes accounted for at least USD 966 billion in healthcare costs (9% of the total adult budget). Diabetes can lead to serious complications, among them vision loss, cardiovascular problems, kidney insufficiency, cerebrovascular accidents, nerve damage and lower-limb amputations. The key measures capable of mitigating this damage include appropriate medication, adapted physical exercise [2] and personalized diet [3–5]. To determine an optimal control and implement these measures, several mathematical control models have been proposed in the literature. Unfortunately, these models do not distinguish between the different components of the diabetic population we wish to control, while certain conflicting characteristics are implicitly embedded in this problem. This results in standard strategies that are adequate for some patients but not appropriate for others [6,7].

This work proposes a new approach to control the dynamics of the diabetic population based on a multi-objective dynamic model, using multi-objective swarm intelligence optimizers and soft clustering algorithms.

Most of the strategies aimed at controlling the dynamics of the diabetic population are based on single-objective mathematical models, which formulate the optimal control using the Pontryagin's maximum principle and decompose the model using various numerical techniques to estimate the control [8–11]. Numerical methods employing Pontryagin's maximum principle result in controls that are prohibitively resource-hungry in terms of both personnel and material. To overcome this problem, the authors of [2] used several heuristic methods to estimate a control by introducing an adequate objective function that makes a compromise between the components of the model proposed in [8]. However, focusing on a single objective leads to controls that heavily minimize certain compartments at the detriment of others. In addition, this kind of solution provides a unique policy that may not be suitable in some contexts or may not be appropriate.

In this work, we introduce a multi-objective strategy to control the dynamics of a diabetic population, implementing three compartments, i.e., pre-diabetics, diabetics, and diabetics with complications, and consisting of the following steps: (a) the introduction of two controls to protect diabetics from developing complications, (b) the introduction of two objectives functions to reduce the size of the two last compartments and the necessary resources to realize such a reduction, (c) the discretization of the obtained model using the Euler–Cauchy method and the trapezoidal rule, (d) the construction of two appropriate objective functions that make a compromise between the components of the model, and (e) employing NSGA-II [12] and MOFA [13] multi-objective optimization algorithms to build the Pareto front that presents the set of control actions to the customers. In step (c), we evaluated the error due to the discretization process. NSGA-II and MOFA produce a set of controls from which the user has to select the appropriate one considering its requirements. To assist the user, we structured the Pareto solutions front using two soft clustering methods: the Gaussian mixture model (GMM) [2] and Fuzzy-CMeans (FCM) [14]. The employed soft computing methods based on fuzzy or probabilistic approaches provide decision system makers with the necessary capabilities to deal with imprecise and incomplete information. A soft clustering method permits us to have an observation that belongs to two or more clusters. The optimal number of groups was selected on the basis of the silhouette criteria [14]. The experimental results show that the proposed multi-objective approach offers several effective and personalized controls that will enable experts to implement diversified group therapies to alleviate the human and material damage of diabetes.

The main contributions of this paper are summarized as follows:

(1) A multi-objective mathematical model for controlling the dynamics of the diabetic population is introduced;
(2) A discretization of the proposed model is realized based on the trapezoidal rule and the Euler–Cauchy method (we demonstrate that this error is bounded);
(3) Two multi-objective optimizers are used to solve the proposed multi-objective model;

(4) As a first postprocessing phase, FFT convolution is used to clean the noise from the control;
(5) As a second post-processing phase, two soft clustering methods are used to structure the Pareto front.

The rest of the paper is organized as follows: Section 2 presents some related works. Section 3 provides a methodology overview. Section 4 presents the proposed multi-objective mathematical model. Section 5 presents the employed computational intelligence algorithms (NSGA-II, MOFA, FCM and GMM). Section 6 discusses the experimental results obtained using our approach. Finally, Section 7 gives some conclusions and future directions.

2. Related Works

Faced with a large population of diabetics, a long-term strategy (group therapy) is needed that involves dividing this population into a reasonable number of compartments, then estimating the magic combinations comprising diet, exercise and medication for each compartment.

To meet these needs, a number of mathematical models have been developed, in particular dynamic models of the diabetic population. Various compartments have been considered, i.e., pre-diabetic people, diabetic people without complications, people with complications, healthy people and other types [15–19]. All the dynamic models of diabetes proposed in the literature focus on a single objective function and follow the same steps, with slight differences depending on the types of diabetic groups studied. At first, research was performed only on two types of compartments, the compartment of people with diabetes and prediabetic people. In 2014, the first controlled population dynamics model of diabetes was put forward [20]. It considered three types of populations: pre-diabetic individuals, diabetic individuals without complications and diabetic individuals with complications [20]. The authors proposed this controlled diabetic population model to reduce the negative effects of this disease and used the Gumel numerical method [21] to solve the system. They also proposed a control strategy over a period of 120 months. Subsequently, another controlled dynamic model of the diabetic population, based on a new control strategy, was proposed in [22]; this time, they thought of dividing the population into two types of compartments, which were the uncomplicated diabetic individuals and the diabetic individuals with complication. Unfortunately, this approach excludes many diabetes-related groups and fails to specify how to control a very large population size with the same control strategy. In 2018, the authors of [23] introduced a population control model for diabetics, implementing five compartments, but the proposed model takes into account people with disabilities, which is not the most prevalent general case. Consequently, this control cannot be generalized.

An alternative study, which has some overlap with our study, examines the fact of being pregnant [24]. Unfortunately, this study focuses only on women, and more specifically on pregnant women. Moreover, the control phase has not yet been carried out. In [25], authors suggested a reduced monitoring framework using the time-discrete method, which models the progression of prediabetes to diabetes with and without complications and the impact of the lifestyle context. Anther work [26], by the same authors, provided a mathematical model of the diabetic population split into six compartments considering other aspects such as the effects of genetics and behavioral factors. The authors of the study suggest that in order to decrease the proportion of diabetics, three controls could be implemented: an awareness program through education and media, therapy and psychological assistance with follow-up. In the end, several strategies were proposed in [25,26], but they lead to confusion for doctors, and the proposed controls are difficult for non-mathematicians to understand. In [8], the authors investigated a model that outlines the evolution of the population, as well as the pain of diabetic subjects with the socio-environmental effect depending on the age category; the authors suggested an optimized monitoring plan to protect patients from the negative influence of a lifestyle that causes them to develop complications. Numerical methods, employing Pontryagin's principle, result in controls that are prohibitively expensive in terms of human and material resources. Recently, in [2], we used heuristic methods to estimate a control based on a fitness function that achieves a compromise between constraints and the model's objective function. However, implementing conflicting criteria in a single function results in non-personalized standard controls. This paper introduces a new approach based on a multi-objective model, which we solve with two multi-objective swarm optimizers. In order to help the user select a suitable control from the Pareto front, we also employ two soft clustering methods as mentioned in the previous section.

3. Methodology Overview

Notations

Time variables: we denote by T the control period and by h the time step.

Compartments: E, I, D and C are the number of pre-diabetics, the incidence of being prediabetic, the number of diabetics and the number of diabetics with complications.

Parameters: In the system described in Equation (1), μ is the biological death ratio, β_1 is the risk of acquiring diabetes mellitus (DM), β_2 is the estimate of the likelihood that an individual with DM will develop a complication, β_3 is the risk of developing complicated DM, γ is the success ratio for complications, ν is the degree to which severely disabled individuals become seriously handicapped and δ is the death ratio caused by medical complications.

Controls: u_1 and u_2 are the functions introduced to control the compartments of diabetics with and without complications.

Model

Variables: we introduced two functions (u_1 and u_2), in model (1), to control the compartments D and C during the period T; see system (2).

First objective function: this function was introduced to minimize the number of diabetic D and the resources required to realize this objective (i.e., u_1).

Second objective function: this function was introduced to minimize the number of diabetics with complications C and the resources required to realize this objective (i.e., u_2).

Constraints: ordinary differential equation that governs the exchange between different compartments in the presence of the controls.

Discretization

Objective functions: The first and the second objective functions implement integral operators to consider the number of diabetics with and without complications and the resources required to control these compartments during the period T. To transform these integral to a discrete sum, we used the trapezoidal rule based on time step h.

Constraints: to transform the differential equations of the proposed model into a combinatorial system, we used the Euler–Cauchy method because of its simplicity.

Error estimation: in lemma 1 and proposition 1, we demonstrate that the error due to the discretization is still bounded with a cubic function that implements all the outputs of the multi-objective model.

Smart local search

To estimate the controls u_1 and u_2, we used two multi-objective local search methods, namely the NSGA [12,27] and MOFA [13] algorithms; the configurations of these methods were experimentally chosen. Here $[T/h] + 1$ represents the integer par of T/h.

Post-processing

Features extraction: to avoid high-dimensional vectors ($[T/h]$), when structuring the control space, we extracted the relevant information by adopting certain criteria (control fluctuation, control cost, quality of compartments and spatial characteristics).

Structuring of the Pareto front: NSGA-II and MOFA produce a set of controls that are difficult to exploit, so we used two soft clustering methods, GMM [14] and FCM [28], to summarize the Pareto front.

Fluctuation corrections: To correct fluctuations caused by successive approximations, we used the FFT convolution operator and test several masks of different sizes

$(4 \times 4, \ldots, 10 \times 10)$. In this sense, given a function, represented by the matrix U, whose fluctuations we want to eliminate, we chose a suitable kernel function K, represented by a matrix of ones modified by a suitable weight w, and the convolution formula is given by $\forall j, k\ C(i,j) = \sum_{p,q} U(p,q) * K(j-p+1, k-q+1)$.

The size of the kernel K and w were experimentally chosen.

Performance evaluation

The control quality was measured using three criteria: (a) the rate of growth of E: the faster it is, the better the control; (b) the rate of decay of D and C: the faster the decay, the better the control; and (c) the values taken by u_1 and u_2: the smaller they are, the better the control.

4. Multi-Objective Diabetic Control Model

4.1. Single-Goal Control Model

Let $E(t)$, $D(t)$ and $C(t)$ be the compartment of pre-diabetic individuals, the compartment of diabetic individuals without complication and the compartment of diabetic individuals with complication, respectively. Derouich et al. [20] introduced the following mathematical model:

$$\begin{cases} \frac{dE(t)}{dt} = I - (\mu + (\beta_3 + \beta_1)(1-u(t)))E(t) \\ \frac{dD(t)}{dt} = \beta_1(1-u(t))E(t) - (\mu + \beta_2(1-u(t)))D(t) + \gamma C(t) \\ \frac{dC(t)}{dt} = \beta_3(1-u(t))E(t) + \beta_2(1-u(t))D(t) - (\mu + \gamma + \nu + \delta)C(t) \end{cases} \quad (1)$$

u represents the intervention of the endocrinologist [medication potency level (1 mg, ..., 10 mg), diet level (the glycemic load not to be exceeded), type of exercise (the number of minutes of walking, type of running...)], I is the effect of the presence of pre-DM, μ is the biological death ratio, β_1 is the risk of acquiring DM, β_2 is the estimate of the likelihood that an individual with DM will develop a complication, β_3 is the risk of developing complicated DM, γ is the success ratio for complications, ν is the degree to which severely disabled individuals become seriously handicapped and δ is the death ratio caused by medical complications.

The function they sought to minimize in [20] is of the following form:

$$\Gamma(u) = \int_0^T \left(D(t) + C(t) + Ku^2(t) \right) dt$$

$K \in \mathbb{R}^+$ and all the feasible controls of one goal model form a set denoted U. u is said to be feasible if it is measurable and the system (1) has at least one solution. A feasible control is expressed in percent, i.e., $0 \leq u(t) \leq 1, \forall t \in [0, T]$.

Problems:

For a given decision u, the $\int_0^T (D(t) + C(t) + Ku^2(t))dt$ and $\int_0^T D(t)dt$ can be minimal and $\int_0^T C(t)dt$ is very large, as the terms in Γ are in conflict with each other.

In practice, it is difficult to set up a tradeoff between D, C and u via a penalty constant. In fact, to evacuate compartment D, there are three possibilities: move patients from D to E or move patients from D to C, or both. To evacuate compartment C, there is only one possibility: move patients from C to D. So, the two compartments D and C are in conflict. A possible scenario: if the number of diabetic patients with complications is very small, compared to D, a wrong choice of aggregation parameters can give wrong information to the optimization methods and the number of patients with complications can receive elements from D while the objective function is minimal (for a given local solution). When dealing with conflicting cost functions, a single solution is not reasonable because a solution that may be appropriate in one context may not be appropriate in another. The characteristics of patients in D are not similar to the ones of C. Therefore, the regulation of the system requires individualized policies that take this difference into account.

The Pontryagin principle implies very complex mathematical formulas; moreover, it leads to an expansive strategy that consumes all existing resources.

4.2. Multi-Objective Control Model

Decision variables: To act on E and D with two separate strategies, we introduced two controls, u_1 to control E and u_2 to control D. u_1 and u_2 are two measurable functions defined on $[0, T]$ and take their values in $[0, 1]$. Practically speaking, it is impossible to estimate the decision functions at each instant t, and that is why we estimated these functions at d instants $t_1, \ldots, t_d$ from $[0, T]$ such that $t_{i+1} - t_i = \text{constant} = h$ and $\frac{T}{h} = d$, and let us denote by $u_{11}, \ldots, u_{1d}$ and $u_{21}, \ldots, u_{2d}$ the obtained values.

Constraints: By introducing u_1 and u_2 in the system (1), we prevented $(1 - u_1(t))\%$ of prediabetic people and $(1 - u_2(t))\%$ of people with diabetes from joining the upper compartments; thus, we obtained the following system:

$$\begin{cases} \frac{dE(t)}{dt} = I - (\mu + (\beta_3 + \beta_1)(1 - u_1(t)))E(t) \\ \frac{dD(t)}{dt} = \beta_1(1 - u_1(t))E(t) - (\mu + \beta_2(1 - u_2(t)))D(t) + \gamma C(t) \\ \frac{dC(t)}{dt} = \beta_3(1 - u_1(t))E(t) + \beta_2(1 - u_2(t))D(t) - (\mu + \gamma + \nu + \delta)C(t) \end{cases} \quad (2)$$

We discretized the system (2) using the time step h defined before, and we obtained the following system:

$$\begin{cases} E_{i+1} = Ih - (\mu + (\beta_3 + \beta_1)(1 - u_{1,i}))hE_i + E_i \\ D_{i+1} = \beta_1(1 - u_{1,i})hE_i - (\mu + \beta_2(1 - u_{2,i}))hD_i + \gamma hC_i + D_i \quad i = 0, \ldots, d-1 \\ C_{i+1} = \beta_3(1 - u_{1,i})hE_i + \beta_2(1 - u_{2,i})hD_i - (\mu + \gamma + \nu + \delta)hC_i + C_i \end{cases} \quad (3)$$

where $\widetilde{u}_p = (u_{p,0}, \ldots, u_{p,d})$, $p = 1, 2$.

Objective functions: We introduced the following two objective functions:

$$\Gamma_1(u_1, u_2) = \int_0^T \left(C(t) + Ku_1^2(t)\right)dt \text{ and } \Gamma_2(u_1, u_2) = \int_0^T \left(D(t) + Ku_2^2(t)\right)dt$$

We used the trapezoidal rule to estimate Γ_1 and Γ_2 and then obtained the following objective functions:

$$\Gamma_1(u_1, u_2) \approx \widetilde{\Gamma}_1([\widetilde{u}_1, \widetilde{u}_2]) = \frac{h}{2}\sum_{i=0}^{d-1}(C_{i+1} + C_i) + \frac{Kh}{2}\sum_{i=0}^{d-1}\left(u_{1,i+1}^2 + u_{1,i}^2\right)$$

$$\Gamma_2(u_1, u_2) \approx \widetilde{\Gamma}_2([\widetilde{u}_1, \widetilde{u}_2]) = \frac{h}{2}\sum_{i=0}^{d-1}(D_{i+1} + D_i) + \frac{Kh}{2}\sum_{i=0}^{d-1}\left(u_{2,i+1}^2 + u_{2,i}^2\right)$$

We denote by $U_{1,2}$ the set of the pair feasible control.
The multi-objective problem we propose in this work is given by:

$$\begin{cases} \text{Min } \widetilde{\Gamma}_1([\widetilde{u}_1, \widetilde{u}_2]) \quad \text{Min } \widetilde{\Gamma}_2([\widetilde{u}_1, \widetilde{u}_2]) \\ \text{Subject to} \\ [\widetilde{u}_1, \widetilde{u}_2] \in [0, 1]^{2d} \\ \left(\widetilde{E}, \widetilde{D}, \widetilde{C}\right) \text{ solution of (3)} \end{cases} \quad (4)$$

Convexity of the problem (4): We have $\forall i = 0, \ldots, d-1$ and $\forall j = 1, 2$; the function $[\widetilde{u}_1, \widetilde{u}_2] \to u_{j,i+1}^2 + u_{j,i}^2$ is convex.

Then $\forall j = 1, 2$, and the function $[\widetilde{u}_1, \widetilde{u}_2] \to \sum_{i=0}^{d-1}\left(u_{j,i+1}^2 + u_{j,i}^2\right)$ is convex;

As $\frac{Kh}{2} > 0$, then $\forall j = 1, 2$; the function $[\widetilde{u}_1, \widetilde{u}_2] \to \widetilde{\Gamma}_j([\widetilde{u}_1, \widetilde{u}_2])$ is convex.

We have $\forall i = 0, \ldots, d-1$, the function

$$F\left(\left[\widetilde{u}_1, \widetilde{u}_2\right]\right) = \begin{bmatrix} Ih - (\mu + (\beta_3 + \beta_1)(1 - u_{1,i}))hE_i + E_i - E_{i+1} \\ \beta_1(1 - u_{1,i})hE_i - (\mu + \beta_2(1 - u_{2,i}))hD_i + \gamma hC_i + D_i - D_{i+1} \\ \beta_3(1 - u_{1,i})hE_i + \beta_2(1 - u_{2,i})hD_i - (\mu + \gamma + \nu + \delta)hC_i + C_i - C_{i+1} \end{bmatrix}$$

is linear; thus, F is convex.

Finally, the problem (4) is convex.

As we have shown that the problem (4) is convex, we used gradient methods to solve it:

(a) The primary methods used were gradient descent algorithms [29];
(b) Dual methods, which exploit convexity to calculate the gradient of the dual max–min, were also used [30];
(c) The substitution and the decomposition Lagrange methods that introduce a copy variable to decompose the initial problem to two sub-problems [31]: the first one does not have any constraints (for which we can use gradient descent, among others) and the second one does not have objective functions (for which we can use back tracking methods, among others).

In this sense, we have to transform the multi-objective functions to a single-objective function using aggregation weights; as a result, we find ourselves in front of a system of $3d$ constraints plus $2d$ positivity constraints. In this work, we preferred to use heuristic methods that only meet the $0 \leq \widetilde{u}_{j,i} \leq 1 \; \forall i = 0, \ldots, d-1$ and $\forall j = 1, 2$ positivity constraints, i.e., $2d$, which are easily introduced in Matlab's "gamultiobj" as bounds. To avoid the $3d$ constraints, given an estimation of the control, at the iteration k, we used an interpolation method (for example the spline method) to transform the discrete control $\widetilde{u}$ to the continuous control u; then, we called for the Euler–Cauchy method to approximate the compartments C and D; after that, we estimated the value of the function $\widetilde{\Gamma}_j\left(\left[\widetilde{u}_1, \widetilde{u}_2\right]\right)$ ($\forall j = 1, 2$).

Error of approximation: In this section, we estimate the error related to the approximation of the integral using the trapezoidal rule. We introduced the following functions:

$f_1(t) = C + Ku_1^2$ and $f_2(t) = D + Ku_2^2$

$\forall i \in \{1, 2\}$, we set $M'_i = \max\limits_{t \in [0, T]} |u'_i(t)|$ and $M''_i = \max\limits_{t \in [0, T]} |u''_i(t)|$

Lemma 1. *Considering the system (2), we have:*

$$\|f''_1\| \leq \left(\alpha M'_1 + \alpha M'_2 + 31\alpha^2 + \alpha\right)P + 2KM''_1 + 2K\left(M'_1\right)^2$$

$$\|f''_2\| \leq \left(\alpha M'_1 + \alpha M'_2 + 17\alpha^2 + \alpha\right)P + 2KM''_2 + 2K\left(M'_2\right)^2$$

where P is the size of the population and $\alpha = \max\{\mu, \gamma, \nu, \delta, \beta_1, \beta_2, \beta_3\}$.

Proof of Lemma 1.

We have $f'_1 = C' + 2Ku'_1u_1$

Thus $f'_1 = \beta_3(1 - u_1)E + \beta_2(1 - u_2)D - (\mu + \gamma + \nu + \delta)C + 2Ku'_1u_1$

Then $f''_1(t) = -\beta_3u'_1E + \beta_3(1 - u_1)E' - \beta_2u'_2D + \beta_2(1 - u_2)D' - (\mu + \gamma + \nu + \delta)C'$
$+ 2Ku''_1u_1 + 2K(u'_1)^2$

Thus $f''_1 = -\beta_3u'_1E + \beta_3(1 - u_1)[I - (\mu + (\beta_3 + \beta_1)(1 - u_1))E] - \beta_2u'_2D + \beta_2(1 - u_2)$
$[\beta_1(1 - u_1)E - (\mu + \beta_2(1 - u_2))D + \gamma C] - (\mu + \gamma + \nu + \delta)[\beta_3(1 - u_1)E + \beta_2(1 - u_2)D(t)$
$- (\mu + \gamma + \nu + \delta)C] + 2Ku''_1u_1 + 2K(u'_1)^2$

Then $f''_1 = [-\beta_3u'_1 - \beta_3(1 - u_1)(\mu + (\beta_3 + \beta_1)(1 - u_1)) + \beta_2\beta_1(1 - u_1)(1 - u_2)$
$-(\mu + \gamma + \nu + \delta)\beta_3(1 - u_1)]E + [-\beta_2u'_2 - \beta_2(1 - u_2)(\mu + \beta_2(1 - u_2)) - (\mu + \gamma + \nu + \delta)$
$\beta_2(1 - u_2)]D + [\beta_2(1 - u_2)\gamma + (\mu + \gamma + \nu + \delta)^2]C + 2Ku''_1u_1 + 2K(u'_1)^2 + \beta_3(1 - u_1)I$

Thus $\|f_1''\| \leq (\alpha M_1' + \alpha M_2' + 31\alpha^2 + \alpha)P + 2KM_1'' + 2K(M_1')^2$

Following the same steps, we find $\|f_2''\| \leq (\alpha M_1' + \alpha M_2' + 17\alpha^2 + \alpha)P + 2KM_2'' + 2K(M_2')^2$. □

Proposition 1. *Let (u_1, u_2) be the optimal control of the problem (P). The error associated with the trapezoidal rule discretization is bounded by the number*

$$E = \left(\frac{Th^2}{12}\right)[(\alpha M_1' + \alpha M_2' + 31\alpha^2 + \alpha)P + 2K(M + M^2)]$$

where $M = \max\{M\prime_1, M\prime_2, M''_1, M''_2\}$.

Proof of the Proposition 1. Consider the integral $Integ = \int_a^b f(t)dt$; Rahman Qzi et al. showed that the error of the trapezoidal rule is estimated by [32]:

$$Error = -\frac{(b-a)^3}{12N^2}f''(\xi)$$

where N is the number of the used points in $[a, b]$.

In our case, $a = 0$ and $b = T$, and $N = d = T/h$ (h is the step of the discretization). However, we demonstrated before, in Lemma 1, that
$\|f_1''\| \leq (\alpha M_1' + \alpha M_2' + 31\alpha^2 + \alpha)P + 2KM_1'' + 2K(M_1')^2$ and $\|f_2''\| \leq (\alpha M_1' + \alpha M_2' + 17\alpha^2 + \alpha)P + 2KM_2'' + 2K(M_2')^2$

Thus $\|f_i''\| \leq (\alpha M_1' + \alpha M_2' + 31\alpha^2 + \alpha)P + 2KM_1'' + 2K(M_1')^2$

Thus, the discretization error is given by

$$E = \left|\frac{T^3}{12N^2}f_i''(\xi)\right| \leq \frac{Th^2}{12}[(\alpha M_1' + \alpha M_2' + 31\alpha^2 + \alpha)P + 2K(M + M^2)]$$

The reason why our approach is of practical rather than conceptual importance is that it offers more freedom to fragment the control: the first is focused on E and the second is about D. We will not elaborate too much in the proofs of the results relating to the invariance and the existence of the solution of the proposed multi-objective model for every pair of controls (u_1, u_2).

Invariance: based on the Gronwall inequality (applied to the model (2)) [15], we demonstrate that $E(t) > 0$, $D(t) > 0$, $C(t) > 0$, and $N(t) = E(t) + D(t) + C(t) \leq I/\mu$.

Existence: By adopting the same procedures as in [15], we proved that the right-hand members of the Equation (2) are Lipschitzian. The difference between the proofs given in [15] were the considered bounds. In our situation, these bounds implemented the controls u_1 and u_2, and since $(u_1, u_2) \in U_1 \times U_2$, these results are always true. □

Theorem 1. *Let us consider the next problem:*

$$\begin{cases} \text{Min } \Gamma_1(u_1, u_2) \quad \text{Min } \Gamma_2(u_1, u_2) \\ \text{Subject to} \\ (u_1, u_2) \in U_1 \times U_2 \\ (E, D, C) \text{ solution of } (2) \end{cases}$$

A Pareto front of non-dominant optimal decisions exists (called the Pareto optimal control).

Proof of the Theorem 1: Without going into details, the existence of an optimal front can be proved by applying the finding of Fleming and Rishel [33] and by following the steps mentioned below:

$\Gamma_1(u_1, u_2) = \int_0^T \left(C(t) + K u_1^2(t) \right) dt$ and $\Gamma_2(u_1, u_2) = \int_0^T \left(D(t) + K u_2^2(t) \right) dt$ are convex in (u_1, u_2);

$U_1 \times U_2$ is convex.

The right-hand sides of the system Equation (2) are continous, bounded and can be written as a linear function of u_1 and u_2 with coefficients depending on time and state;

The integral of the objective functionals, $C(t) + K u_1^2(t)$ and $D(t) + K u_2^2(t)$, are clearly convex on $U_1 \times U_2$;

We have $C(t) + K u_1^2(t) \geq \alpha_{1,1} + \alpha_{1,2} \|u_1\|^2$ and $D(t) + K u_2^2(t) \geq \alpha_{2,1} + \alpha_{2,2} \|u_2\|^2$ where $\alpha_{1,1} = \inf_{t \in [0,T]} D(t)$ and $\alpha_{2,1} = \inf_{t \in [0,T]} C(t)$ and $\alpha_{1,2} = \alpha_{2,2} = K$. □

4.3. Pareto Controls Characterization

The Pareto curve is formed by non-dominated controls (we denote by m the size of the pareto set). We estimated these controls on several points, say d, of the control interval $[0, 10]$ *years*; generally, d is very large compared to m, which affects the quality of the grouping. In this part, we extracted the most important features that describe the controls basing on four criterions: the controls fluctuation, the cost of the controls, the spatial characteristics of the decisions and the quality of the controls.

Fluctuation characteristics: First, we removed the linear trend from the controls u_1 and u_2 using FFT processing [34], and we obtained the corrected controls u_1' and u_2'. Two fluctuation features were extracted in this phase: $norm(u_1 - u_1')$ and $norm(u_2 - u_2')$.

Control cost: The total resources mobilized to control the compartments D and C; these were estimated by the sum of the controls on the control duration. In this sense, we have: $sum(u_1(t)/t \in [0, T])$ and $sum(u_2(t)/t \in [0, T])$.

Spatial characteristics: the spatial characteristics were measured based on the coefficients of the polynomial interpolation of the controls.

Quality of Compartments: The adequacy of the different compartments was measured by DIST = distance (compartment without control, compartment with control). We obtained three compartment features: $dist(Ew, E), dist(Dw, D)$ and $dist(Cw, C)$. The greater the DIST, the better the control.

5. Smart Algorithms

5.1. Swarm Intelligence Optimizers

The metaheuristic algorithms were designed based on simulating natural phenomena and laws, which have better global search ability. The most important concept of multi-objective optimization is the Pareto front; it is the curve of the non-dominated points considering the different objective functions at the same time.

Figure 1 illustrates the notion of a Pareto front. Considering the blue line, if we prefer one objective function, then we disadvantage the other objective function.

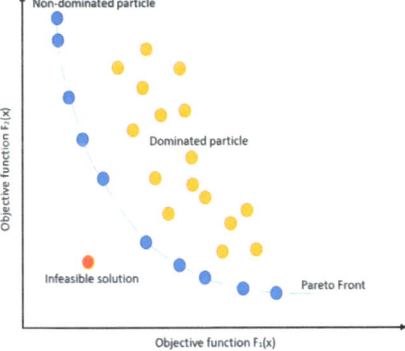

Figure 1. Pareto front concept.

In this section, we provide the main ideas of two well-known swarm intelligence optimizers that produce local Pareto fronts when solving a multi-objective optimization: NSGA-II and MOFA. As the appropriate crossover of local search methods allows us to overcome the shortcomings of the parent methods, it was possible to use some recent hybrid methods to solve the proposed model, such as the hybrid firefly genetic [3] approach, the cuckoo search-based metaheuristic approach [35] and the hybrid marine predator sine–cosine [36] algorithms.

5.1.1. Non-Dominated Sorting Genetic Algorithm II

NSGA-II tends to encourage higher-order chromosomes to appear in the future population [12,27]. The controlled evolutionary selection also encourages the chromosomes to participate in a reasonable diversity of the population in spite of their current weakness. The Pareto fraction limits the number of chromosomes in the solution set. The distance function ensures diversity on the front while promoting chromosomes with an acceptable distance from the front.

The different phases of NSGA-II are shown in Figure 2. The initialization is based on the constraints of the studied problem. Then, the procedure called the non-dominated sorting process about the dominance notion is started. After that, the chromosomes are selected based on two criteria: rank and crowding distance. Then, selection via the tournaments method implementing the crowded-comparison operator is used to select the individuals. The crossover operator and the mutation operator are used to produce chromosomes. The new production is filled by each of the successive borders until the current population size exceeds the tolerated size.

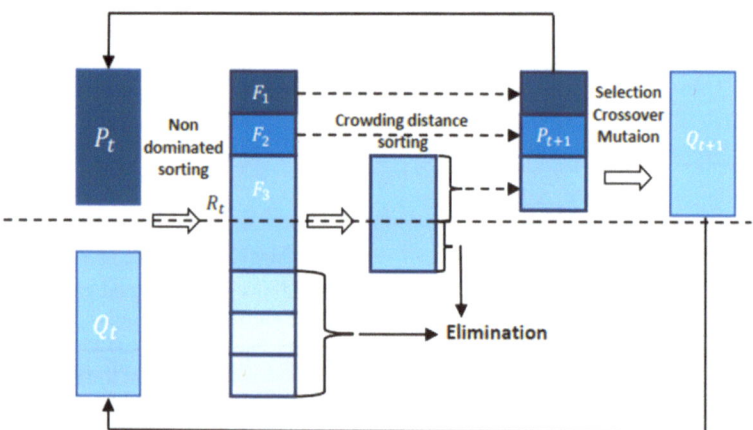

Figure 2. NSGA-II algorithm schematic diagram.

NSGA-II has the advantages of simple coding and excellent performance, and has been successfully applied for solving real-world problems, e.g, solving the flow-shop scheduling problem [37], optimal diet problem [4], data classification [38], Louver configuration [39], 3D laser scanning scheme for engineering structures [40], UAV path planning [41], optimal lane change path planning [42], green pepper detection [43], optimal configuration of landscape storage in public buildings [44], etc.

5.1.2. Multi-Objective Firefly Algorithm

MOFA is inspired by the behavior of fireflies [13]. The basic one-objective version is based on the following rules [27]:

(a) Fireflies have the ability to attract other fireflies, no matter which sex they are.
(b) The attraction is positively proportional to the brightness. If all the fireflies have nearly the same degree of brightness, then one or more fireflies are moving.

(c) The luminosity of a firefly is calculated from the cost function.

If d_{ij} is the distance between two fireflies i and j, then the variability of attractiveness δ_{ij} is estimated by:

$$\delta_{ij} = \delta_0 exp\left(-\sigma d_{ij}^2\right) \quad (5)$$

The parameters δ_0 and σ are chosen by the user.

If x_i^t and x_j^t are the present position of the fireflies i and j, respectively, the FA algorithm uses the following equation to calculate the next position of the ith firefly:

$$x_i^{t+1} = x_i^t + \delta_{ij}\left(x_j^t - x_i^t\right) + \alpha_t \varepsilon_i^t \quad (6)$$

α_t and ε_i^t are the global and local random series corresponding to the ith firefly. An extension of the basic ideas of FA leads to MOFA, presented in Figure 3.

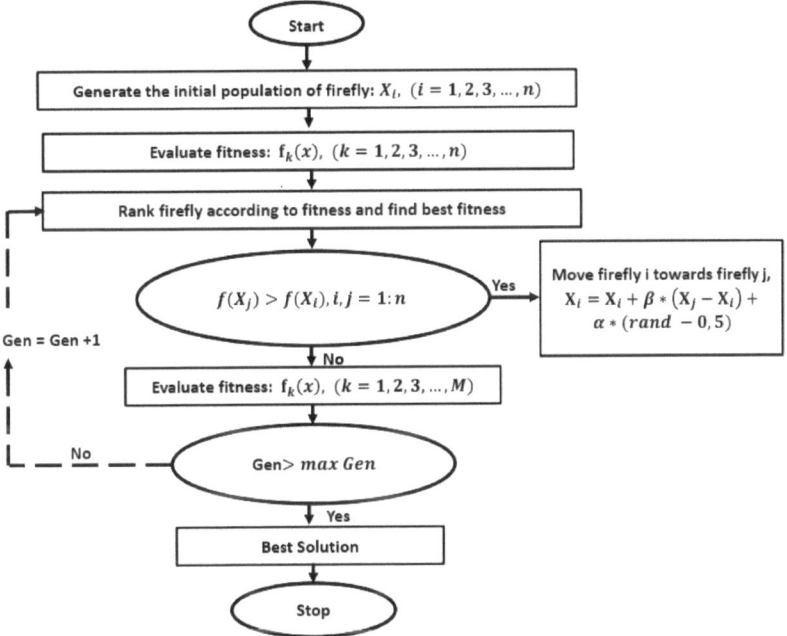

Figure 3. The MOFA algorithm schematic diagram.

Based on the cost functions and the constraints of the problem to be solved, we defined appropriate objective functions. A swarm of fireflies was uniformly chosen from the research space to ensure the same chances to different regions to be explored. The basic cycle starts by measuring the brightness of all individuals in the swarm, and each pair of elements is measured against each other. Then, a smart pairing is carried out on the base of a convex weight matrix. The non-dominated solutions are then forwarded to the following step. Upon termination of a given number of repetitions, n non-dominated solutions are achieved to approach the correct Pareto front.

In this context, MOFA has the advantages of simple coding and a smaller number of parameters to use and has been successfully applied recently for solving interesting problems, such as an energy-efficiency problem [45], automatic EEG channel selection problem [46], reference point reconstruction problem [47], integrated process planning and scheduling problem [48], stochastic techno-economic–environmental optimization [49], optimization of concentric circular antenna array [50], VLSI Floorplanning problem [51], configuration of power quality control device problem [52], identification of compound gear-bearing faults [53], etc.

5.2. Soft Clustering Algorithms

Let us consider the set of not-labeled observations $\mathcal{B} = \{z_1, \ldots, z_N\} \subset \mathbb{R}^n$. The clustering issue consists of partitioning this set into $K \ll N$ groups $P_1, \ldots, P_K$ and representing each group k by a vector w_k called the reference.

Soft computing methods based on fuzzy (based on the degree of membership) or probabilistic (based on the frequency of events) approaches provide decision system makers with the necessary capabilities to deal with imprecise and incomplete information. In fact, a soft clustering method allows an observation to belong to two or more clusters. In this work, we use the well-known and basic soft clustering algorithms Gaussian mixture model (GMM) and Fuzzy-CMeans (FCM) to structure the Pareto front.

5.2.1. Gaussian Mixture Model (GMM)

When we randomly draw an element from the set $\mathcal{B}$, the probability that it is an element of group k is of α_k (statistically estimated). Since the group k is assumed to be sufficiently compact, it makes sense to simulate the distribution of information within this subgroup to a normal distribution of mean w_k and reduced covariance $\sigma^2 I$, where I is the identity matrix of dimension $n \times n$ et $\sigma > 0$. In this way, the probability that an observation is an element of $\mathcal{B}$ is approximated by the following mixture density:
$p(z) = \sum_{k=1}^{k} \alpha_k f_k(z)$, where $\sum_{k=1}^{K} \alpha_k = 1$ and f_k is defined by

$$f_k(z) = \frac{1}{(2\pi)^{\frac{n}{2}} \sigma^n} exp\left(-\frac{\|z - w_k\|^2}{2\sigma^2}\right)$$

Practically, the prior probabilities α_k are all equal to $1/K$. Moreover, the log-likelihood measuring the fact that all observations are generated by p is given by [54]:

$$V(W, \sigma, P) = \frac{1}{2\sigma^2} I(W, P) + Nn\ln(\sigma) + constant$$

$I(W, P)$ is the squared error associated with $w_1, \ldots, w_K$ and $P_1, \ldots, P_K$. The minimization of this function is performed in an iterative way, each of which is divided into two steps: fixing one variable and updating the other (the two variables involved are W and σ).

At the iteration $iter$, if we suppose the references $w_1^{iter}, \ldots, w_k^{iter}$ are known, then σ^{iter} is given by the equation $\sigma^{iter} = \sqrt{nN/I(W^{iter}, P^{iter})}$;

At the end, the observation z_i is allocated to the group $j*$ given by the equation $j^* = argmax(f_k(z_i), k = 1, \ldots, K)$

5.2.2. Fuzzy C-Means (FCM)

FCM is a method of clustering that allows to one observation to be in two or more clusters at the same time [30]. The cost function that this method tends to minimize is given by:

$$I(W, \mu) = \sum_{k=1}^{K} \sum_{i=1}^{N} \mu_{ik}^m \|z_i - w_k\|^2, 1 \leq m < \infty$$

where μ_{ik} is the degree of membership of zi in the cluster k. The function $I(W, \mu)$ is the minimization of this function, performed in an iterative way based on the equations:

$$\mu_{ik}^{-1} = \sum_{j=1}^{K} \left(\frac{\|z_i - w_k\|}{\|z_i - w_j\|}\right)^{\frac{2}{m-1}} \text{ and } w_k = \frac{\sum_{i=1}^{N} \mu_{ik}^m z_i}{\sum_{i=1}^{N} \mu_{ik}^m}.$$

At the end, the observation z_i is allocated to the group j^* given by the equation $j^* = argmax(\mu_{ik}, k = 1, \ldots, K)$.

6. Experimental Results and Discussion

In this section, we used two multi-objective heuristic methods, NSGA-II [12,27] and MOFA [13], to estimate the optimal Pareto front of the problem (4). The configurations of these two algorithms were performed experimentally, i.e., several configurations were performed and the ones producing better results were retained.

To structure the obtained fronts, we used two soft clustering methods, FCM and GMM. The choice of the number of clusters was based on the silhouette criterion. The number of clusters was chosen on the basis of the silhouette criterion. To this end, we tested GMM and FCM for different values of K (number of classes) and evaluated the silhouette of the resulting partitions; the best K is the one corresponding to the highest silhouette value. Figure A5 gives different silhouette values for different numbers of clusters (1 to 6); in our case, the silhouette was maximal when K = 4. In this sense, we did not consider all the approximation points of the controls to structure the control space, but we instead based our analysis on the characterization shown above (Section 4.3). In addition, in order not to clutter the paper with many figures, we introduce and analyze the results obtained via FCM and we put, in the Appendix A, those obtained via GMM.

In addition, convolution filters, such as 9×9 kernels, were used to eliminate the fluctuations, intrinsic to each approximation, in order to obtain reasonable strategies that are easy to implement.

6.1. NSGA-II Combined with Soft Clustering Methods

In this section, we use the NSGA-II method to estimate the elements of the Pareto front at several points by adopting the configuration described in Table 1. This configuration was chosen using the traditional approach of running a number of pilot tests; for example, we noticed that after 60 iterations, the fitness function remained constant, so we set the maximum number of iterations at 60. In addition, the adaptive mutation ratio was used to explore other regions to avoid early convergence to poor local minima. It should be noted that the best parameters of NSGA-II are the ones that optimize the criteria given in Section 3 in the sub-section entitled "Performance measures".

Table 1. NSGA-II configuration.

Option [12]	Configuration
Crossover operator	New_indiv = indiv1 + rand × atio × (indiv2 − indiv1)
Crossover ratio	0.8
Number of iterations	60
Mutation ratio	adaptive feasible

Figure 4 shows the Pareto front obtained by the NSGA-II method. The shape and the richness of this front show that this algorithm offers several customized control strategies.

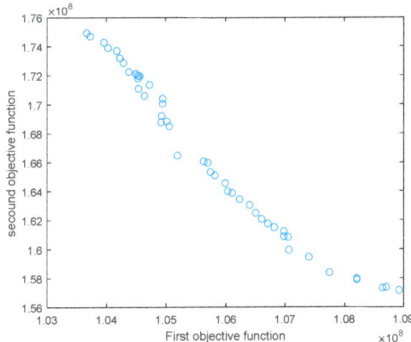

Figure 4. Pareto front produced via NSGA-II applied to the model (4).

Figure 5a,b give the two sets of sub-controls offered by the Pareto front produced via NSGA-II applied to the problem (4). We notice that at each instant of the control interval, the two sub-controls do not have the same value, so the controller of the population dynamics under study did not adopt the same strategy for the two compartments C and D. This justifies, experimentally, the fact of associating different controls to the different compartments in the mathematical model (2).

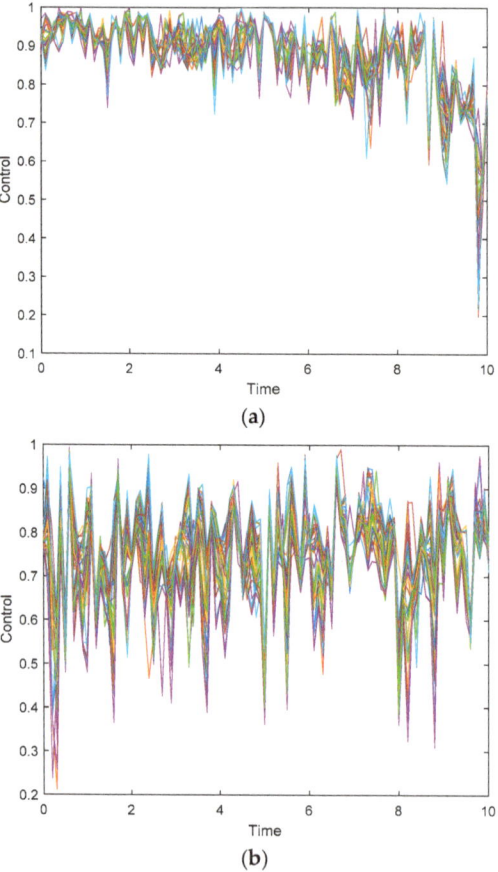

Figure 5. (**a**) The set of controls u_1 extracted from the Pareto front produced via NSGA-II applied to model (4). (**b**) The set of controls u_2 extracted from the Pareto front produced via NSGA-II applied to model (4).

On the one hand, the richness of the solution space provides several possibilities, but on the other hand, it is difficult to exploit these strategies directly because they are numerous and an expert in the medical field needs assistance to choose what suits them. For this reason, we used two soft clustering methods to structure the two subspaces of controls: FCM and GMM. Based on the silhouette criterion, the best K is 4.

Figure 6 gives the pair controls obtained via FCM applied to the sets of subcontrols extracted from the Pareto front produced via NSGA-II applied to the problem (4). We noticed that during the first 6 years, the core 1 controls consumed all the resources and then this effort resulted in a saving of a good portion of the resources. Concerning the second control, the strategies proposed by the two core 2 controls of the different clusters are moderate, except for the third cluster, which always requires the exhaustion of more than 90% of the resources.

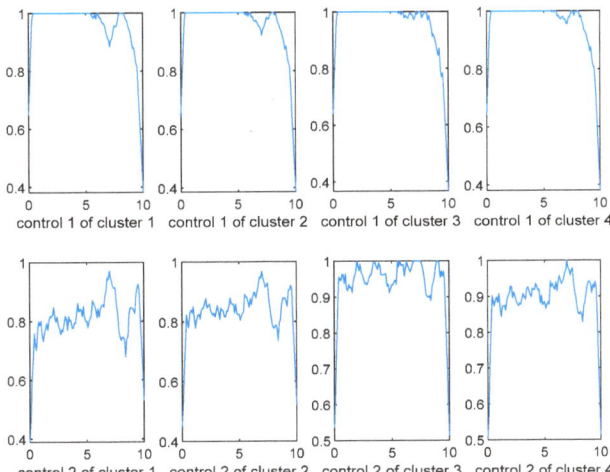

Figure 6. Control pairs, over 10 years, obtained via FCM applied to the Pareto front produced via NSGA-II.

Figure 7 shows the behavior of the different compartments obtained by different controls produced via FCM applied to the Pareto front produced via NSGA-II applied to the problem (4). It can be seen that even if all the resources were not consumed, the population studied can be controlled and the desired behavior was obtained, except that at the end of the control period, we noticed a slight growth in the two compartments. This phenomenon is almost absent when we apply the strategy offered by cluster 1.

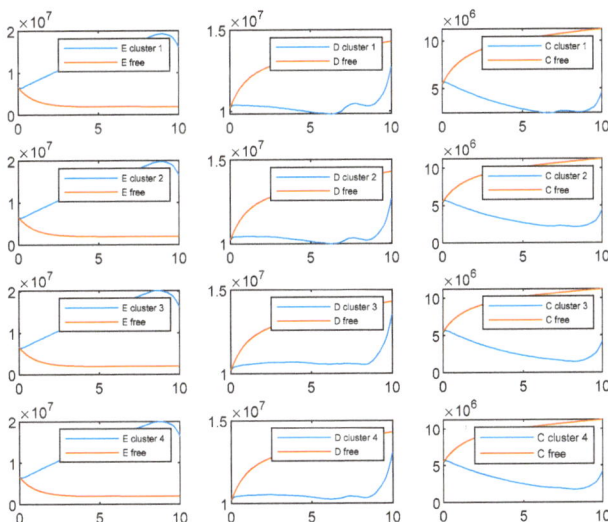

Figure 7. Compartments obtained when introducing different controls, produced via FCM applied to the Pareto front produced via NSGA-II, in model (4).

6.2. MOFA Combined with Soft Clustering Methods

In this section, we used the MOFA method to estimate the elements of the Pareto front at several points by adopting the configuration described in Table 2. This configuration was chosen using the traditional approach of running a number of pilot tests. It should be noted that the best parameters of MOFA are the ones that optimize the criteria given in Section 3 in the sub-section entitled "performance measures".

Table 2. MOFA configuration.

Option [13]	Configuration
Maximum number of iterations	1000
Swarm size	25
Light absorption coefficient	1
Attraction coefficient base value	2
Mutation coefficient	0.2
Mutation coefficient damping ratio	0.98

Figure 8 gives the Pareto front obtained using the MOFA optimizer applied to the problem (4). The shape of the front shows that this algorithm provides a few diversified choices compared to NSGA-II.

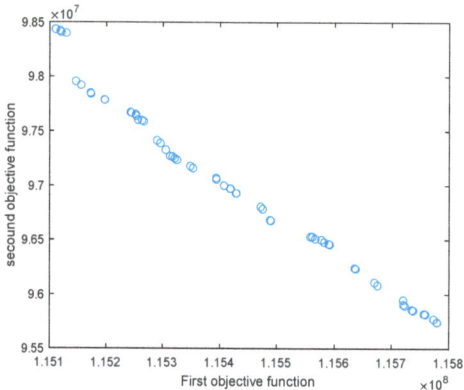

Figure 8. Pareto front produced via MOFA applied to the model (4).

Figure 9a,b give the sets of subcontrols extracted from the Pareto front produced via MOFA applied to the problem (4). We notice that at each time of the control interval, the two subcontrols do not have the same value, so the controller of the studied population dynamics does not adopt the same strategy for the two compartments C and D. This justifies, experimentally, the act of associating different controls to the different compartments in the mathematical model (2).

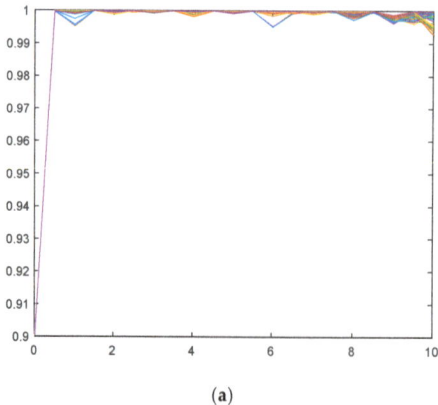

(a)

Figure 9. *Cont.*

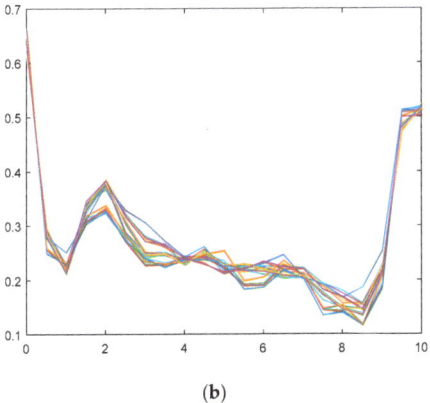

(b)

Figure 9. (a) The set of controls u_1 extracted from the Pareto front produced via MOFA applied to the model (4). (b) The set of controls u_2 extracted from the Pareto front produced via MOFA applied to the model (4).

Similar to the previous subsection, and in order to assist medical experts in the choice of strategies compatible with their requirements, we used two soft clustering methods to structure the two control subspaces: FCM and GMM.

Figure 10 gives the controls pair obtained via FCM applied to a Pareto front produced via MOFA applied to the problem (4). We notice that the sub-controls associated with compartment D have the form of a trapezoid using all the resources on the time scales of [1 years, 7 years]. Concerning the second sub-control, MOFA manages to control the compartment C with few resources (between 25% and 40%). We always notice that the strategy followed to control compartment D is totally different from the one adopted for compartment C, which justifies the use of two decision functions in the model (2).

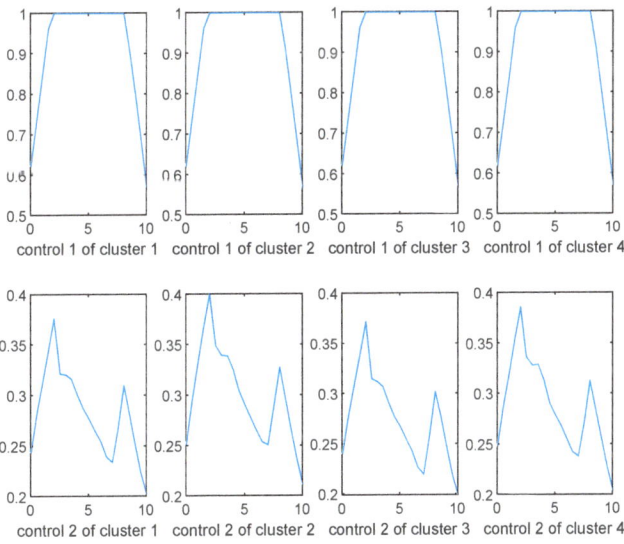

Figure 10. Controls pair obtained via FCM applied to the Pareto front produced via MOFA.

Figure 11 shows the behavior of the different compartments obtained using different controls produced via FCM applied to the Pareto front produced via MOFA when applied to the problem (4). It can be seen that even if all the resources were not consumed, the population studied could be controlled and the desired behavior was obtained, except for at the end of the control period, when we notice a slight growth of the two compartments with nearly the same size and in the same way in all the clusters.

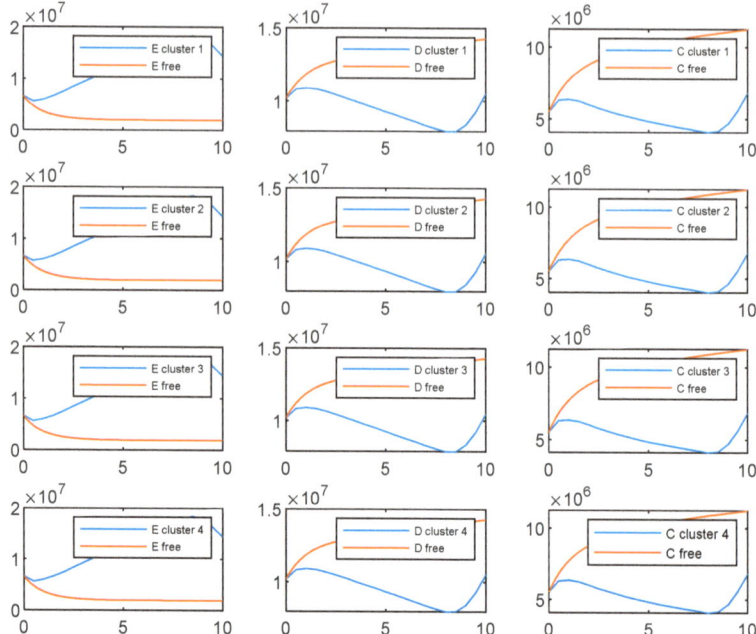

Figure 11. Compartments obtained when introducing different controls, produced via FCM applied to the Pareto front produced via MOFA, in the model (4).

To measure the percentage of the resources saved by the proposed multi-objective strategy compared to a mono-objective strategy, we use the following equation:

$$SavedResources = 1 - sum(MultiObjControl)/sum(SingObjControl)$$

Table 3 gives the percentage of the resources saved by using the proposed method. In this regard, we find that the multi-objective strategy saved up to 4% of resources for the control of compartment D. In addition, it saved up to 18% of resources for the control of compartment C.

Table 3. Multi-objective control strategy vs. single-objective control strategy.

		Cluster 1	Cluster 2	Cluster 3	Cluster 4
Multi-objective vs. single-objective	u_1	3%	4%	4%	4%
	u_2	14%	6%	18%	11%

Notes:

(a) Considering the experimental results shown in the Figures 1–4, we notice that the two soft clustering methods FCM and GMM give approximately the same groups (considering a simple permutation), so we extended the same remarks, conclusions, and recommendations to the other method.
(b) The shapes of the front given in Figure 5 and those given in Figure 9 show that NSGA-II offers several highly non-dominated customized control strategies compared to MOFA.

(c) According to Tables 1 and 2, compared with NSGA-II, MOFA requires a large number of generations to achieve feasible and acceptable controls (60 generations for NSGA-II versus 1000 generations for MOFA).

We remark that MOFA-Soft-Cluster produces controls completely different to the ones produced via NSGA-II-Soft-Cluster. The sub-controls associated with compartment D, produced via MOFA-Soft-Cluster, are a symmetrical trapezoid of the support [2 years, 8 years], and the sub-controls associated with compartment C have very low cost along the control duration (they do not even consume 40% of the resources). So, what is wasted on one side is recovered on the other side. The sub-controls, associated with D, produced via NSGA-II-Soft-Cluster are a non-symmetrical trapezoid and they start by being too expensive and end up with a very low cost. Meanwhile, the sub-controls associated with C, produced via NSGA-II-Soft-Cluster, are too expensive from the beginning to the end (they consume between 80% and 100% of the resources). We cannot talk about the processing time because the size of the population and the size of the swarm influence the time complexity and we cannot establish a mathematical relation between the two sizes.

6.3. Single-Objective vs. Multi-Objective on the Control of the Dynamics of the Diabetic Population Problem (C2D2P)

In [19], the authors modeled the problem C2D2P in terms of the single-objective dynamic mathematical model. To solve these models, they used the Gumel method based on Pantriagin's principle; the obtained control is represented in red in Figure A6. Since a control mobilizes human and material resources, it would be better if this control took as small values as possible. In this sense, Gumel was rejected because it takes very high values (these controls consume practically all resources). In our previous work [4], we used the bees algorithm (BA), firefly algorithm (FA), particle swarm algorithm (PSO), genetic algorithm (GA), moth swarm algorithm (MSA), stochastic fractal search (SFS), wind-driven optimization (WDO) and probabilistic bees algorithm (PBA); see Figure A6. The stochastic fractal search (SFS) method has shown an unprecedented ability to produce continuous, economical controls capable of alleviating socio-economic damage on a reasonable budget. Compared to a multi-objective strategy (introduced in this work), PSO, FA, GA, AWD, SBA, PBA, MWA and SFS propose the same strategies to control compartments C and D. When dealing with conflicting cost functions, a single solution is not reasonable because a solution that may be appropriate in one context may not be appropriate in another. The characteristics of patients in D are not similar to the ones of C. For example, 40 min of running combined with 2 g of medication may regulate the blood sugar of a compartment D patient, but this solution may not be suitable for a compartment C patient (whose complications prevent them from running), and 40 min of walking plus 6 g of medication may prove more appropriate.

6.4. Sensitivity of the Proposed System

To study the sensitivity of the controls obtained using the proposed model + NSGA-II + FCM, we applied Gaussian perturbations to the controls obtained using this system; this noise was generated between 0.001 and 0.3.

Figures A7–A12 show the comparisons obtained for certain noise values; see Appendix B as well. For Gaussian perturbations between 0.001 and 0.1, there was almost no change in the compartments. For Gaussian perturbations between 0.12 and 0.3, changes were noted in compartment D and small changes were noted in compartment C (the number of diabetics increases very rapidly compared with the case of optimal control developed in this work). This is normal, since the values taken by the control are between 0 and 1, and when almost 30% change is applied, the control changes; subsequently, the behavior of the compartments also changes.

In the end, we can say that the proposed approach is consistent when it comes to small Gaussian noise.

At the end of this section, we prefer not to make technical comparisons between the two types of strategies offered by the four systems because a choice may be appropriate in one context while it may be bad in another context. When there is a multi-objective problem, the role of the data scientist is the modeling of the studied phenomenon, the numerical simulation and the structuring of the control space. Then, it is left to the medical experts to choose what suits them according to their requirements and availability.

7. Conclusions

Single-objective mathematical modeling was previously used for controlling the dynamics of populations having one or more chronic diseases, in particular controlling the dynamics of the diabetic population. Unfortunately, the controls obtained make no distinction between the different compartments. In this work, we introduce a multi-objective approach that implements a multi-objective optimization model, population-based metaheuristics (NSGA-II and MOFA), flexible unsupervised learning (FCM and GMM), polynomial interpolation and FFT convolution for the problem of controlling the diabetic population. The optimization model implements two objective functions: one for diabetics without complications and another for diabetics with complications. To avoid solving large constraint systems, our approach calls the Euler–Cauchy method once we have a premature approximation of the control. The parameters of the heuristic methods are chosen experimentally. To clean up the resulting controls from noise due to successive approximations, we used a fast Fourier transform with a kernel size of 9×9, chosen experimentally. Since it is difficult for a diabetes specialist to choose the right control for their use, from the Pareto front, we used two soft clustering methods to structure the solution space, where the optimal number of clusters was selected on the basis of the silhouette criterion. The controls produced enabled the evacuation of the compartments of diabetics with and without complications, except that towards the end of the control period, we noticed a small increase in these compartments, a problem we can solve by adding more control approximation points. In addition, the controls produced are customized because the required resources to control the diabetics without complications are totally different from the required resources to control the diabetics with complications. In addition, the multi-objective strategy permits us to save a good number of resources. In the future, we will use variational Bayes techniques to estimate the parameters of the multi-objective model in order to remedy sampling-related problems. To improve the control quality, we will use hybrid metaheuristics (MFOA + NSGA) while introducing the notion of attractiveness during crossover or mutation. In addition, we will introduce the fractional version of the multi-objective model to handle more information about the dynamics of the diabetic population.

Author Contributions: Conceptualization, K.E.M.; Methodology, K.E.M. and V.P.; Validation, A.C.; Investigation, K.E.M. and A.C.; Data curation, K.E.M., H.B., S.C. and M.C.; Writing – original draft, K.E.M. and A.E.O.; Writing – review & editing, V.P., A.C. and A.O. All authors have read and agreed to the published version of the manuscript.

Funding: This research received no external funding.

Data Availability Statement: Data are available under request.

Acknowledgments: This work was supported by the Ministry of National Education, Professional Training, Higher Education and Scientific Research (MENFPESRS), the Digital Development Agency (DDA) and the CNRST of Morocco (Nos. Alkhawarizmi/2020/23).

Conflicts of Interest: The authors declare no conflict of interest.

Appendix A

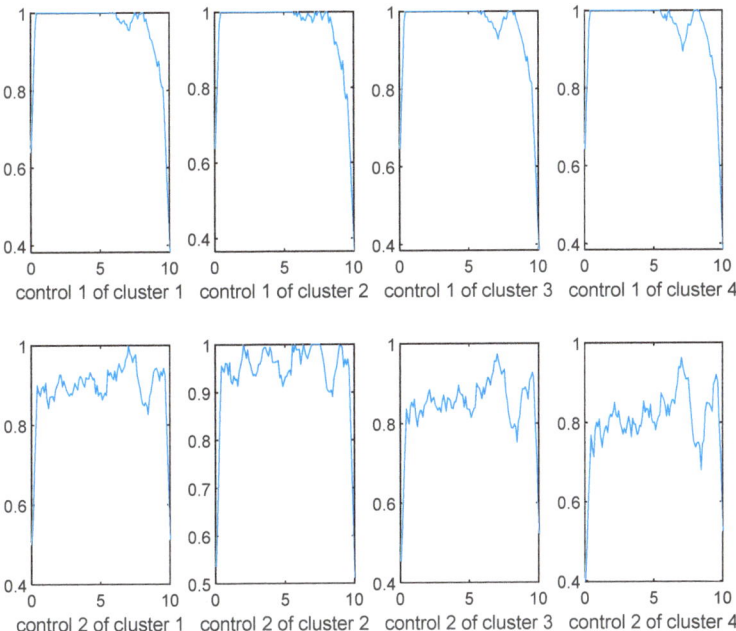

Figure A1. Controls pair obtained via GMM applied to the Pareto front produced via NSGA-II.

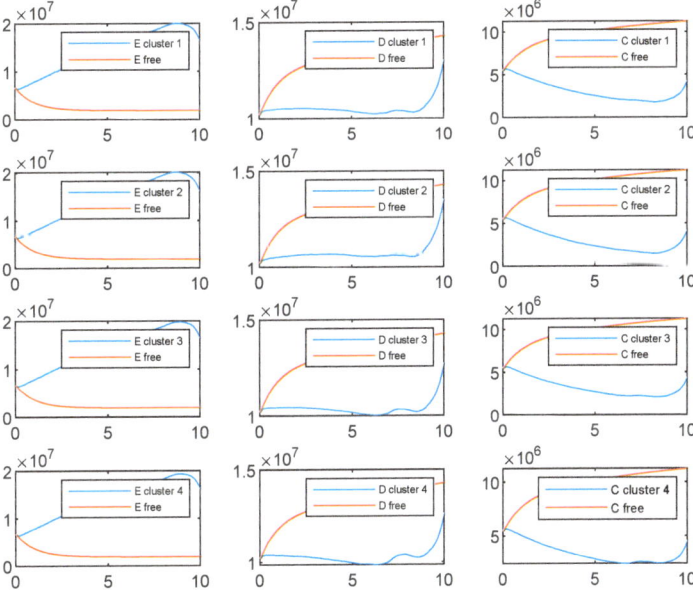

Figure A2. Compartments obtained using different controls produced via GMM applied to the Pareto front produced via NSGA-II.

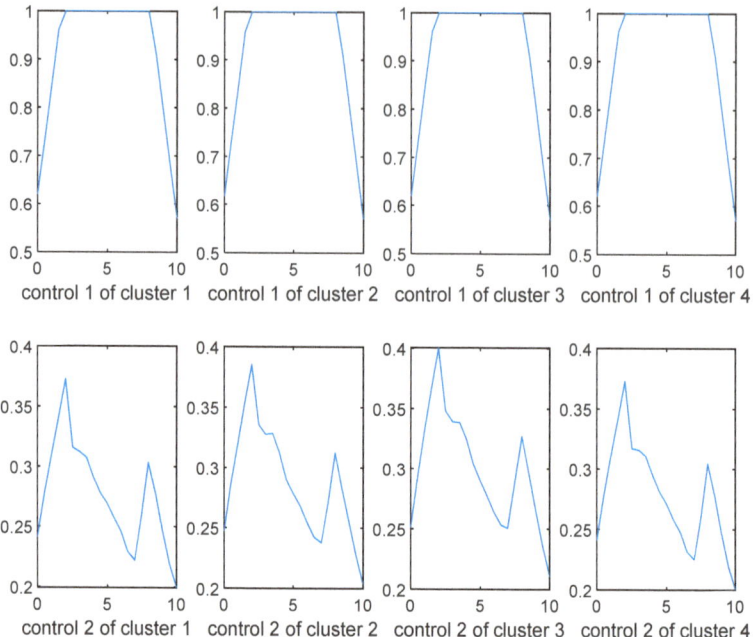

Figure A3. Controls pair obtained via GMM applied to the Pareto front produced via MOFA.

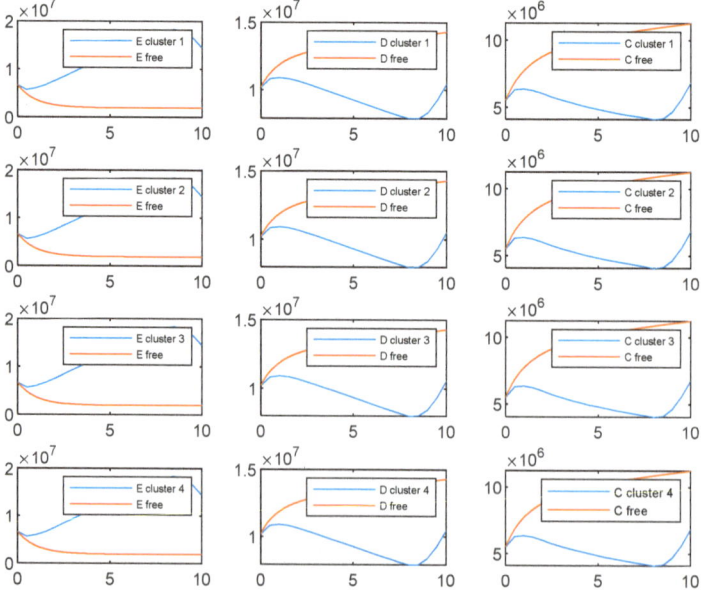

Figure A4. Compartments obtained when introducing different controls, produced via GMM applied to the Pareto front produced via MOFA applied to the model (4).

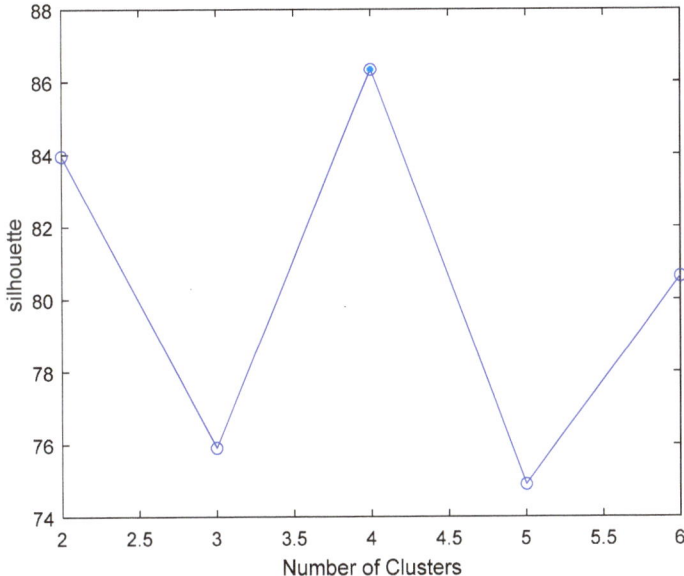

Figure A5. Selection of the optimal number of clusters based on the silhouette criteria of groups obtained via FCM.

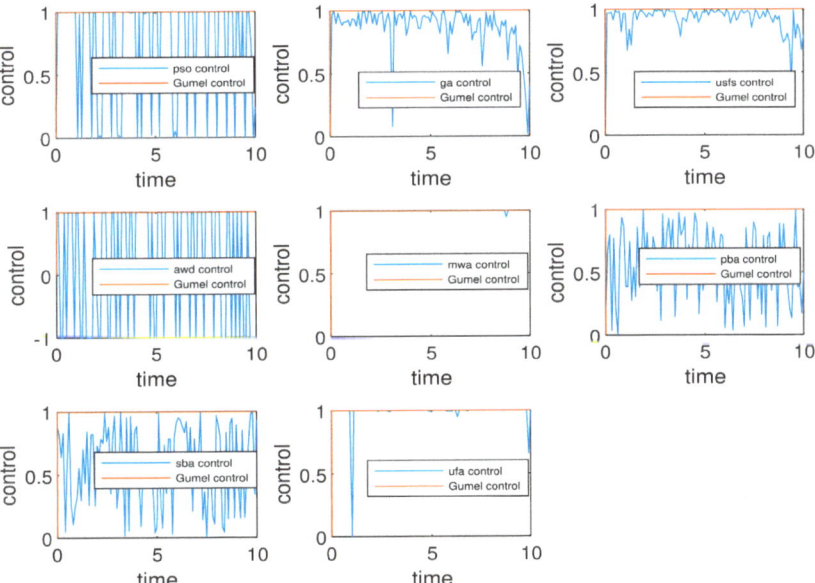

Figure A6. Gumel control and meta-heuristics controls for the Bouteyeb model.

Appendix B

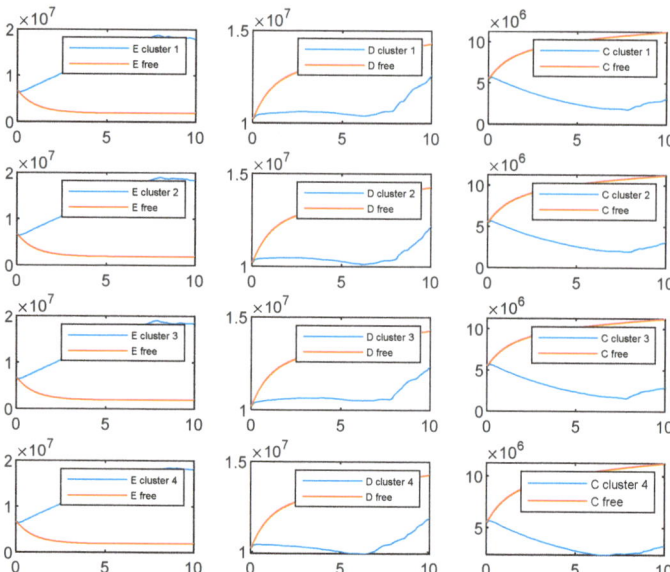

Figure A7. Compartments obtained when introducing different controls, produced via FCM, applied to the Pareto front produced via NSGA-II, for which we added Gaussian noise from [0, 0.05].

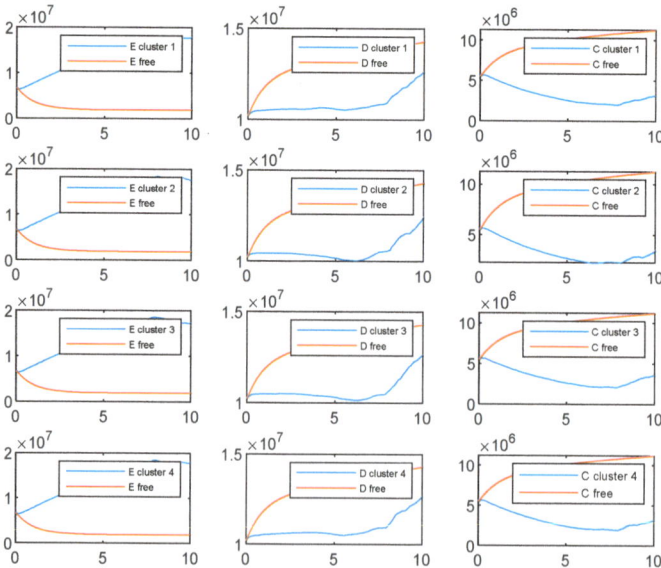

Figure A8. Compartments obtained when introducing different controls, produced via FCM, applied to the Pareto front produced via NSGA-II, for which we added Gaussian noise from [0, 0.1].

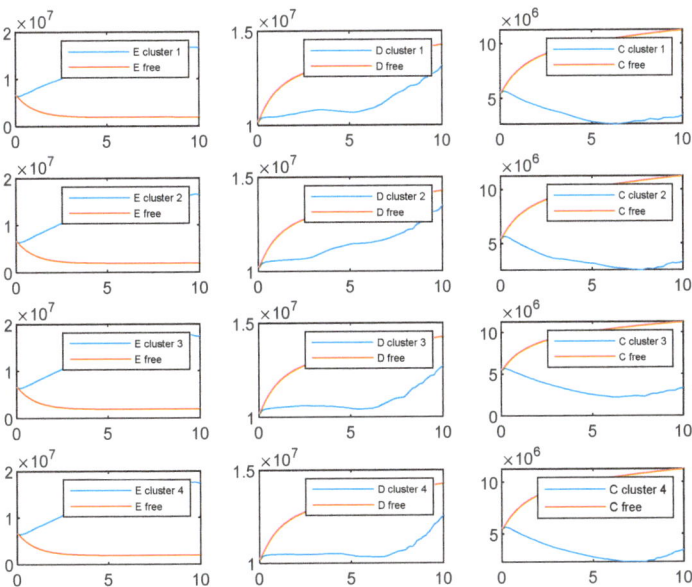

Figure A9. Compartments obtained when introducing different controls, produced via FCM, applied to Pareto front produced via NSGA-II, for which we added Gaussian noise from [0, 0.15].

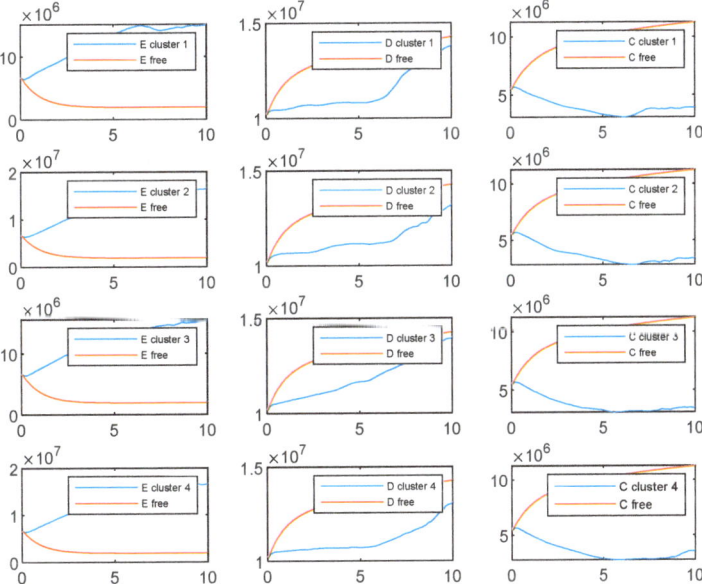

Figure A10. Compartments obtained when introducing different controls, produced via FCM applied to the Pareto front produced via NSGA-II, for which we added Gaussian noise from [0, 0.2].

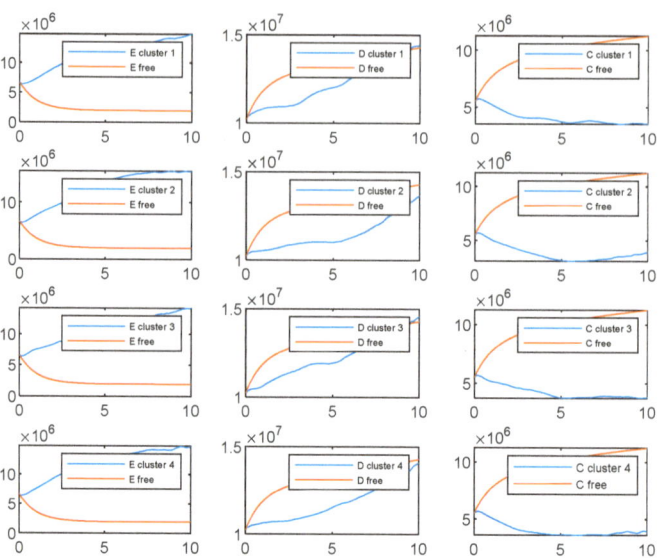

Figure A11. Compartments obtained when introducing different controls, produced via FCM applied to the Pareto front produced via NSGA-II, for which we added Gaussian noise from [0, 0.25].

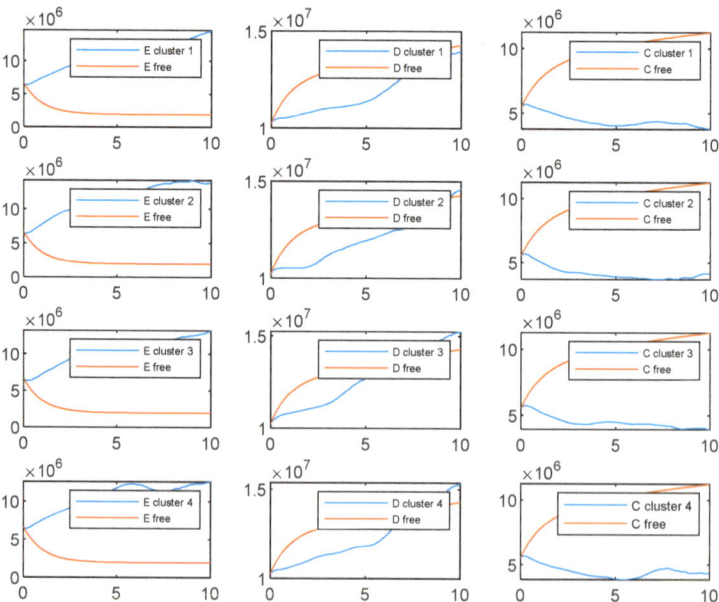

Figure A12. Compartments obtained when introducing different controls, produced via FCM, applied to the Pareto front produced via NSGA-II, for which we added Gaussian noise from [0, 0.3].

References

1. International Diabetes Federation (IDF). About—Diabetes. Available online: https://idf.org/about-diabetes/facts-figures (accessed on 1 May 2023).
2. Abdellatif, E.O.; Karim, E.M.; Hicham, B.; Saliha, C. Intelligent local search for an optimal control of diabetic population dynamics. *Math. Model. Comput. Simul.* **2022**, *14*, 1051–1071. [CrossRef]
3. El Moutaouakil, K.; Ahourag, A.; Chakir, S.; Kabbaj, Z.; Chellack, S.; Cheggour, M.; Baizri, H. Hybrid firefly genetic algorithm and integral fuzzy quadratic programming to an optimal Moroccan diet. *Math. Model. Comput.* **2023**, *10*, 338–350. [CrossRef]

4. Ahourag, A.; El Moutaouakil, K.; Cheggour, M.; Chellak, S.; Baizri, H. Multiobjective optimization to optimal moroccan diet using genetic algorithm. *Int. J. Eng. Model.* **2023**, *36*, 67–79.
5. Ahourag, A.; Chellak, S.; Cheggour, M.; Baizri, H.; Bahri, A. Quadratic programming and triangular numbers ranking to an optimal moroccan diet with minimal glycemic load. *Stat. Optim. Inf. Comput.* **2023**, *11*, 85–94.
6. World Health Organisation. *Definition and Diagnosis of Diabetes Mellitus and Intermediate Hyperglycemia*; WHO: Geneva, Switzerland, 2016.
7. *IDF Diabetes Atlas*, 9th ed.; International Diabetes Federation (IDF): Brussels, Belgium, 2019.
8. Kouidere, A.; Khajji, B.; Balatif, O.; Rachik, M. A multi-age mathematical modeling of the dynamics of population diabetics with effect of lifestyle using optimal control. *J. Appl. Math. Comput.* **2021**, *67*, 375–403. [CrossRef]
9. Abdellatif, E.O.; Karim, E.M.; Saliha, C.; Hicham, B. Genetic algorithms for optimal control of a continuous model of a diabetic population. In Proceedings of the 2022 IEEE 3rd International Conference on Electronics, Control, Optimization and Computer Science (ICECOCS), Fez, Morocco, 1–2 December 2022; pp. 1–5.
10. Li, H.; Peng, R.; Wang, Z. On a diffusive susceptible-infected-susceptible epidemic model with mass action mechanism and birth-death effect: Analysis, simulations, and comparison with other mechanisms. *SIAM J. Appl. Math.* **2018**, *78*, 2129–2153. [CrossRef]
11. Jin, H.Y.; Wang, Z.A. Global stabilization of the full attraction-repulsion keller-segel system. *Discret. Contin. Dyn. Syst.—Ser. A* **2020**, *40*, 3509–3527. [CrossRef]
12. Yuan, M.; Li, Y.; Zhang, L.; Pei, F. Research on intelligent workshop resource scheduling method based on improved NSGA-II algorithm. *Robot. Comput. Integr. Manuf.* **2021**, *71*, 102141. [CrossRef]
13. Ranganathan, S.; Surya Kalavathi, M.; Asir Rajan, C.C. Self-adaptive firefly algorithm based multi-objectives for multi-type FACTS placement. *IET Gener. Transm. Distrib.* **2016**, *10*, 2576–2584. [CrossRef]
14. El Moutaouakil, K.; Yahyaouy, A.; Chellak, S.; Baizri, H. An optimized gradient dynamic-neuro-weighted-fuzzy clustering method: Application in the nutrition field. *Int. J. Fuzzy Syst.* **2022**, *24*, 3731–3744. [CrossRef]
15. Boutayeb, A.; Chetouani, A. A population model of diabetes and pre-diabetes. *Int. J. Comput. Math.* **2007**, *84*, 57–66. [CrossRef]
16. Mahata, A.; Mondal, S.P.; Alam, S.; Chakraborty, A.; De, S.K.; Goswami, A. Mathematical model for diabetes in fuzzy environment with stability analysis. *J. Intell. Fuzzy Syst.* **2019**, *36*, 2923–2932. [CrossRef]
17. Ollerton, R.L. Application of optimal control theory to diabetes mellitus. *Int. J. Control* **1989**, *50*, 2503–2522. [CrossRef]
18. Swan, G.W. An optimal control model of diabetes mellitus. *Bull. Math. Biol.* **1982**, *44*, 793–808. [CrossRef] [PubMed]
19. Makroglou, A.; Karaoustas, I.; Li, J.; Kuang, Y. Delay differential equation models in diabetes modeling. *Theor. Biol. Med. Model.* 2009. Available online: https://d1wqtxts1xzle7.cloudfront.net/39776544/Delay_differential_equation_models_in_di20151107-11553-9sa4j7-libre.pdf?1446919153=&response-content-disposition=inline%3B+filename%3DDelay_differential_equation_models_in_di.pdf&Expires=1688109253&Signature=gzNPLMm9mZ3KYeZD9wfLLcKorB-7z3XPMW8kUnqEXooVDMVVRyQqbvUD1timDez8PEcjfkgNsLfYgASjLAJ~LP~rY5M7aihIVP5~wu4y5GR29sMBYBgTyszEQSG5g10Gt~LYiWgqbmPcrBXBP7Lcv5rkQkORQzTOxJhWoiRYadd8Hw6kBVlr4mjVPEMHxnQkgp6QEW-qlqF1FUKKG8pxI338xA~bkZDSiDKmGgzvjiEBBcBj3W1LCGQMZh1maPlnmVqaldj8n33dkCEXny5bM-yvEz64JsAJ6my3qC59kctUZR4YwI2rVjKTWhEN4dlD3ogdKkBRNA2bal4vQPpRCA__&Key-Pair-Id=APKAJLOHF5GGSLRBV4ZA (accessed on 1 May 2023).
20. Derouich, M.; Boutayeb, A.; Boutayeb, W.; Lamlili, M. Optimal control approach to the dynamics of a population of diabetics. *Appl. Math. Sci.* **2014**, *8*, 2773–2782. [CrossRef]
21. Gumel, A.B.; Shivakumar, P.N.; Sahai, B.M. A mathematical model for the dynamics of HIV-1 during the typical course of infection. *Nonlinear Anal. Theory Methods Appl.* **2001**, *47*, 1773–1783. [CrossRef]
22. Yusuf, T.T. Optimal control of incidence of medical complications in a diabetic patients' population. *FUTA J. Res. Sci.* **2015**, *11*, 180–189.
23. Permatasari, A.H.; Tjahjana, R.H.; Udjiani, T. Existence and characterization of optimal control in mathematics model of diabetics population. *J. Phys. Conf. Ser.* **2018**, *983*, 012069. [CrossRef]
24. Daud, A.A.M.; Toh, C.Q.; Saidun, S. Development and analysis of a mathematical model for the population dynamics of Diabetes Mellitus during pregnancy. *Math. Model. Comput. Simul.* **2020**, *12*, 620–630. [CrossRef]
25. Kouidere, A.; Balatif, O.; Ferjouchia, H.; Boutayeb, A.; Rachik, M. Optimal control strategy for a discrete time to the dynamics of a population of diabetics with highlighting the impact of living environment. *Discret. Dyn. Nat. Soc.* **2019**, *2019*, 6342169. [CrossRef]
26. Kouidere, A.; Labzai, A.; Ferjouchia, H.; Balatif, O.; Rachik, M. A new mathematical modeling with optimal control strategy for the dynamics of population of diabetics and its complications with effect of behavioral factors. *J. Appl. Math.* **2020**, *2020*, 1943410. [CrossRef]
27. Ahourag, A.; El Moutaouakil, K.; Chellak, S.; Baizri, H.; Cheggour, M. Multi-criteria optimization for optimal nutrition of Moroccan diabetics. In Proceedings of the 2022 International Conference on Intelligent Systems and Computer Vision (ISCV), Fez, Morocco, 18–20 May 2022; pp. 1–6.
28. El Moutaouakil, K.; Palade, V.; Safouan, S.; Charroud, A. FP-Conv-CM: Fuzzy probabilistic convolution C-means. *Mathematics* **2023**, *11*, 1931. [CrossRef]
29. Bolduc, E.; Knee, G.C.; Gauger, E.M.; Leach, J. Projected gradient descent algorithms for quantum state tomography. *npj Quantum Inf.* **2017**, *3*, 44. [CrossRef]

30. Auslender, A.; Teboulle, M. Lagrangian duality and related multiplier methods for variational inequality problems. *SIAM J. Optim.* **2000**, *10*, 1097–1115. [CrossRef]
31. Föllmer, H.; Kabanov, Y.M. Optional decomposition and Lagrange multipliers. *Financ. Stoch.* **1997**, *2*, 69–81. [CrossRef]
32. Rahman, Q.I.; Schmeisser, G. Characterization of the speed of convergence of the trapezoidal rule. *Numer. Math.* **1990**, *57*, 123–138. [CrossRef]
33. Fleming, W.H.; Rishel, R.W. *Deterministic and Stochastic Optimal Control*; Springer: New York, NY, USA, 1975.
34. Frigo, M.; Johnson, S.G. FFTW: An adaptive software architecture for the FFT. In Proceedings of the International Conference on Acoustics, Speech, Signal Processing, Seattle, WA, USA, 15 May 1998; Volume 3, pp. 1381–1384.
35. Jawad, K.; Mahto, R.; Das, A.; Ahmed, S.U.; Aziz, R.M.; Kumar, P. Novel cuckoo search-based metaheuristic approach for deep learning prediction of depression. *Appl. Sci.* **2023**, *13*, 5322. [CrossRef]
36. Ali, S.; Bhargava, A.; Saxena, A.; Kumar, P. A hybrid marine predator sine cosine algorithm for parameter selection of hybrid active power filter. *Mathematics* **2023**, *11*, 598. [CrossRef]
37. Wang, Y.J.; Wang, G.G.; Tian, F.M.; Gong, D.W.; Pedrycz, W. Solving energy-efficient fuzzy hybrid flow-shop scheduling problem at a variable machine speed using an extended NSGA-II. *Eng. Appl. Artif. Intell.* **2023**, *121*, 105977. [CrossRef]
38. Yazdinejad, A.; Dehghantanha, A.; Parizi, R.M.; Epiphaniou, G. An optimized fuzzy deep learning model for data classification based on NSGA-II. *Neurocomputing* **2023**, *522*, 116–128. [CrossRef]
39. Rafati, N.; Hazbei, M.; Eicker, U. Louver configuration comparison in three Canadian cities utilizing NSGA-II. *Build. Environ.* **2023**, *229*, 109939. [CrossRef]
40. Chen, H.; Feng, Z.; Liu, Y.; Chen, B.; Deng, T.; Qin, Y.; Xu, W. Multiobjective optimization of a 3D laser scanning scheme for engineering structures based on RF-NSGA-II. *J. Constr. Eng. Manag.* **2023**, *149*, 04022169. [CrossRef]
41. Singh, M.K.; Choudhary, A.; Gulia, S.; Verma, A. Multi-objective NSGA-II optimization framework for UAV path planning in an UAV-assisted WSN. *J. Supercomput.* **2023**, *79*, 832–866. [CrossRef]
42. Wang, D.; Wang, G.; Wang, H. Optimal lane change path planning based on the NSGA-II and TOPSIS algorithms. *Appl. Sci.* **2023**, *13*, 1149. [CrossRef]
43. Nan, Y.; Zhang, H.; Zeng, Y.; Zheng, J.; Ge, Y. Faster and accurate green pepper detection using NSGA-II-based pruned YOLOv5l in the field environment. *Comput. Electron. Agric.* **2023**, *205*, 107563. [CrossRef]
44. Li, S.; Zhou, H.; Xu, G. Research on optimal configuration of landscape storage in public buildings based on improved NSGA-II. *Sustainability* **2023**, *15*, 1460. [CrossRef]
45. Wang, Z.; Shen, L.; Li, X.; Gao, L. An improved multi-objective firefly algorithm for energy-efficient hybrid flowshop rescheduling problem. *J. Clean. Prod.* **2023**, *385*, 135738. [CrossRef]
46. Tiwari, A.; Chaturvedi, A. Automatic EEG channel selection for multiclass brain-computer interface classification using multiobjective improved firefly algorithm. *Multimed. Tools Appl.* **2023**, *82*, 5405–5433. [CrossRef]
47. He, Y.; Peng, H.; Deng, C.; Dong, X.; Wu, Z.; Guo, Z. Reference point reconstruction-based firefly algorithm for irregular multi-objective optimization. *Appl. Intell.* **2023**, *53*, 962–983. [CrossRef]
48. Ri, K.W.; Mun, K.H. Firefly algorithm hybridized with genetic algorithm for multi-objective integrated process planning and scheduling. *Res. Sq.* **2023**. [CrossRef]
49. Ahmadi, S.E.; Kazemi-Razi, S.M.; Marzband, M.; Ikpehai, A.; Abusorrah, A. Multi-objective stochastic techno-economic-environmental optimization of distribution networks with G2V and V2G systems. *Electr. Power Syst. Res.* **2023**, *218*, 109195. [CrossRef]
50. Li, J.; Sun, G.; Wang, A.; Zheng, X.; Chen, Z.; Liang, S.; Liu, Y. Multi-objective sparse synthesis optimization of concentric circular antenna array via hybrid evolutionary computation approach. *Expert Syst. Appl.* **2023**, *231*, 120771. [CrossRef]
51. Srinivasan, B.; Venkatesan, R.; Aljafari, B.; Kotecha, K.; Indragandhi, V.; Vairavasundaram, S. A novel multicriteria optimization technique for VLSI floorplanning based on hybridized firefly and ant colony systems. *IEEE Access* **2023**, *11*, 14677–14692. [CrossRef]
52. Shou, S.; Luo, H.; Wang, X.; Li, Y.; Hu, J.; Su, L. Optimal configuration of power quality control device for new distribution network based on firefly algorithm. In Proceedings of the 2023 Panda Forum on Power and Energy (PandaFPE), Chengdu, China, 27–30 April 2023; pp. 872–876.
53. Athisayam, A.; Kondal, M. Fault feature selection for the identification of compound gear-bearing faults using firefly algorithm. *Int. J. Adv. Manuf. Technol.* **2023**, *125*, 1777–1788. [CrossRef]
54. El Moutaouakil, K.; Touhafi, A. A new recurrent neural network fuzzy mean square clustering method. In Proceedings of the 2020 5th International Conference on Cloud Computing and Artificial Intelligence: Technologies and Applications (CloudTech), Marrakesh, Morocco, 24–26 November 2020; pp. 1–5. [CrossRef]

Disclaimer/Publisher's Note: The statements, opinions and data contained in all publications are solely those of the individual author(s) and contributor(s) and not of MDPI and/or the editor(s). MDPI and/or the editor(s) disclaim responsibility for any injury to people or property resulting from any ideas, methods, instructions or products referred to in the content.

Article

The Optimal Consumption, Investment and Life Insurance for Wage Earners under Inside Information and Inflation

Rui Jiao [1], Wei Liu [1,*] and Yijun Hu [2]

1. College of Mathematics and System Science, Xinjiang University, Urumqi 830017, China; jiaorui@stu.xju.edu.cn
2. School of Mathematics and Statistics, Wuhan University, Wuhan 430072, China; yjhu.math@whu.edu.cn
* Correspondence: liuwei.math@xju.edu.cn

Abstract: This paper studies the dynamically optimal consumption, investment and life-insurance strategies for a wage earners under inside information and inflation. Assume that the wage earner can invest in a risk-free asset, a risky asset and an inflation-indexed bond and that the wage earner can obtain some additional information on the risky asset from the financial market. By maximizing the expected utility of the wage earner's consumption, inheritance and terminal wealth, we obtain the dynamically optimal consumption, investment and life-insurance strategies for the wage earner. The method of this paper is mainly based on (dynamical) stochastic control theory and the technique of enlargement of filtrations. Moreover, sensitivity analysis is carried out, which reveals that a wage earner with inside information tends to increase his/her consumption and investment, while reducing his/her purchase of life insurance.

Keywords: investment; consumption; life insurance; inside information; inflation

MSC: 93E30; 97M30

Citation: Jiao, R.; Liu, W.; Hu, Y. The Optimal Consumption, Investment and Life Insurance for Wage Earners under Inside Information and Inflation. *Mathematics* **2023**, *11*, 3415. https://doi.org/10.3390/math11153415

Academic Editors: Adrian Olaru, Gabriel Frumusanu and Catalin Alexandru

Received: 28 June 2023
Revised: 30 July 2023
Accepted: 1 August 2023
Published: 5 August 2023

Copyright: © 2023 by the authors. Licensee MDPI, Basel, Switzerland. This article is an open access article distributed under the terms and conditions of the Creative Commons Attribution (CC BY) license (https://creativecommons.org/licenses/by/4.0/).

1. Introduction

Since Merton's seminal work [1], investment and consumption problems have been extensively studied. Karatzas et al. [2] used the dynamic programming method to explicitly propose a solution to the consumption-portfolio problem under a general utility function and general rates of return. Fleming and Pang [3] obtained the optimal investment and consumption strategy for investors under the fluctuation of interest rates. Chang and Chang [4] solved the investment-consumption problem under the Vasicek model and Hyperbolic Absolute Risk Aversion (HARA) utility. As financial markets have become more sophisticated, investors are no longer limited to purchasing stocks, bonds and other products in the securities markets to earn investment returns. Instead, they can choose products from a broader range of investment products. With the booming insurance industry, more and more people are investing their money in insurance, and life insurance is one of the most interesting insurance products. Besides being widely accepted as a new type of investment product, life insurance is also used by individuals or families to protect themselves against risk. According to Campbell [5], uncertainty about a wage earner's future age of death leads to uncertainty about the family's financial situation. Many wage earners purchase life insurance to protect their families against the death risk. Based on Merton's elegant theoretical framework, many investment-consumption problems with life insurance have been studied in the literature. Richard [6] was the first to study the individual's portfolio–consumption–life insurance problem under the maximization of the expected utility, considering that the investor's lifetime follows a random but known distribution. Subsequently, Pliska and Ye [7] studied the optimization problem by maximizing the expected utility and analyzed the demand for life insurance using numerical experiments. Under the HARA utility, Huang and Milevsky [8] investigated the portfolio-selection problem, where life insurance is involved. Considering that stocks have a mean-reverting drift term, the optimal strategies under Constant Relative Risk Aversion (CRRA) utility were studied by Pirvu and Zhang [9]. Zeng et al. [10] solved the optimization problem under the no-borrowing restriction. They used the duality method to determine optimal strategies and indicated that the optimal strategies are influenced by no-borrowing restrictions. So for individuals, they are buying life insurance both as a more popular way to manage their finances and to provide financial security for their families. In addition, Wei et al. [11] provided the optimal strategies for lifetime correlation couples. They used copula and common-shock to model the mortality dependence and thus measured correlated longevity. Considering a household in the context of a continuous two-generation period, the robust optimal strategies were studied by Wang et al. [12], which assumed that the income growth rate is unknown. They indicated that wealth does not influence investment strategy, but higher wealth levels contribute to lower life insurance and higher consumption. Therefore, life insurance provides the necessary protection for the economic stability of individuals and families in real life. Based on the investment and consumption problem, the study of the optimal strategy of life insurance is a current hotspot and has high theoretical value for enriching the application of stochastic optimal control theory.

Most of the references mentioned above use individual life insurance. In reality, however, the insurance market exists and insurance companies offer different insurance contracts, and wage earners face a variety of choices in the insurance market. Therefore, it is more relevant and promising to consider insurance consisting of multiple life-insurance policies. The optimal strategies were obtained by Mousa et al. [13] in the case of multidimensional life insurance, which assumed that a life-insurance market consists of different life-insurance contracts from a finite number of insurers. Hoshiea et al. [14] took both a social welfare system and multiple life-insurance policies into account to study the optimal strategies. Considering multidimensional life insurance, Mousa et al. [15] introduced an economic indicator represented by a stochastic process that affects the financial assets and studied an optimal asset-allocation problem of a wage earner.

In addition to the risk of death, the increased level of inflation should not be ignored. The purchasing power of wage earners can be significantly affected by inflation. Kwak and Lim [16] studied a family's optimal asset allocation under inflation risk and discovered inflation's impact on life insurance premiums. Han and Hung [17] considered risks of interest rate and inflation to investigate the optimal economic decisions of a wage earner. They discovered that fluctuation in inflation would discourage people from buying life insurance. Liang and Zhao [18] took into account the inflation risk and studied the optimal strategies including life insurance under a mean-variance utility. Quite recently, inflation risk and consumption habits were considered by Shi et al. [19] and their effects on optimal consumption–investment–life-insurance strategies were analyzed.

In reality, most common people could access public information published by companies and/or regulators. Professional investors would most like to investigate private markets to obtain additional information about the financial market. This leads to the so-called inside-information issue. For example, Kyle [20] first pointed out that insiders in the market make positive profits by exploiting their monopoly power and that the existence of noise trading makes insider trading undetectable to market makers. Pikovsky and Karatzas [21], based on Kyle's research, pointed out that inside-information situations are real and involve an investor in possession of some information about the future and possessing relevant mathematical models. This could affect the investment strategy and wealth levels of wage earners and hence life-insurance and consumption strategies. Therefore, inside information should be considered with respect to optimal asset-allocation problems for wage earners. The existence of inside information gives the wage earner access to a much larger filtration than that generated by the market, which requires solving the optimality problem under a new filtration. A common approach to modeling the behavior of wage earners in possession of inside information is the enlargement of filtration techniques. Early studies of inside information focused on investors in financial markets. The impact of inside information on investment strategies and welfare was studied by Pikovsky and Karatzas [21]. Imkeller et al. [22] considered the problem of possible arbitrage opportunities. The problems of non-life insurance with inside information have been studied, where the insurers may have some inside information about their claim process; see Baltas et al. [23], Cao et al. [24], Peng et al. [25]. Assuming that the claims process and the risky assets of insurers are related to jump–diffusion processes, Peng and Wang [26] took into account inside information in both financial and insurance markets and provided the optimal risk-management and investment strategies for insurance companies. Peng and Chen [27] studied the problem of asset-liability management under inside information. Nevertheless, the study on individual asset allocation with inside information leaves much to be explored.

In this paper, we investigate the dynamically optimal consumption, investment and life-insurance strategies for a wage earner under inside information and inflation. The wage earner is allowed to invest in a portfolio consisting of risk-free assets, risky assets and inflation-linked bonds. Assume that the wage earner has access to inside information in the stock market. Correspondingly, we develop a dynamic control system in which the state equation consists of a wealth process and an income process. The control variables are the proportion of investment in risky assets, the proportion of investment in inflation-indexed bonds, consumption and life insurance premium rate. The objective is to maximize the expected utility of consumption, inheritance and final wealth. For this stochastic control problem, the optimal solution is obtained by applying the dynamic programming method and solving the corresponding HJB equation. The main contributions of this paper are as follows:

(i) Solving the asset-allocation strategies for a wage earner under inside information, and analyzing the impact of inside information on asset-allocation strategies.
(ii) Taking multidimensional life insurance in the insurance market into consideration.
(iii) Solving the optimal inflation-indexed bond strategy. By addressing these key aspects, we aim to shed light on the intricate dynamics of consumption, investment and life-insurance decisions when individuals have access to inside information and are navigating the complexities associated with inflation.

The remainder of the paper has the following structure. A model that includes the wealth process and the performance function is presented in Section 2. Section 3 identifies the optimal decisions and value function. Section 4 provides numerical analyses and explanations of the economic significance of the optimal strategies. The conclusions of the paper are presented in Section 5.

2. Model

Let $(\Omega, \mathbb{F}, \mathbb{P})$ be a complete probability space and filtration $\mathcal{F}_{t \in [0,T]}$ generated by two standard one-dimension Brownian motions $B_S(t)$ and $B_I(t)$. $T > 0$ is the terminal time, considered to be the wage earner's retirement time.

2.1. The Financial Market

As is common in the literature, assume that the price process of the risk-free asset is

$$dS_1(t) = r_1 S_1(t) dt,$$

where the risk-free interest rate $r_1 > 0$. The price process of the risky asset (stock) $S_2(t)$ can be given as

$$dS_2(t) = \lambda S_2(t) dt + \sigma_S S_2(t) dB_S(t),$$

and $\lambda > 0$ is the instantaneous expected return rate. σ_S represents the volatility rate. To measure inflation, the commodity-price-index process is expressed as

$$dI(t) = \lambda_I I(t) dt + \sigma_I I(t) dB_I(t),$$

where the constant $\lambda_I \in (0, \Lambda_I]$ stands for the expected inflation rate and the constant Λ_I means the possible maximum value for the inflation rate. $\sigma_I > 0$ expresses the price index's volatility rate. The price dynamic of an inflation-indexed bond $p(t)$ is as follows

$$dp(t) = r_2 p(t) dt + p(t) \frac{dI(t)}{I(t)} = (r_2 + \lambda_I) p(t) dt + \sigma_I p(t) dB_I(t),$$

where r_2 is the real interest rate. $r_2 + \lambda_I$ is the expected return rate of the bond.

2.2. The Income and the Insurance Market

The nominal income process $L_N(t)$ is described as

$$dL_N(t) = \lambda_L L_N(t) dt + \sigma_L L_N(t) dB_I(t),$$

where λ_L denotes the expected return rate of nominal income. σ_L is the volatility of nominal income.

Suppose the investor is alive at time t. Let τ stand for the the investor's lifetime. Assume the insurance market includes K life insurances from K insurance companies. The life-insurance premium rate of the kth company is $\theta_k(t), k \in \{1, 2, \cdots, K\}$. $\eta_k : [0, T] \to R^+$ can be called the premium–insurance ratio.

Assumption 1. *For each $k \in 1, \cdots, K$, $\eta_k(t)$ is a deterministic continuous function. Furthermore, the kth insurer considered here is assumed to offer a different set of contracts, i.e., $\eta_{k_1} \neq \eta_{k_2}$ for each $k_1 \neq k_2$ and $t \in [0, T]$. Once the wage earner dies at time t, the kth insurance company will pay $\theta_k(t)/\eta_k(t)$. Therefore, the legacy W at death time τ is expressed as*

$$W(\tau) = X(\tau) + \sum_{k=1}^{K} \frac{\theta_k(\tau)}{\eta_k(\tau)}.$$

Let $\pi_1(t)$ and $\pi_2(t)$ denote the proportion of assets in stocks and inflation-indexed bonds, respectively. $\theta_{N,k}(t)$ represents the nominal premium rate of the kth life insurance company and $C_N(t)$ is the nominal consumption for the wage earner. Denote the control variables as $\phi = (\pi_1(t), \pi_2(t), \theta_{N,k}(t), C_N(t))$. The nominal wealth process under ϕ is as follows

$$dX_N(t) = \left[X_N(t) r_1 + X_N(t) \pi_1(t)(\lambda - r_1) + X_N(t) \pi_2(t)(r_2 + \lambda_I - r_1) + L_N(t) - C_N(t) \right. \tag{1}$$
$$\left. - \sum_{k=1}^{K} \theta_{N,k}(t) \right] dt + X_N(t) \pi_1(t) \sigma_S dB_S(t) + X_N(t) \pi_2(t) \sigma_I dB_I(t).$$

2.3. Inside Information

We assume that a wage earner can obtain inside information in the risky asset. Specifically, let $\mathcal{L} = B_S(T_0)$ denote the wage earner's inside information, with $T_0 > T$. The filtration of the wage earner would be as follows

$$\mathcal{G}_t = \mathcal{F}_t \vee \sigma(B_S(T_0)),$$

and the relationship between $\mathcal{G}_t$ and $\mathcal{F}_t$ is

$$\mathcal{G}_t \supset \mathcal{F}_t, \forall t \in [0, T].$$

The following lemma is from Theorem 3.1 of Baltas et al. [23].

Lemma 1. *The process $\{B_S(t), t \geq 0\}$ is a semimartingale with respect to $\mathbb{G} = \{\mathcal{G}_t, t \geq 0\}$. Its semimartingale decomposition is as follows*

$$B_S(t) = \widetilde{B}_S(t) + \int_0^t \kappa(s)ds,$$

where

$$\kappa(t) = \frac{B_S(T_0) - B_S(t)}{T_0 - t}, 0 \leq t < T_0,$$

and $\widetilde{B}_S(t)$ is a $(\mathbb{G}, \mathbb{P})$ Brownian motion.

Considering the inside information $B_S(T_0)$, the nominal wealth can be described as

$$dX_N(t) = \left[X_N(t)r_1 + X_N(t)\pi_1(t)(\lambda - r_1 + \sigma_S \kappa_0 - \sigma_S M(t)) + X_N(t)\pi_2(t)(r_2 + \lambda_I - r_1) \right.$$

$$\left. + L_N(t) - C_N(t) - \sum_{k=1}^{K} \theta_{N,k}(t) \right] dt + X_N(t)\pi_1(t)\sigma_S d\widetilde{B}_S(t) + X_N(t)\pi_2(t)\sigma_I dB_I(t),$$

where

$$\kappa_0 = \lim_{t \to 0} \kappa(t) = \frac{B_S(T_0)}{T_0}, \quad (2)$$

and

$$M(t) = \int_0^t \frac{1}{T_0 - s} d\widetilde{B}_S(t).$$

2.4. The Stochastic Optimal Control Problem

Let $X(t) = X_N(t)/I(t)$ be the actual wealth, removing the effects of inflation. Actual income, actual consumption and the actual insurance premium rate are denoted by $L(t) = L_N(t)/I(t)$, $C(t) = C_N(t)/I(t)$ and $\theta_k(t) = \theta_{N,k}(t)/I(t)$, respectively. Then the actual wealth and actual income processes can be presented as

$$dX(t) = \left[X(t)(r_1 - \lambda_I + \sigma_I^2) + X(t)\pi_1(t)(\lambda - r_1 + \sigma_S \kappa_0 - \sigma_S M(t)) + X(t)\pi_2(t)(r_2 + \lambda_I - r_1 \right.$$

$$\left. - \sigma_I^2) + L(t) - C(t) - \sum_{k=1}^{K} \theta_k(t) \right] dt + X(t)\pi_1(t)\sigma_S d\widetilde{B}_S(t) + X(t)(\pi_2(t) - 1)\sigma_I dB_I(t),$$

and

$$dL(t) = L(t)(\lambda_L - \lambda_I + \sigma_I^2 - \sigma_I \sigma_L)dt + L(t)(\sigma_L - \sigma_I)dB_I(t).$$

The performance function can be expressed as

$$J(t, x, m, l; \phi) = E_{t,x} \left[\int_t^{T \wedge \tau} U(s, C(s))ds + Y(\tau, W(\tau))\mathbf{1}_{\{\tau \leq T\}} + \Gamma(X(T))\mathbf{1}_{\{\tau > T\}} \right], \quad (3)$$

where $U(x,y)$, $Y(x,y)$ and $\Gamma(x)$ are utility functions.

From the results of Pliska and Ye [7], we have

$$J(t, x, m, l; \phi) = E_{t,x} \left[\int_t^{T} f(s,t)U(s, C(s)) + \bar{F}(s,t)Y(s, W(s))ds + \bar{F}(T,t)\Gamma(X(T)) \right].$$

where $f(s,t)$ and $\bar{F}(s,t)$ are the conditional probability density and conditional survival probability, respectively. Let $\mu(t)$ denote the hazard function, then

$$f(s,t) = \mu(t) \exp\left\{ -\int_t^s \mu(u)du \right\}, \quad \bar{F}(s,t) = \exp\left\{ -\int_t^s \mu(u)du \right\}.$$

Then define the value function as

$$V(t,x,m,l) := \sup_{\phi \in \mathcal{A}} J(t,x,m,l;\phi). \tag{4}$$

Definition 1. *The strategies $\phi = (\pi_1(t), \pi_2(t), \theta_k(t), C(t))$ are called admissible strategies if they satisfy the following conditions. The admissible-strategies set is denoted as $\mathcal{A}$.*

(i) *The life-insurance purchase $\theta_k(t)$ is $\mathcal{F}_{t\in[0,T]}$-measurable and satisfies*

$$\int_0^T \theta_k(s)ds < \infty, \quad k = 1, \ldots, K.$$

(ii) *The consumption $C(t)$ is $\mathcal{F}_{t\in[0,T]}$-measurable and satisfies*

$$\int_0^T C(s)ds < \infty \quad \text{a.s..}$$

(iii) *The investment strategies $\pi_1(t)$ and $\pi_2(t)$ are $\mathcal{F}_{t\in[0,T]}$-measurable processes and comply with*

$$\int_0^T \|\pi_2(t)\|^2 dt < \infty \quad \text{a.s.,}$$
$$\int_0^T \|\pi_1(t)\|^2 dt < \infty \quad \text{a.s.,}$$
$$E\left\{\exp\left[-\int_0^T \pi_1(s)d\widetilde{B}_S(s) - \frac{1}{2}\int_0^T \|\pi_1(s)\|^2 ds\right]\right\} = 1.$$

3. Solution to the Stochastic Optimal Control Problem

This section derives the optimal strategies and corresponding value function.

Theorem 1 (Verification Theorem). *If there exists a function $Z(t,x,m,l)$ that satisfies the following HJB equation*

$$\max_{\phi \in \mathcal{A}}\{U(s,C(s)) + \mu(t)Y(s,W(s)) - \mu(t)Z(t,x,m,l) + \Phi(t,x,m,l;\phi)\} = 0,$$

with the boundary condition

$$Z(T,x,m,l) = \Gamma(X(T)),$$

where the infinitesimal generator

$$\Phi(t,x,m,l;\phi) = Z_t(t,x,m,l) + \left[x\left(r_1 - \lambda_I + \sigma_I^2\right) + x\pi_1(t)(\lambda - r_1 + \sigma_S\kappa_0 - \sigma_S m) + x\pi_2(t)\right.$$
$$\times \left(r_2 + \lambda_I - r_1 - \sigma_I^2\right) + l - C - \sum_{k=1}^K \theta_k\right]Z_x(t,x,m,l) + (\lambda_L - \lambda_I + \sigma_I^2 - \sigma_I\sigma_L)$$
$$\times lZ_l(t,x,m,l) + \frac{1}{2}\left(\pi_1^2\sigma_S^2 + (\pi_2 - 1)^2\sigma_I^2\right)x^2 Z_{xx}(t,x,m,l) + \frac{1}{2}\left(\frac{1}{T_0 - t}\right)^2$$
$$\times Z_{mm}(t,x,m,l) + \frac{1}{2}(\sigma_L - \sigma_I)^2 l^2 Z_{ll}(t,x,m,l) + \frac{x\pi_1\sigma_S}{T_0 - t}Z_{xm}(t,x,m,l)$$
$$+ xl\sigma_L\sigma_I(\pi_2 - 1)Z_{xl}(t,x,m,l).$$

and

$$\phi^* = \arg\max_{\phi \in \mathcal{A}}\{U(s,C(s)) + \mu(t)Y(s,W(s)) - \mu(t)Z(t,x,m,l) + \Phi(t,x,m,l;\phi)\},$$

then the value function $V(t,x,m,l) = Z(t,x,m,l)$.

The proof of the verification theorem can be found in Fleming and Soner [28] and Ye [29].

Let $U_y(x,y)$ and $Y_y(x,y)$ represent the derivative of $U(x,y)$ and $Y(x,y)$ about its second variable. $U(x,y)$ and $Y(x,y)$ are strictly concave to their second variable; thus, $U_y(x,y)$ and $Y_y(x,y)$ are invertible. Therefore, $\Theta : [0,T] \times R_0^+ \to R_0^+$ is defined as the function complying with

$$\Theta_1(x, U_y(x,y)) = y, \quad U_y(x, \Theta_1(x,y)) = y,$$
$$\Theta_2(x, Y_y(x,y)) = y, \quad Y_y(x, \Theta_2(x,y)) = y.$$

Theorem 2. *The value function reaches its maximum under* $\phi^* = (\pi_1^*(t), \pi_2^*(t), \theta_k^*(t), C^*(t)) \in \mathcal{A}$. *The optimal strategies are*

$$\pi_1^*(t,x) = -\frac{\frac{\sigma_S}{T_0-t} V_{xm}(t,x,m,l) + (\lambda - r_1 + \sigma_S \kappa_0 - \sigma_S m) V_x(t,x,m,l)}{x \sigma_S^2 V_{xx}(t,x,m,l)},$$

$$\pi_2^*(t,x) = 1 - \frac{L \sigma_L \sigma_I V_{xl}(t,x,m,l) + (r_2 + \lambda_I - r_1 - \sigma_I^2) V_x(t,x,m,l)}{x \sigma_I^2 V_{xx}(t,x,m,l)},$$

$$\theta_k^*(t,x) = \begin{cases} \left[\Theta_2\left(t, \frac{\eta_k(t) V_x(t,x,m,l)}{\mu(t)}\right) - x \right] \eta_k(t), & k = k^*(t), \\ 0, & \text{others}, \end{cases}$$

$$C^*(t,x) = \Theta_1(t, V_x(t,x,m,l)),$$

where $k^*(t) = \underset{k \in \{1,2,\cdots,K\}}{\arg\min} \{\eta_k(t)\}$.

Proof. Please see Mousa et al. [13] for the proof. □

We consider that wage earners use the same discounted CARA utility function for household consumption, inheritance and terminal wealth. These utility functions are

$$U(x,y) = -\frac{1}{\gamma} e^{-\rho x} \exp\{-\gamma y\}, Y(x,y) = -\frac{1}{\gamma} e^{-\rho x} \exp\{-\gamma y\}, \Gamma(x) = -\frac{1}{\gamma} e^{-\rho T} \exp\{-\gamma x\},$$

where $\rho > 0$ is the discount rate, γ ($\gamma < 1$, $\gamma \neq 0$) is the risk-aversion parameter. If $\phi = (\pi_1(t), \pi_2(t), \theta_{k^*}(t), C(t))$, we obtain the following HJB equation

$$\max_{\phi \in \mathcal{A}} \left\{ -\frac{1}{\gamma} e^{-\rho t} \exp\{-\gamma C\} - \frac{\mu(t)}{\gamma} e^{-\rho t} \exp\left\{-\gamma \left(x + \frac{\theta_{k^*}}{\eta_{k^*}}\right)\right\} - \mu(t) V(t,x,m,l) \right. \quad (5)$$
$$\left. + \Phi(t,x,m,l;\phi) \right\} = 0,$$

where

$$\Phi(t,x,m,l;\phi) = V_t(t,x,m,l) + \left[x \left(r_1 - \lambda_I + \sigma_I^2 \right) + x \pi_1(t)(\lambda - r_1 + \sigma_S \kappa_0 - \sigma_S m) + x \pi_2(t) \right.$$
$$\times \left. \left(r_2 + \lambda_I - r_1 - \sigma_I^2 \right) + l - C - \theta_{k^*} \right] V_x(t,x,m,l) + \left(\lambda_L - \lambda_I + \sigma_I^2 - \sigma_I \sigma_L \right)$$
$$\times l V_l(t,x,m,l) + \frac{1}{2} \left(\pi_2^2 \sigma_S^2 + (\pi_2 - 1)^2 \sigma_I^2 \right) x^2 V_{xx}(t,x,m,l) + \frac{1}{2} \left(\frac{1}{T_0 - t} \right)^2$$
$$\times V_{mm}(t,x,m,l) + \frac{1}{2} (\sigma_L - \sigma_I)^2 l^2 V_{ll}(t,x,m,l) + \frac{x \pi_1 \sigma_S}{T_0 - t} V_{xm}(t,x,m,l)$$
$$+ x l \sigma_L \sigma_I (\pi_2 - 1) V_{xl}(t,x,m,l).$$

Theorem 3. *The value function can be obtained as*

$$V(t,x,m,l) = -\frac{1}{\gamma} \exp\left\{ -\gamma \left[A(t) x + D_1(t) m^2 + D_2(t) m + Q(t) l + H(t) \right] \right\}.$$

The optimal strategies are

$$\pi_1^*(t) = -\frac{\frac{\sigma_s}{T_0-t}[2D_1(t)m + D_2(t)] - (\lambda - r_1 + \sigma_s \kappa_0 - \sigma_s m)}{x\sigma_s^2 \gamma A(t)},$$

$$\pi_2^*(t) = 1 - \frac{l\sigma_L \sigma_I Q(t) - (r_2 + \lambda_I - r_1 - \sigma_I^2)}{x\sigma_I^2 \gamma A(t)},$$

$$\theta_{k^*}^*(t) = -\frac{\eta_{k^*}}{\gamma}\left[\ln \frac{\eta_{k^*} A(t)}{\mu(t)} + \rho t + \gamma x\right] + \eta_{k^*}\left[A(t)x + D_1(t)m^2 + D_2(t)m + Q(t)l + H(t)\right],$$

$$C^*(t) = -\frac{1}{\gamma}[\ln A(t) + \rho t] + \left[A(t)x + D_1(t)m^2 + D_2(t)m + Q(t)l + H(t)\right].$$

where

$$A(t) = \frac{r_2 + \eta_{k^*}}{e^{-(r_2+\eta_{k^*})(T-t)}\left[(\eta_{k^*}+1)e^{(r_2+\eta_{k^*})(T-t)} + r_2 - 1\right]},$$

$$D_1(t) = -\frac{1}{2\gamma}(T_0 - T)^2 e^{(\eta_{k^*}+1)\int_t^T A(s)ds} \int_t^T \frac{e^{-(\eta_{k^*}+1)\int_t^s A(u)du}}{(T_0 - s)^2} ds,$$

$$D_2(t) = (T_0 - t)e^{(\eta_{k^*}+1)\int_t^T A(s)ds} \int_t^T \left(\frac{2(\lambda - r_1 + \sigma_s \kappa_0)D_1(s)}{(T_0 - s)\sigma_s} + \frac{\lambda - r_1 + \sigma_s \kappa_0}{\gamma \sigma_s}\right)$$
$$\times \frac{e^{-(\eta_{k^*}+1)\int_t^s A(u)du}}{T_0 - s} ds,$$

$$Q(t) = -e^{\left[-\lambda_L + \lambda_I - \sigma_I^2 + \sigma_I \sigma_L + \frac{\sigma_L}{\sigma_I}(r_2 + \lambda_I - r_1 - \sigma_I^2)\right](T-t) + (\eta_{k^*}+1)\int_t^T A(s)ds}$$
$$\times \int_t^T A(s)e^{\left[-\lambda_L + \lambda_I - \sigma_I^2 + \sigma_I \sigma_L + \frac{\sigma_L}{\sigma_I}(r_2 + \lambda_I - r_1 - \sigma_I^2)\right](s-t) - (\eta_{k^*}+1)\int_t^s A(u)du} ds,$$

$$H(t) = -e^{(\eta_{k^*}+1)\int_t^T A(s)ds}\left[\int_t^T G(s)e^{-(\eta_{k^*}+1)\int_t^s A(u)du} ds + \frac{\rho T}{\gamma}\right],$$

$$G(t) = -\frac{\eta_{k^*}+1}{\gamma}A(t) + \frac{\eta_{k^*}+1}{\gamma}\rho t A(t) + \frac{A(t)}{\gamma}\ln A(t) + \frac{\eta_{k^*} A(t)}{\gamma}\ln \frac{\eta_{k^*} A(t)}{\mu(t)} + \frac{\mu(t)}{\gamma}$$
$$+ \frac{D_1(t)}{(T_0-t)^2} - \frac{(\lambda - r_1 + \sigma_s \kappa_0)D_2(t)}{(T_0-t)\sigma_s} + \frac{(\lambda_I + r_2 - r_1 - \sigma_I^2)^2}{2\gamma \sigma_I^2} + \frac{(\lambda - r_1 + \sigma_s \kappa_0)^2}{2\gamma \sigma_s^2}.$$

Proof. Please see Appendix A. □

Proposition 1 (No inflation case). *When there is no inflation in the model, the optimal value function and optimal strategies are expressed as*

$$V(t, x, m, l) = -\frac{1}{\gamma}\exp\left\{-\gamma\left[A(t)x + D_1(t)m^2 + D_2(t)m + Q(t)l + H(t)\right]\right\},$$

$$\pi_1^*(t) = -\frac{\frac{\sigma_s}{T_0-t}[2D_1(t)m + D_2(t)] - (\lambda - r_1 + \sigma_s \kappa_0 - \sigma_s m)}{x\sigma_s^2 \gamma A(t)},$$

$$\theta_{k^*}^*(t) = -\frac{\eta_{k^*}}{\gamma}\left[\ln \frac{\eta_{k^*} A(t)}{\mu(t)} + \rho t + \gamma x\right] + \eta_{k^*}\left[A(t)x + D_1(t)m^2 + D_2(t)m + Q(t)l + H(t)\right],$$

$$C^*(t) = -\frac{1}{\gamma}[\ln A(t) + \rho t] + [A(t)x + D_1(t)m^2 + D_2(t)m + Q(t)l + H(t)],$$

where

$$A(t) = \frac{r_1 + \eta_{k^*}}{e^{-(r_2+\eta_{k^*})(T-t)}\left[(\eta_{k^*}+1)e^{(r_2+\eta_{k^*})(T-t)} + r_2 - 1\right]},$$

$$D_1(t) = -\frac{1}{2\gamma}(T_0 - T)^2 e^{(\eta_{k^*}+1)\int_t^T A(s)ds} \int_t^T \frac{e^{-(\eta_{k^*}+1)\int_t^s A(u)du}}{(T_0 - s)^2} ds,$$

$$D_2(t) = (T_0 - t)e^{(\eta_{k^*}+1)\int_t^T A(s)ds} \int_t^T \left(\frac{2(\lambda - r + \sigma_S \kappa_0)D_1(s)}{(T_0 - s)\sigma_S} + \frac{\lambda - r + \sigma_S \kappa_0}{\gamma \sigma_S}\right)$$
$$\frac{e^{-(\eta_{k^*}+1)\int_t^s A(u)du}}{T_0 - s} ds,$$

$$Q(t) = -e^{-\lambda_L(T-t) + (\eta_{k^*}+1)\int_t^T A(s)ds} \int_t^T A(s)e^{\lambda_L(s-t) - (\eta_{k^*}+1)\int_t^s A(u)du} ds,$$

$$H(t) = -e^{(\eta_{k^*}+1)\int_t^T A(s)ds}\left[\int_t^T G(s)e^{-(\eta_{k^*}+1)\int_t^s A(u)du} ds + \frac{\rho T}{\gamma}\right],$$

$$G(t) = -\frac{\eta_{k^*}+1}{\gamma}A(t) + \frac{\eta_{k^*}+1}{\gamma}\rho t A(t) + \frac{A(t)}{\gamma}\ln A(t) + \frac{\eta_{k^*}A(t)}{\gamma}\ln\frac{\eta_{k^*}A(t)}{\mu(t)} + \frac{\mu(t)}{\gamma}$$
$$+ \frac{D_1(t)}{(T_0 - t)^2} - \frac{(\lambda - r_1 + \sigma_S \kappa_0)D_2(t)}{(T_0 - t)\sigma_S} + \frac{(\lambda - r_1 + \sigma_S \kappa_0)^2}{2\gamma \sigma_S^2}.$$

Proposition 2 (No inside information case). *When there is no inside information in the model, we solve the optimization problem under filtration $\mathbb{F}$. The optimal value function and optimal strategies are expressed as*

$$V(t,x,l) = -\frac{1}{\gamma}\exp\{-\gamma[A(t)x + Q(t)l + H(t)]\},$$

$$\pi_1^*(t) = \frac{\lambda - r_1}{x\sigma_S^2 \gamma A(t)},$$

$$\pi_2^*(t) = 1 - \frac{l\sigma_L\sigma_I Q(t) - (r_2 + \lambda_I - r_1 - \sigma_I^2)}{x\sigma_I^2 \gamma A(t)},$$

$$\theta_{k^*}^*(t) = -\frac{\eta_{k^*}}{\gamma}\left[\ln\frac{\eta_{k^*}A(t)}{\lambda(t)} + \rho t + \gamma x\right] + \eta_{k^*}[A(t)x + Q(t)l + H(t)],$$

$$C^*(t) = -\frac{1}{\gamma}[\ln A(t) + \rho t] + [A(t)x + Q(t)l + H(t)],$$

where

$$A(t) = \frac{r_2 + \eta_{k^*}}{e^{-(r_2+\eta_{k^*})(T-t)}\left[(\eta_{k^*}+1)e^{(r_2+\eta_{k^*})(T-t)} + r_2 - 1\right]},$$

$$Q(t) = -e^{-\lambda_L(T-t) + (\eta_{k^*}+1)\int_t^T A(s)ds} \int_t^T A(s)e^{\lambda_L(s-t) - (\eta_{k^*}+1)\int_t^s A(u)du} ds,$$

$$H(t) = -e^{(\eta_{k^*}+1)\int_t^T A(s)ds}\left[\int_t^T G(s)e^{-(\eta_{k^*}+1)\int_t^s A(u)du} ds + \frac{\rho T}{\gamma}\right],$$

$$G(t) = -\frac{\eta_{k^*}+1}{\gamma}A(t) + \frac{\eta_{k^*}+1}{\gamma}\rho t A(t) + \frac{A(t)}{\gamma}\ln A(t) + \frac{\eta_{k^*}A(t)}{\gamma}\ln\frac{\eta_{k^*}A(t)}{\mu(t)} + \frac{\mu(t)}{\gamma}$$
$$+ \frac{(\lambda_I + r_2 - r_1 - \sigma_I^2)^2}{2\gamma \sigma_I^2} + \frac{(\lambda - r_1 + \sigma_S)^2}{2\gamma \sigma_S^2}.$$

Remark 1. *The proofs of Propositions 1 and 2 are similar to that of Theorem 3.*

(i) Note that when there is no inflation in consideration, the wealth process under inside information can be expressed as

$$dX(t) = \left[X(t)r_1 + X(t)\pi_1(t)(\lambda - r_1 + \sigma_S \kappa_0 - \sigma_S M(t)) + L(t) - C(t) - \sum_{k=1}^{K}\theta_k(t)\right]dt$$
$$+ X(t)\pi_1(t)\sigma_S d\widetilde{B}_S(t).$$

The optimal value function V under strategy $\phi = (\pi_1(t), \theta_k(t), C(t))$ satisfies the following HJB equation

$$\max_{\phi \in \mathcal{A}}\{U(s, C(s)) + \mu(t)Y(s, W(s)) - \mu(t)V(t, x, m, l) + \Phi(t, x, m, l; \phi)\} = 0,$$

where infinitesimal generator

$$\Phi(t, x, m, l; \phi) = V_t(t, x, m, l) + \left[xr_1 + x\pi_1(t)(\lambda - r_1 + \sigma_S \kappa_0 - \sigma_S m) + l - C - \sum_{k=1}^{K}\theta_k\right]$$
$$\times V_x(t, x, m, l) + \lambda_L l V_l(t, x, m, l) + \frac{1}{2}\pi_1^2 \sigma_S^2 x^2 V_{xx}(t, x, m, l)$$
$$+ \frac{1}{2}\left(\frac{1}{T_0 - t}\right)^2 V_{mm}(t, x, m, l) + \frac{1}{2}\sigma_L^2 l^2 V_{ll}(t, x, m, l) + \frac{x\pi_1 \sigma_S}{T_0 - t}V_{xm}(t, x, m, l).$$

(ii) On the other hand, when there is no inside information, the real wealth process under $\mathbb{F}$ can be obtained as

$$dX(t) = \left[X(t)(r_1 - \lambda_I + \sigma_I^2) + X(t)\pi_1(t)(\lambda - r_1) + X(t)\pi_2(t)(r_2 + \lambda_I - r_1 - \sigma_I^2) + L(t)\right.$$
$$\left. - C(t) - \sum_{k=1}^{K}\theta_k(t)\right]dt + X(t)\pi_1(t)\sigma_S dB_S(t) + X(t)(\pi_2(t) - 1)\sigma_I dB_I(t).$$

The HJB equation corresponding to problem (4) under filtration $\mathbb{F}$ is

$$\max_{\phi \in \mathcal{A}}\{U(s, C(s)) + \lambda(t)Y(s, W(s)) - \mu(t)V(t, x, l) + \Phi(t, x, l; \phi)\} = 0,$$

where $\phi = (\pi_1(t), \pi_2(t), \theta_k(t), C(t))$ and infinitesimal generator

$$\Phi(t, x, l; \phi) = V_t(t, x, l) + \left[x\left(r_1 - \lambda_I + \sigma_I^2\right) + x\pi_1(t)(\lambda - r_1) + x\pi_2(t)\left(r_2 + \lambda_I - r_1\right.\right.$$
$$\left.\left. - \sigma_I^2\right) + l - C - \sum_{k=1}^{K}\theta_k\right]V_x(t, x, l) + \left(\lambda_L - \lambda_I + \sigma_I^2 - \sigma_I \sigma_L\right)lV_l(t, x, l)$$
$$+ \frac{1}{2}\left(\pi_1^2\sigma_S^2 + (\pi_2 - 1)^2\sigma_I^2\right)x^2 V_{xx}(t, x, l) + \frac{1}{2}(\sigma_L - \sigma_I)^2 l^2 V_{ll}(t, x, l)$$
$$+ xl\sigma_L \sigma_I(\pi_2 - 1)V_{xl}(t, x, l).$$

4. Numerical Illustrations

This section discusses the impact of important parameters on the optimal strategies of the wage earner. The results in this section are obtained by applying MATLAB software (version R2016a, MathWorks Inc., Natick, MA, USA) for numerical analysis. Suppose the hazard function is described using the Gompertz parameter form

$$\lambda(t) = \frac{1}{10}e^{\frac{t-40}{10}}.$$

According to Baltas et al. [23], we discuss the optimal strategies under $x = 100$, $l = 10, T = 65, r_1 = 0.1, r_2 = 0.08, m = 0.5, T_0 = 70, \rho = 0.1, \sigma_S = 0.4, \sigma_I = 0.2$, $\sigma_L = 0.1, \lambda = 0.8, \lambda_I = 0.7, \lambda_L = 0.5, \gamma = 0.3$.

4.1. Optimal Investment Strategy

Figure 1 plots the impact of the risk-aversion coefficient γ on optimal investment strategies π_1^* and π_2^*. It is observed that as γ increases, the optimal investment strategies π_1^* and π_2^* both decrease. The risk-aversion coefficient γ increases, implying that the wage earner is more risk averse, and therefore accepts a lower investment risk. Thus, it is reasonable for investors with higher levels of risk aversion to adopt a more cautious investment strategy.

In Figure 2, the effect of the expected return rate λ and volatility rate σ_S on the optimal stock strategy π_1^* is explored. Figure 2a shows that the optimal stock strategy π_1^* increases as the expected rate of return λ increases. However, as shown in Figure 2b, when the value of the volatility rate σ_S increases, the optimal stock strategy π_1^* decreases. It is common sense that as the expected rate of return λ increases, investors will gain more from the stock market. However, an increase in instantaneous volatility σ_S will lead to an increase in investment risk. Thus, as returns increase and volatility decreases, wage earners will invest more in risky assets, which is consistent with the general conclusion of the investment problem.

The impact of the expected inflation rate λ_I and price-index volatility rate σ_I on the optimal inflation-indexed bond strategy π_2^* is shown in Figure 3. What can be seen is that the optimal bond strategy π_2^* is larger when expected inflation λ_I is larger and price index volatility σ_I is lower. This observation implies that the wage earner stands to invest in the inflation-index bond if the expected inflation rate is higher and volatility is lower. This is because larger expected inflation implies larger inflation-indexed expected returns, while increased volatility in price indices implies increased uncertainty in investing in inflation-indexed bonds.

The effect of inside information on the optimal stock strategy π_1^* is shown in Figure 4. According to Equation (2), the average rate at which the wage earner obtains inside information is captured by the parameter κ_0. We can observe that the optimal stock strategy π_1^* is an increasing function of κ_0. This is a reasonable result from the assumption that the wage earner has a priori knowledge of the random variable $\mathcal{L} = B_S(T_0), T_0 > T$ and $\kappa(t)$ is the drift induced by this random variable. This implies that the wage earner is taking advantage of additional information as an insider, and having access to inside stock information will encourage the wage earner to be bolder when investing in stocks. Moreover, since the inside information in our model only affects the stock process and not the inflation-index bond process, it is also quite natural that the optimal percentage invested in the inflation-index bond is not affected by the information drift.

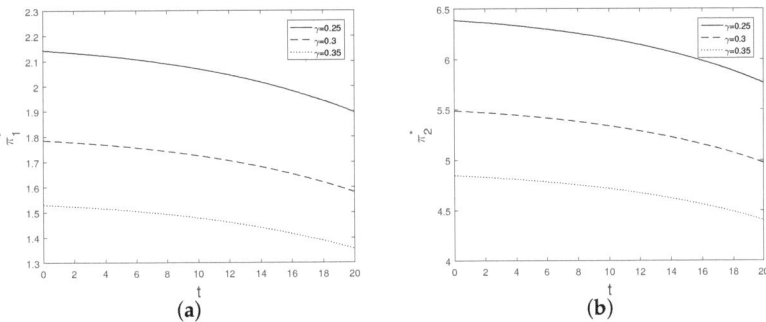

Figure 1. The effect of the risk-aversion parameter on the optimal investment strategies. (**a**) The effect of the risk-aversion parameter γ on the optimal stock strategy π_1^*. (**b**) The effect of the risk-aversion parameter γ on the optimal bond strategy π_2^*.

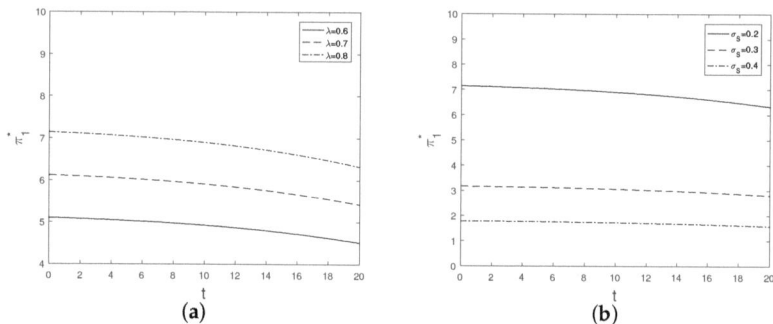

Figure 2. The effect of the expected return rate and volatility rate on the optimal stock strategy π_1^*. (a) The effect of the expected return rate λ on the optimal stock strategy π_1^*. (b) The effect of the volatility rate σ_S on the optimal stock strategy π_1^*.

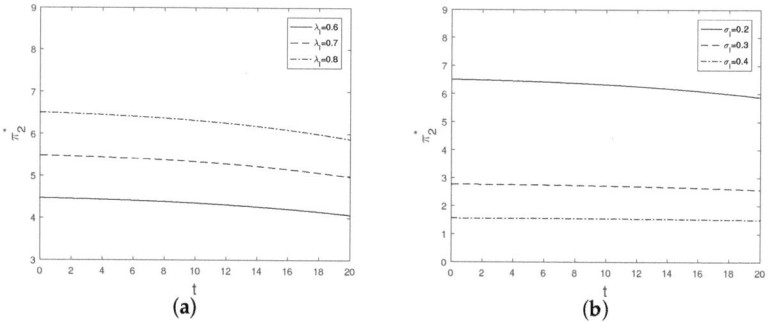

Figure 3. The effect of the expected inflation rate and volatility rate of price index on the optimal bond strategy π_2^*. (a) The effect of the expected inflation rate λ_I of price index on the optimal bond strategy. (b) The effect of the volatility rate σ_I of price index on the optimal bond strategy π_2^*.

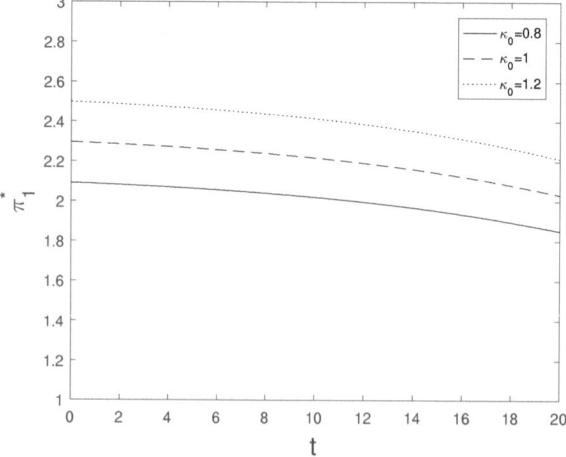

Figure 4. The effect of the inside information on the optimal stock strategy π_1^*.

4.2. Optimal Life Insurance Strategy

In Figure 5, the influence of the risk-aversion coefficient γ on the optimal life-insurance strategy θ^* is illustrated. As depicted in the graph, the spending on buying life insurance increases as γ increases, indicating that the wage earner is inclined to purchase life insurance at an optimal level when he/she is more risk averse. Moreover, it shows that the optimal life-insurance strategy θ^* increases over time t before decreasing. This is because the risk of death increases as the wage earner gets older; thus, naturally, life-insurance purchases will increase. However, after a certain age, the wage earner has a higher mortality rate and it would cost more to purchase life insurance at this time. Combined with the fact that they may have accumulated some wealth, it is more cost-effective to take other economic actions than to purchase life insurance.

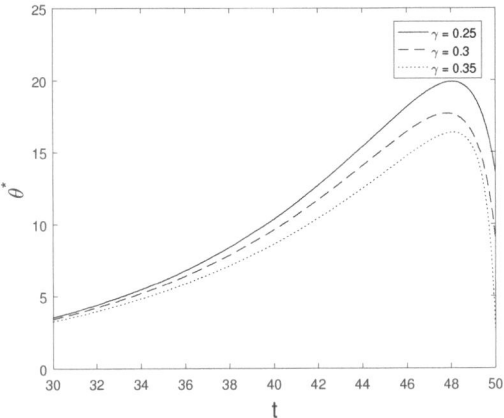

Figure 5. The effect of the risk-aversion parameter γ on the optimal life-insurance strategy θ^*.

Figure 6 shows the impact of inside information on the optimal life-insurance strategy θ^*. As the image illustrates, a higher value of κ_0 is associated with a larger optimal life-insurance strategy θ^*. This may be because when more inside information about the stock is available, the wage earner is more certain about investing in stock and thus is more inclined to invest in risky markets to gain wealth. Therefore, the increase in investment and expected increase in wealth makes the wage earner less inclined to purchase life insurance to protect against financial risk.

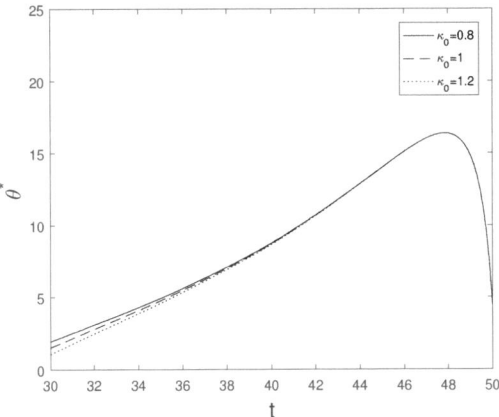

Figure 6. The effect of the inside information on the optimal life-insurance strategy θ^*.

4.3. Optimal Consumption Strategy

Figure 7 plots the influence of the risk-aversion parameter γ on the optimal consumption strategy C^*. The graph shows that as the value of the risk-aversion coefficient γ rises, the optimal consumption strategy C^* declines. This indicates that the higher the risk aversion of the wage earner, the more cautious the consumption, which is consistent with common sense.

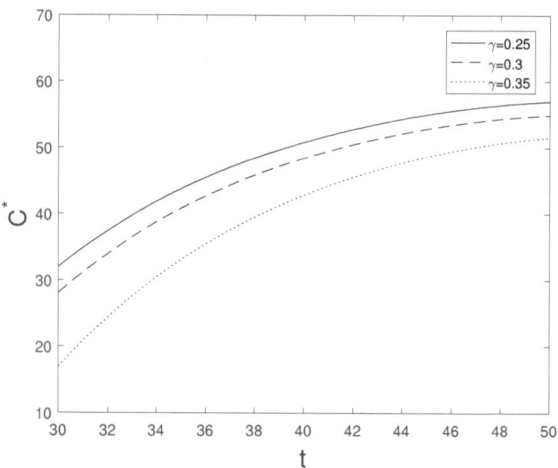

Figure 7. The effect of the risk-aversion parameter γ on the optimal consumption strategy C^*.

Figure 8 shows the effect of inside information on optimal consumption strategy C^*. As can be seen, the larger the value of κ_0, the larger the corresponding optimal consumption strategy C^*. This implies that inside information about the stock will have a positive effect on the optimal consumption. This can be interpreted as follows: when the wage earner has more inside information about stock, he/she is more inclined to invest in risky assets; thus, an increase in wealth is expected. As a result, wage earners tend to spend more on consumption at the optimal level.

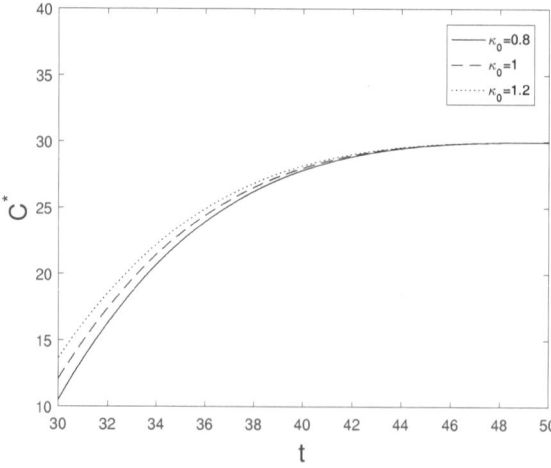

Figure 8. The effect of inside information on the optimal consumption strategy C^*.

5. Conclusions

Due to the wide range of applications of life insurance in reality and the fact that it has become a current research hotspot, the factors influencing a wage earner's investment, consumption and demand for life insurance under certain conditions need to be studied and justified. The economics of wage earners' behavior under these conditions, including investment, consumption and life-insurance purchases also need to be analyzed. In this paper, we study the dynamically optimal consumption, investment and life-insurance strategies for a wage earner in the presence of inside information and inflation. Specifically, the dynamically optimal strategies for consumption, investment in risky assets, investment in inflation-indexed bonds and life insurance are obtained by maximizing the expected utility of consumption, inheritance and final wealth. To provide a comprehensive analysis, we consider alternative scenarios as well, including an inflation-free model and a model without the presence of inside information. Finally, in order to study the factors affecting the investment, consumption and life-insurance demand of the wage earner under conditions of inside information and inflation, sensitivity analysis is provided through numerical studies. The dynamically optimal strategies and value function properties suggest that the dynamic financial behavior of the wage earner are as follows: (i) Inside information leads to an increase in investment and consumption but a decrease in life-insurance purchases. (ii) As expected inflation increases and volatility decreases, the purchase of inflation-indexed bonds should be increased to protect against inflation risk. (iii) If wage earners are more risk-averse, they will invest more money in life insurance while reducing consumption and spending on investments in risky markets.

However, due to the complexity of life insurance as a product, the reality of life insurance is susceptible to a number of factors, such as the economic situation, the health status of the wage earner, the status of family members and the wage earner's own subjective awareness of life insurance. The measurement of these real-world factors requires more sophisticated modeling to represent them. Therefore, the model in this paper explores more from a mathematical perspective, which has some limitations for the real situation. Consequently, more realistic models about investment, consumption and life insurance are expected.

Author Contributions: Writing—original draft, R.J.; writing—review & editing, W.L. and Y.H. All authors have read and agreed to the published version of the manuscript.

Funding: This research was supported by the National Natural Science Foundation of China (No. 11961064).

Data Availability Statement: Not applicable.

Acknowledgments: The authors are very grateful to the editors and the anonymous referees for their constructive and valuable comments and suggestions, which led to the present greatly improved version of our manuscript.

Conflicts of Interest: The authors declare no conflict of interest.

Appendix A

Proof. Applying the first-order optimality condition for the HJB equation, the investment, consumption and life-insurance strategies can be expressed as

$$
\begin{aligned}
\pi_1^*(t) &= -\frac{\frac{\sigma_s}{T_0-t} V_{xm}(t,x,m,l) + (\lambda - r_1 + \sigma_s \kappa_0 - \sigma_s m) V_x(t,x,m,l)}{x \sigma_s^2 V_{xx}(t,x,m,l)}, \\
\pi_2^*(t) &= 1 - \frac{L \sigma_L \sigma_I V_{xl}(t,x,m,l) + (r_2 + \lambda_I - r_1 - \sigma_I^2) V_x(t,x,m,l)}{x \sigma_I^2 V_{xx}(t,x,m,l)}, \\
\theta_{k^*}^*(t) &= -\frac{\eta_{k^*}}{\gamma} \left[\ln \frac{\eta_{k^*} V_x(t,x,m,l)}{\mu(t)} + \rho t + \gamma x \right], \\
C^*(t) &= -\frac{1}{\gamma} [\ln V_x(t,x,m,l) + \rho t].
\end{aligned} \quad \text{(A1)}
$$

Substituting Equation (A1) into Equation (5) yields

$$-\frac{1}{\gamma}V_x - \frac{\eta_{k^*}}{\gamma}V_x - \mu(t)V + V_t + x\left(r_1 - \lambda_I + \sigma_I^2\right)V_x + lV_x - \frac{\frac{\sigma_s}{T_0-t}V_{xm} + (\lambda - r_1 + \sigma_s\kappa_0 - \sigma_s m)V_x}{\sigma_s^2 V_{xx}}$$

$$\times (\lambda - r_1 + \sigma_s\kappa_0 - \sigma_s m)V_x + x\left(r_2 + \lambda_I - r_1 - \sigma_I^2\right)V_x - \frac{l\sigma_L\sigma_I V_{xl} + (\lambda_I + r_2 - r_1 - \sigma_I^2)V_x}{\sigma_I^2 V_{xx}}$$

$$\times \left(r_2 + \lambda_I - r_1 - \sigma_I^2\right)V_x + \frac{1}{\gamma}(\ln V_x + \rho t)V_x - \frac{\eta_{k^*} V_x}{\gamma}\left[\ln \frac{\eta_{k^*} V_x}{\mu(t)} + \rho t + \gamma x\right] + \left(\lambda_L - \lambda_I + \sigma_I^2\right)$$

$$-\sigma_I\sigma_L\Big)lV_l + \frac{\frac{\sigma_s^2}{(T_0-t)^2}V_{xm}^2 + (\lambda - r_1 + \sigma_s\kappa_0 - \sigma_s m)^2 V_x^2 + \frac{2\sigma_s}{(T_0-t)}(\lambda - r_1 + \sigma_s\kappa_0 - \sigma_s m)V_x V_{xm}}{2\sigma_s^2 V_{xx}} \quad (A2)$$

$$+\frac{l^2\sigma_L^2\sigma_I^2 V_{xl}^2 + (\lambda_I + r_2 - r_1 - \sigma_I^2)^2 V_x^2 + 2(\lambda_I + r_2 - r_1 - \sigma_I^2)l\sigma_L\sigma_I V_{xl}V_x}{2\sigma_I^2 V_{xx}} + \frac{1}{2}\left(\frac{1}{T_0-t}\right)^2 V_{mm}$$

$$-\frac{\frac{\sigma_s}{T_0-t}V_{xm}^2 + (\lambda - r_1 + \sigma_s\kappa_0 - \sigma_s m)V_x V_{xm}}{(T_0-t)\sigma_s V_{xx}} - \frac{l^2\sigma_L^2\sigma_I V_{xl}^2 + l\sigma_L(\lambda_I + r_2 - r_1 - \sigma_I^2)V_x V_{xl}}{\sigma_I V_{xx}}$$

$$+\frac{1}{2}(\sigma_L - \sigma_I)^2 l^2 V_{ll} = 0.$$

The value function can be conjectured to be of the following form

$$V(t,x,m,l) = -\frac{1}{\gamma}\exp\left\{-\gamma\left[A(t)x + D_1(t)m^2 + D_2(t)m + Q(t)l + H(t)\right]\right\}.$$

Then

$$\begin{aligned}
V_t &= \left[A'(t)x + D_1'(t)m^2 + D_2'(t)m + Q'(t)l + H'(t)\right]\Psi, \\
V_x &= A(t)\Psi, \\
V_l &= Q(t)\Psi, \\
V_m &= [2D_1(t)m + D_2(t)]\Psi, \\
V_{xx} &= -\gamma A^2(t)\Psi, \\
V_{ll} &= -\gamma Q^2(t)\Psi, \\
V_{mm} &= -\gamma[2D_1(t)m + D_2(t)]^2\Psi + 2D_1(t)\Psi, \\
V_{xm} &= -\gamma A(t)[2D_1(t)m + D_2(t)]\Psi, \\
V_{xl} &= -\gamma A(t)Q(t)\Psi,
\end{aligned} \quad (A3)$$

where $\Psi = \exp\left\{-\gamma\left[A(t)x + D_1(t)m^2 + D_2(t)m + Q(t)l + H(t)\right]\right\}$.

Thus

$$\pi_1^*(t) = -\frac{\frac{\sigma_s}{T_0-t}[2D_1(t)m + D_2(t)] - (\lambda - r_1 + \sigma_s\kappa_0 - \sigma_s m)}{x\sigma_s^2 \gamma A(t)},$$

$$\pi_2^*(t) = 1 - \frac{l\sigma_L\sigma_I Q(t) - (r_2 + \lambda_I - r_1 - \sigma_I^2)}{x\sigma_I^2 \gamma A(t)},$$

$$\theta^*(t) = -\frac{\eta_{k^*}}{\gamma}\left[\ln \frac{\eta_{k^*} A(t)}{\mu(t)} + \rho t + \gamma x\right] + \eta_{k^*}\left[A(t)x + D_1(t)m^2 + D_2(t)m + Q(t)l + H(t)\right],$$

$$C^*(t) = -\frac{1}{\gamma}[\ln A(t) + \rho t] + \left[A(t)x + D_1(t)m^2 + D_2(t)m + Q(t)l + H(t)\right].$$

Substitute Equation (A3) into Equation (A2), we have

$$-\frac{\eta_{k^*}+1}{\gamma}A(t)+\frac{\mu(t)}{\gamma}+\left[A'(t)x+D_1'(t)m^2+D_2'(t)m+Q'(t)l+H'(t)\right]+xr_2A(t)+lA(t)$$
$$-\frac{(\lambda-r_1+\sigma_S\kappa_0-\sigma_Sm)[2D_1(t)m+D_2(t)]}{(T_0-t)\sigma_S}+\frac{(\lambda-r_1+\sigma_S\kappa_0-\sigma_Sm)^2}{2\gamma\sigma_S^2}+\frac{(\lambda_I+r_2-r_1-\sigma_I^2)^2}{2\gamma\sigma_I^2}$$
$$+\frac{A(t)}{\gamma}\ln A(t)+\frac{\eta_{k^*}A(t)}{\gamma}\ln\frac{\eta_{k^*}A(t)}{\mu}+\frac{\eta_{k^*}+1}{\gamma}\rho tA(t)+\eta_{k^*}xA(t)+\left(\lambda_L-\lambda_I+\sigma_I^2-\sigma_I\sigma_L\right)lQ(t)$$
$$+\frac{D_1(t)}{(T_0-t)^2}+\sigma_L\sigma_Il^2\gamma Q^2(t)-\frac{1}{2}\gamma l^2\sigma_I^2 Q^2(t)-\frac{\sigma_L}{\sigma_I}\left(r_2+\lambda_I-r_1-\sigma_I^2\right)lQ(t)$$
$$-(\eta_{k^*}+1)A(t)\left[A(t)x+D_1(t)m^2+D_2(t)m+Q(t)l+H(t)\right]=0.$$

Let the coefficients of x, m and l equals zero respectively, we obtain the following differential equation

$$A'(t)-(\eta_{k^*}+1)A^2(t)+(r+\eta_{k^*})A(t)=0,$$
$$D_1'(t)+\left[\frac{2}{T_0-t}-(\eta_{k^*}+1)A(t)\right]D_1(t)+\frac{1}{2\gamma}=0,$$
$$D_2'(t)+\left[\frac{1}{T_0-t}-(\eta_{k^*}+1)A(t)\right]D_2(t)-\frac{2(\lambda-r_1+\sigma_S\kappa_0)D_1(t)}{(T_0-t)\sigma_S}-\frac{\lambda-r_1+\sigma_S\kappa_0}{\gamma\sigma_S}=0,$$
$$Q'(t)+\left[\lambda_L-\lambda_I+\sigma_I^2-\sigma_I\sigma_L-(\eta_{k^*}+1)A(t)-\frac{\sigma_L}{\sigma_I}\left(r_2+\lambda_I-r_1-\sigma_I^2\right)\right]Q(t)+A(t)=0,$$
$$H'(t)-(\eta_{k^*}+1)A(t)H(t)-\frac{\eta_{k^*}+1}{\gamma}A(t)+\frac{\eta_{k^*}+1}{\gamma}\rho tA(t)+\frac{A(t)}{\gamma}\ln A(t)+\frac{\eta_{k^*}A(t)}{\gamma}\ln\frac{\eta_{k^*}A(t)}{\mu(t)}$$
$$+\frac{\mu(t)}{\gamma}+\frac{B(t)}{(T_0-t)^2}-\frac{(\lambda-r_1+\sigma_S\kappa_0)D_2(t)}{(T_0-t)\sigma_S}+\frac{(\lambda_I+r_2-r_1-\sigma_I^2)^2}{2\gamma\sigma_I^2}+\frac{(\lambda-r_1+\sigma_S\kappa_0)^2}{2\gamma\sigma_S^2}=0.$$

According to the boundary conditions $A(T)=1, D_1(T)=D_2(T)=Q(T)=0, H(t)=\frac{\rho T}{\gamma}$, we have

$$A(t)=\frac{r_2+\eta_{k^*}}{e^{-(r_2+\eta_{k^*})(T-t)}\left[(\eta_{k^*}+1)e^{(r_2+\eta_{k^*})(T-t)}+r_2-1\right]},$$
$$D_1(t)=-\frac{1}{2\gamma}(T_0-T)^2 e^{(\eta_{k^*}+1)\int_t^T A(s)ds}\int_t^T\frac{e^{-(\eta_{k^*}+1)\int_t^s A(u)du}}{(T_0-s)^2}ds,$$
$$D_2(t)=(T_0-t)e^{(\eta_{k^*}+1)\int_t^T A(s)ds}\int_t^T\left(\frac{2(\lambda-r_1+\sigma_S\kappa_0)D_1(s)}{(T_0-s)\sigma_S}+\frac{\lambda-r_1+\sigma_S\kappa_0}{\gamma\sigma_S}\right)$$
$$\frac{e^{-(\eta_{k^*}+1)\int_t^s A(u)du}}{T_0-s}ds,$$
$$Q(t)=-e^{\left[-\lambda_L+\lambda_I-\sigma_I^2+\sigma_I\sigma_L+\frac{\sigma_L}{\sigma_I}(r_2+\lambda_I-r_1-\sigma_I^2)\right](T-t)+(\eta_{k^*}+1)\int_t^T A(s)ds}$$
$$\times\int_t^T A(s)e^{\left[-\lambda_L+\lambda_I-\sigma_I^2+\sigma_I\sigma_L+\frac{\sigma_L}{\sigma_I}(r_2+\lambda_I-r_1-\sigma_I^2)\right](s-t)-(\eta_{k^*}+1)\int_t^s A(u)du}ds,$$
$$H(t)=-e^{(\eta_{k^*}+1)\int_t^T A(s)ds}\left[\int_t^T G(s)e^{-(\eta_{k^*}+1)\int_t^s A(u)du}ds+\frac{\rho T}{\gamma}\right],$$

where

$$G(t)=-\frac{\eta_{k^*}+1}{\gamma}A(t)+\frac{\eta_{k^*}+1}{\gamma}\rho tA(t)+\frac{A(t)}{\gamma}\ln A(t)+\frac{\eta_{k^*}A(t)}{\gamma}\ln\frac{\eta_{k^*}A(t)}{\mu(t)}+\frac{\mu(t)}{\gamma}$$
$$+\frac{D_1(t)}{(T_0-t)^2}-\frac{(\lambda-r_1+\sigma_S\kappa_0)D_2(t)}{(T_0-t)\sigma_S}+\frac{(\lambda_I+r_2-r_1-\sigma_I^2)^2}{2\gamma\sigma_I^2}+\frac{(\lambda-r_1+\sigma_S\kappa_0)^2}{2\gamma\sigma_S^2}.$$

□

status through the use of artificial intelligence (AI) and vibration analysis [4]. Substantial research has been conducted to establish algorithms based on a large volume of data that train based on specific moments of failure, through machine learning, to obtain specific failure models [4,5].

This paper presents a method that leverages Fourier spectrum analysis and machine learning-based data extraction techniques for predicting wear in wind turbine operation. The novelty of the applied method lies in its utilization of unlabeled and uncategorized data to infer meaningful results for the predictive maintenance of wind turbines. In this study, functions representing the vibration trends of turbines across certain speed parameters, power levels, and wind flow conditions have been constructed. Furthermore, a density-based data filtering technique drawn from a machine learning-based method, Density Based Support Vector Machines (DBSVMs), has been employed at the data acquisition stages of the experiments.

The research was carried out over a period of about two months. The Fourier spectra were analyzed at different points in time while maintaining regulated and controlled parameters. With the help of at least five points from the Fourier spectra, the objective functions were defined. The evolution over time of these Fourier spectra's maximum points (amplitude–frequency) offers an effective approach to ensuring predictive maintenance. The established objective functions can be utilized to determine the wear evolution in both the low-frequency and high-frequency areas of a wind turbine. As a result of the experiments, the envelope of normal operation and the envelope of the maximum limit of operation are obtained for the gearbox, which is the most vulnerable part of the wind turbine. The envelope of the maximum limit of operation refers to wind turbine operation until the appearance of a defect. These experiments define the frequency–amplitude limits, which allow for the predictive maintenance of turbine components by setting the intervention thresholds without the need for extensive data collection.

The organization of the paper is as follows. Section 1 includes the details of the current scenario of the predictive maintenance of wind turbines and the state-of-the-art methods used for the condition monitoring of wind turbines. The research methods and experiments conducted in this study are discussed in Section 2. The results of the experiments and their interpretations are presented in Section 3, and the conclusion and future work are briefed in Section 4.

1.2. Overview of Wind Turbine Condition Monitoring and Its Need

Wind energy has seen remarkable growth over the past decade and continues to be on an upward trend in the power generation industry [3]. In the current context of the reduction in and even abandonment of conventional energy sources, wind energy emerges as a primary source, along with nuclear energy and hydropower [5]. In these conditions, the reliability and stability of wind turbine operations are crucial to maintaining the production capacity for prolonged periods and with optimal predictability [1].

The monitoring of wind turbine (WT) conditions is defined as a complex process of monitoring the parameters of the state of the machine so that a significant change is detected, which indicates a possible developing fault [6]. This can potentially help in different stages of wear: the early detection of incipient failures, thus reducing the chances of catastrophic failures; accurately assessing the proper functioning of the components and reducing maintenance costs; the analysis of the fundamental causes of the occurrence of defects, which can ensure the optimal determination of the input parameters for an improved operation of the turbine; the establishment of the control strategy and the optimal design of the components [7–10]. In a broad sense, the CMS of a wind turbine can target almost all of its major subsystems, including the blades, nacelle, power transmission, tower, and foundation [9]. This paper presents a method that focuses on the monitoring of wind turbines and can be applied to the different components of the wind turbine: the rotor shaft with main bearings, the gearbox, and the generator. From a CMS perspective, the three major monitored transmission components are the rotor shaft, the gearbox, and the generator. Of these three components, the gearbox causes the longest downtimes [11–13]. For this reason, the gearbox was chosen as the main subsystem targeted in this study. In detail, this paper will cover the typical practices, challenges, and future research opportunities related to CM wind turbine drivetrains [14].

To understand the dynamic behavior of a WT and especially of a planetary gearbox, a number of techniques have been used in research and in the industrial field: vibration analysis, oil condition analysis, thermography, acoustic measurement, boroscopic inspection, electrical parameters effects, the self-diagnostic of sensors, etc. [15]. In order to ensure the optimal conditions for predictive maintenance, a combination of different techniques is needed. Even if the vibration technique has a dominant proportion, it is supported in the decision by the other specific technologies.

However, a vibration analysis on component fault diagnosis in wind turbines is a hard challenge due to the complex mechanical conditions of the power transmission kinematic chain, the variable operating conditions with transient phenomena, and the speed differences between the different elements of the gearbox [15–17]. In the use of vibration transducers specifically, piezoelectric accelerometers are the most used method, with different sensitivities depending on the speed and with a rigid fixation on the structure of the components [7–9]. The repartition of the sensors in the monitoring process of the wind turbine from the actual stage of the research is shown in Figure 1 and Table 1.

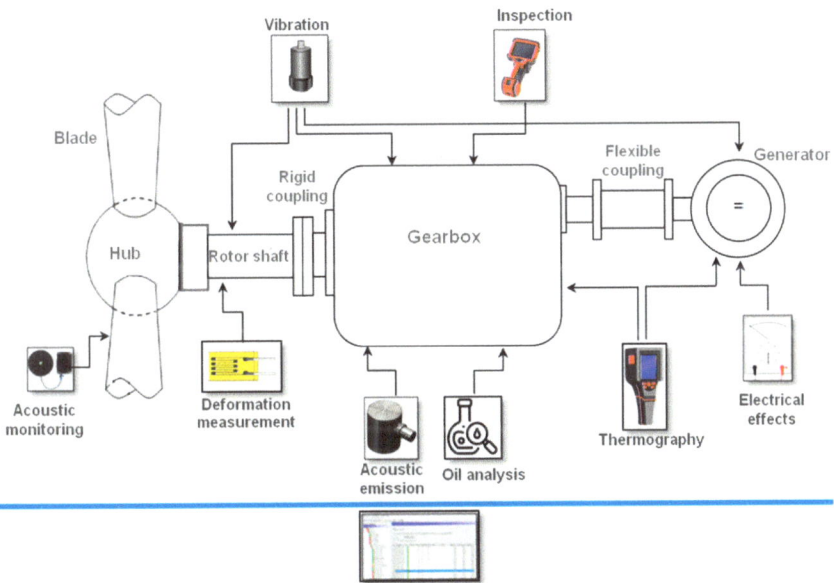

Figure 1. The position of the sensors for the monitoring process.

Table 1. Sensor notation and position on the wind turbine.

Sensor Label	Description
B1-MB-RS	Main bearing accelerometer—rotor side
B2-MB-GS	Main bearing accelerometer—generator side
B3-LSS	Gearbox accelerometer—low-speed shaft
B4-IS	Gearbox accelerometer—intermediary shaft
B5-HSS	Gearbox accelerometer—high-speed shaft
B6-G-DE	Generator accelerometer—drive end side
B7-G-NDE	Generator accelerometer—non-drive end side

In this paper, the focus is on the monitoring of wind turbine drivetrains. The drivetrains consist of the main bearing, main shaft, gearbox, brake, generator shaft, and generator. From a CM perspective, the three major monitored transmission components are the main bearing, the gearbox, and the generator. Of these three components [6], the gearbox causes the longest downtimes. Other research has also shown that the gearbox is the most expensive subsystem to maintain during the 20-year operating life of a turbine [1–7]. For this reason, the gearbox was chosen as the main subsystem targeted in this study.

1.3. State of the Art in Turbine Wear Monitoring and Trend Analysis

Current research has led to the identification of the following monitoring techniques and directions, which can be applied to wind turbines [14,15]: vibration analysis; oil condition analysis; the thermography of important elements in the turbine structure (gearbox); the analysis of the physical condition of the materials; the measurement of elastic yielding and deformation of various components; acoustic measurements in various sensitive areas of the turbine; the measurement of various electrical effects; process parameter measurement; visual inspection; performance monitoring by comparing output sizes for the same input data; the use of self-diagnostic sensors (Figure 1).

(a) *Vibration analysis*—Vibration analysis is the most well-known technology for monitoring working conditions, especially for rotating equipment [15]. The type of sensors used depends on the frequency range used for monitoring, the position of transducers on the transmission chain for the low-frequency range, the velocity sensor in the 5–1000 Hz frequency domain, the accelerometers for the high-frequency range, and the acoustic sensor for gearbox monitoring or blades.

(b) *Oil analysis*—Oil analysis is another evaluation technique, which, coupled with vibration analysis, contributes to decision-making in predictive maintenance. Oil analysis is mostly conducted offline via sampling and also ensuring the quality of the oil. The contamination with dirt from the turbine parts in contact, the moisture, the degradation of additives, and the maintenance of the oil filter are also aspects of this method. However, to protect oil quality, the application of online sensors is used more and more often, especially for particle counters. In addition, protecting the condition of the oil filter is currently mainly applied to both hydraulic oil and lubricating oil. In the case of the excessive pollution of the filter, or a change in the characteristics of the oil, this leads to excessive wear [15].

(c) *Thermography*—Thermography is often applied for the monitoring and fault identification of electrical and electronic components [15]. Hot spots due to component degeneration or poor contact can be identified in a simple and fast way using cameras and diagnostic software. Mainly, they are used in generator and power converter monitoring but also for thermal gear contact.

(d) *Inspection of component condition*—This type of monitoring mainly focuses on detecting and tracking the evolution of wear using a boroscope device. This method is normally offline and is a very important decision criterion for stopping, limiting, or planning a repair [15,16].

(e) *Deformation measurement*—Deformation measurement using manometers is a common technique but is not often applied in the case of wind turbine monitoring. Strain gauges are not robust in the long term [15–17]. For wind turbines, deformation measurement can be very useful for life prediction and stress level protection, especially for blades [18] but also for the main shaft.

(f) *Acoustic monitoring*—Acoustic monitoring is related to vibration monitoring using noise measurement. Acoustic monitoring technology can be used for blade condition monitoring using an acoustic microphone or for bearing and gearbox monitoring using acoustic emission sensors fixed directly to the housing [15].

(g) *Electrical effects*—The electrical parameter monitoring of a generator represents a mandatory condition in based condition maintenance (CBM). The analysis of electrical parameters, such as electrical current, voltage, insulation, power, etc., allows for both the evaluation of the quality of the generated power and the analysis of the potential faults [17].

(h) *Process parameters*—Condition monitoring systems (CMSs) are becoming more sophisticated, and their diagnostic capabilities are improving. However, protection is mostly based on level detection or signal comparison, which directly leads to alarm when the signals exceed predefined threshold values. The integration of machine learning is still in the beginning stages, but in the future, solutions using AI will be sought for large-scale development [15].

(i) *Performance monitoring*—Wind turbine performance is often gauged through the relationship between power, wind speed, rotor speed, and blade angle, and in the case of large deviations, an alarm sounds or a stop is even initiated [15]. The detection of margins is important to prevent false alarms [19]. Similar to process parameter estimation, more sophisticated methods like performance evolution monitoring are still not a common practice.

Thus, to obtain reliable predictive maintenance results, a combination of different techniques is needed. While vibration analysis may hold a predominant role, it is complemented by other specific technologies to perform decision-making accurately (Figure 1).

2. Applied Research Methods

2.1. Condition Monitoring System

In this research, the experimental protocol is based on the Condition Monitoring System (CMS). The data used are part of the online data protocol regarding the wind turbines' state of operation. The recorded data are analyzed using signal evaluation both in the time domain and in the frequency domain. The CMS provides all datasets as originally optimized for all turbines. The data are collected from a wind turbine gearbox. The repartition of the sensors in the monitoring process of the wind turbine from the actual stage of the research is shown in Figure 2.

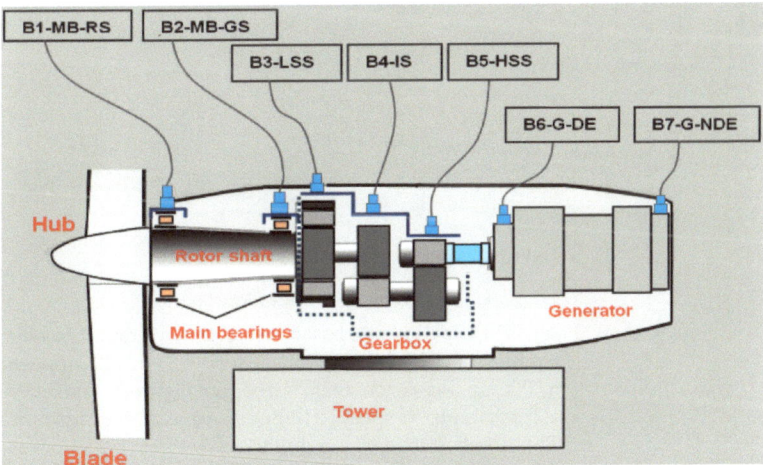

Figure 2. Schema of the experimental stand with the position of the used sensors.

The analysis is centered on the gearbox, examining the vibrations at three specific points of the gearbox: the low-speed shaft (LSS), the intermediary shaft (IS), and the high-speed shaft (HSS). The data acquisition is conducted using vibration sensors fixed on the bearings of the kinematic chain, starting from the input, which is the rotor side, and extending to the output, which is the generator side.

The data transmission and processing chain is illustrated in Figure 3. The online acquisition system allows the data to be recorded according to the original settings, thus capturing signals along with their speed and power readings. In this way, the evolution of vibrations can be determined specific to certain values of speed and power [20]. The system allowed the definition of parameters in the frequency domain both in the acquisition and analysis phases. The selected frequency range is according to ISO 10816-21 standards [21], including the rotor, gearbox, generator, and tower/nacelle. Figure 3 shows the datasets according to CMS, for the gearbox in the 3 entry points: LSS, IS, and HSS.

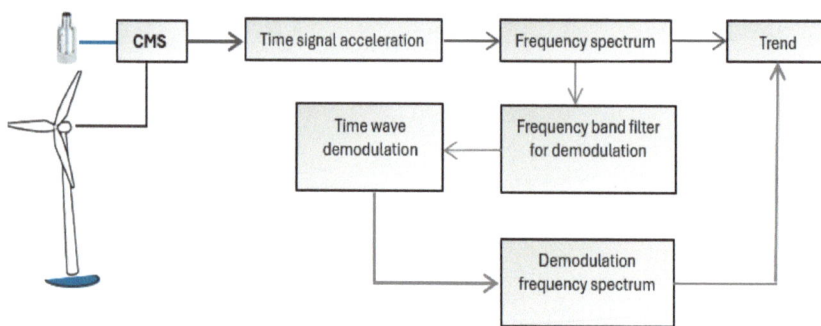

Figure 3. Synopsis of data acquisition and signal processing.

In these experiments, the data from the input of the gearbox, the acceleration in the frequency domain at LSS, and the data of the gearbox output, in the frequency domain at HSS, are taken into account, Figure 4.

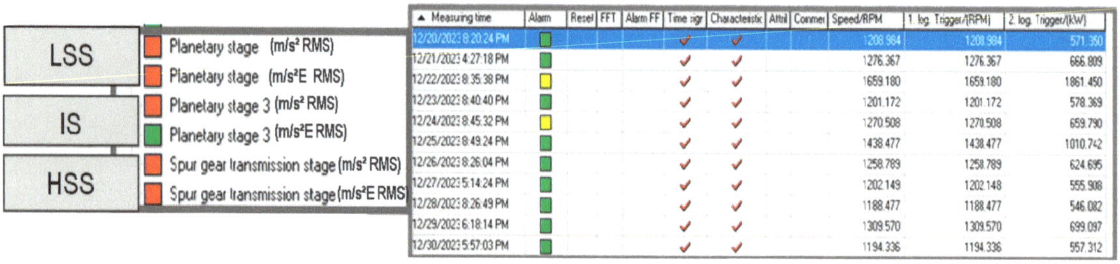

Figure 4. Example of CMS data presentation.

2.2. Signal Processing and Defect Detection

The experiment is based on real-time vibration monitoring, using National Instrument equipment cRIO-9076 (Austin, TX, USA), with 12 input channels, a 24-bit resolution, and a 50 k samples/s/ch. max. speed; see Figure 5. The real-time monitoring data are set on a 25 k samples/s speed, a buffer size of 32,768 samples, and a block size of 10 k samples. The vibration monitoring provides the signal data from the 3 accelerometers fixed on the 3 gearbox points: the LSS with the 1–2 stages, the IS with the 3 stages, and the HSS with the spur gear stage. The accelerometers used have a 100 mV/g sensitivity for the IS and HSS points and a 500 mV/g sensitivity for the LSS point. For a precise synchronization between vibration signals and speed signals, a laser speed sensor fixed at the generator side was used.

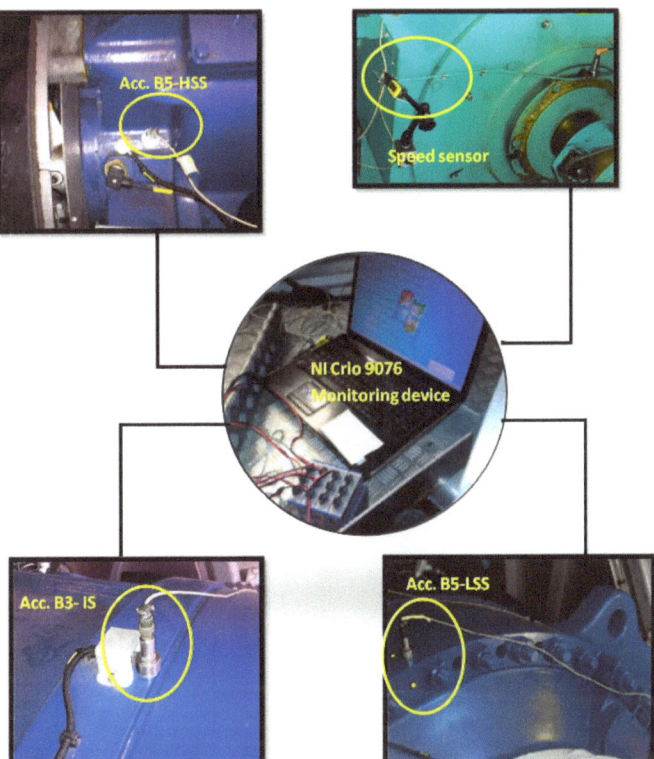

Figure 5. Vibration monitoring devices.

Signal analysis was performed via numerical processing, taking into account the parameters (frequency and amplitude) being monitored. Thus, Figure 6 shows the waveforms obtained with the help of the monitoring software both in the time domain and in the frequency domain; Figures 7 and 8 show the acceleration signal in the case of the gearbox wear. The vibration parameters are set according to the ISO 10816-21 standard, specifying acceleration in m/s^2 RMS, vibration velocity in mm/s RMS, and demodulated acceleration in m/s^{2}E. With the bearing frequency data, the characteristic frequency of the bearing defect can be identified. The structure of the vibration parameters is complex and based on the vibration defect theory [5,22]. The vibration limits for wind turbines, provided by the ISO 10816-21 standard, present an integrated base defining the recommended state of operation [23–25]. Even in this situation, many specific cases of the vibration of the wind turbine components are difficult to classify according to this standard [26]. For this purpose, it is proposed to develop a model that can interpret the state of operation in real operating conditions using data provided via the CMS.

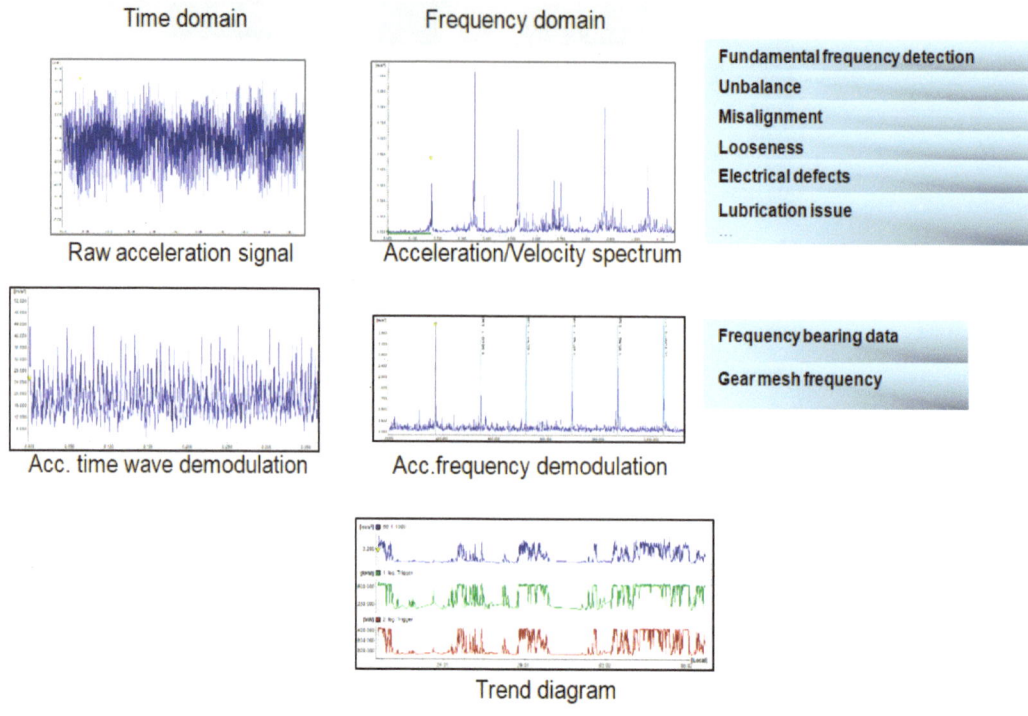

Figure 6. Signal processing and defect detection from the CMS.

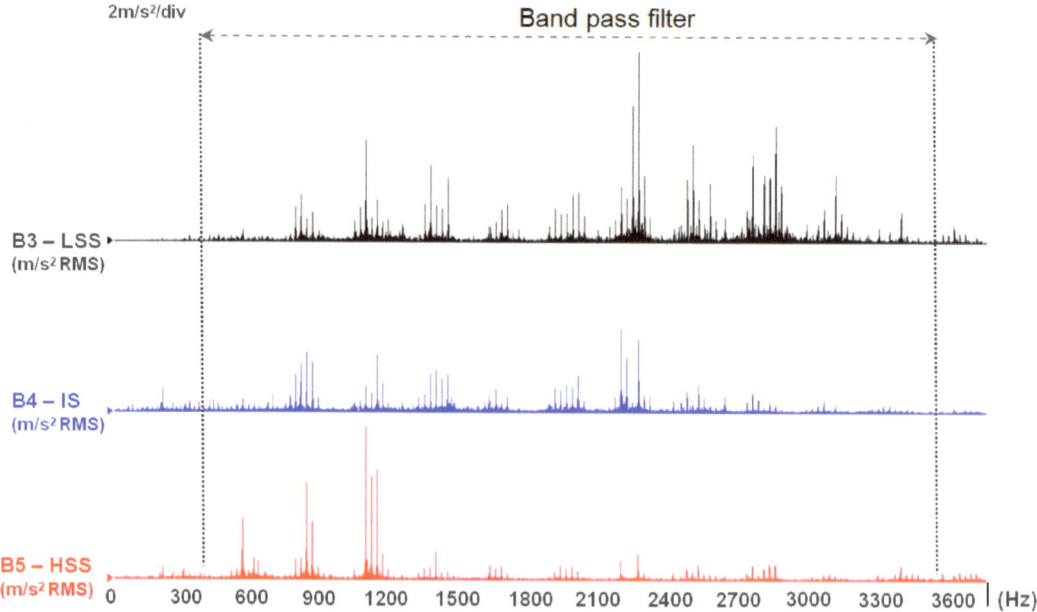

Figure 7. Acceleration signal in the case of gearbox wear.

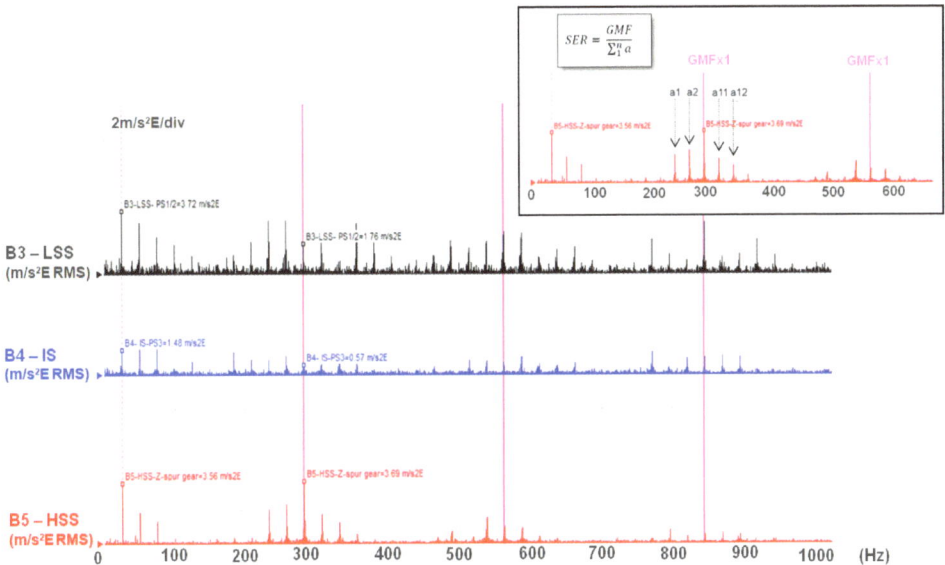

Figure 8. Envelope acceleration in the case of gearbox wear.

The processing data and analysis approach for bearing detection are also applied for gear characterization using the gear mesh frequency data according to the kinematic chain of the gearbox [9–12]. The signal processing and analysis are performed with Fastview software (v300124), which allows for the use of both vibration monitoring and analysis in real time. The software allows for the identification of the specific failure frequencies of the gear and bearings through the method of vibration demodulating using the envelope function [27] with the dynamic filtering of the specific domain frequencies (Figure 7).

A novelty in the evaluation analysis of the gearbox wear condition is the envelope method using the Hilbert transform [27] with sideband energy coefficient integration, called SER coefficient (Sideband Energy Ratio™, a patent-pending algorithm utilized in the General Electric) [28–32], so that the impact energy generated by the defect can be quantified (Figure 8).

Figure 8 shows the spectrum of the acceleration envelope in the case of the gearbox defect. The quantification of the defective condition is evaluated by means of the gear mesh frequency presence (GMF) in relation to the sidebands, as well as its harmonics. According to the quantification of the level of sidebands in relation to the amplitude of the GMF frequency, it can be found that the ratio is less than one, which means that the defect in the HSS stage is present and is in an advanced state.

2.3. Using DBSVM-Based Data Extraction Technique

The Base density of the Support Vector for Machine Learning (*DBSVM*) [30] has been beneficial in establishing the basic data for neural network learning. In any monitoring activity, it is more efficient to train the neural network using DBSVM as it reduces the learning input data (decreasing computational complexity) and determines the resulting weights matrices to identify a mechanical failure without being impacted by the outliers. This study exploits this method to find the most relevant data points and establish the objective function (*FO*).

This data extraction method is based on the filtering of data points based on their population density. The population density of data points refers to the correlation between the population size and the space they occupy. The rationale behind this data filtering is to deal with the data points that are influenced by random noises or gross errors. These data points do not accurately represent the general trend. These points are considered outliers and can affect the accuracy of the established objective functions and, subsequently, the analysis. The densely populated areas in the input space are determined by calculating the Mahalanobis distance (1). The points lying in this region are considered meaningful points while the points lying outside of this region are considered outliers.

The Mahalanobis distance is calculated from the quantity μ which represents the average of the points' distances, to each point. The cov^{-1} represents the inverse covariance matrix. This distance is explained in [33,34]. The Mahalanobis distance takes into account the correlation of the dataset and does not depend on the measurement scale [34–36]. The population variance is calculated with a variance–covariance matrix [35]. The Mahalanobis distance from the point to the mean of the distribution μ can be calculated by (1), and the Mahalanobis distance from one point to another can be calculated by (2):

$$d = \sqrt{(x-\mu)^T cov^{-1}(x-\mu)} \tag{1}$$

$$d = \sqrt{(x-y)^T cov^{-1}(x-y)} \tag{2}$$

where the population variance is calculated with [12]

$$var(x_n) = \frac{\sum_1^n (x-\mu)^2}{n} \tag{3}$$

and population covariance with

$$cov(x_n, y_n) = \frac{\sum_1^n (x_i - \mu_x)(y_i - \mu_y)}{n} \tag{4}$$

If $cov(x_i)$ and $cov(y_i) > 0$
both of them increase or decrease;
If $cov(x_i)$ and $cov(y_i) < 0$
when x_i increases, y_i decreases, or vice versa;
If $cov(x_i)$ and $cov(y_i) = 0$
no relation exists between x_i and y_i;
If $var(x_i) > var(y_i)$
x_i increases or decreases faster than y_i;
End.

The average of d is

$$average\_d = \frac{\sum_1^n \sqrt{(x_i - \mu)^T cov^{-1}(x_i - \mu)}}{n} \tag{5}$$

where d_i is the distance between the points and d is the average of these distances
If $d_i > d$,
the point i is in the outlier group;
Else
the point i will be considered an important (meaningful) point in DBSVM;
End.

2.4. Objective Functions

The optimization function (FO) was proposed as a polynomial function of the fifth order with real coefficients that will be constructed using the data from the acquisition of Fourier spectra of the vibrations:

$$FO = a_1 \times x^5 + a_2 \times x^4 + a_3 \times x^3 + a_4 \times x^2 + a_5 \times x + a_6 \tag{6}$$

where a_i will be determined using the matrix equation:

$$\begin{pmatrix} a_1 \\ a_2 \\ \cdots \\ a_6 \end{pmatrix} = \left\{ \begin{bmatrix} x_1^5 & \cdots & x_1 & 1 \\ \vdots & \ddots & \vdots \\ x_5^5 & \cdots & x_5 & 1 \end{bmatrix} \begin{bmatrix} x_1^5 & \cdots & x_1 & 1 \\ \vdots & \ddots & \vdots \\ x_5^5 & \cdots & x_5 & 1 \end{bmatrix}^T \right\}^{-1} \begin{pmatrix} FO_1 \\ \cdots \\ FO_5 \end{pmatrix} \quad (7)$$

with the following constraints:

- $x_i > 0$;
- x_i must be meaningful points, $x_i \in$ group 1;
- $x_i \in DBSVM$;

where FO_i is the amplitude of the vibration evolution in time where the defect will appear and x_i is the frequency in time. To define the FO, 5 boundary points $(x_i, FO_i) \in DBSVM$ will be used for each moment of time vs. frequency points but under the same conditions of forced vibration and for the same wind turbine. The DBSVM points must strictly adhere to the condition of belonging to DBSVM, which is that

$$d_i < average\_d. \quad (8)$$

The boundary of the FO will be the limit of the optimal functioning of the wind turbine. In this way, the moment of time for the intervention on the gearbox will be determined to eliminate the danger of an imminent defect.

2.5. The Used Proper LabView Virtual Instrumentation for FO

To solve the objective function FO, proper LabView virtual instrumentation was used, and the block schemas are shown in Figures 9–11.

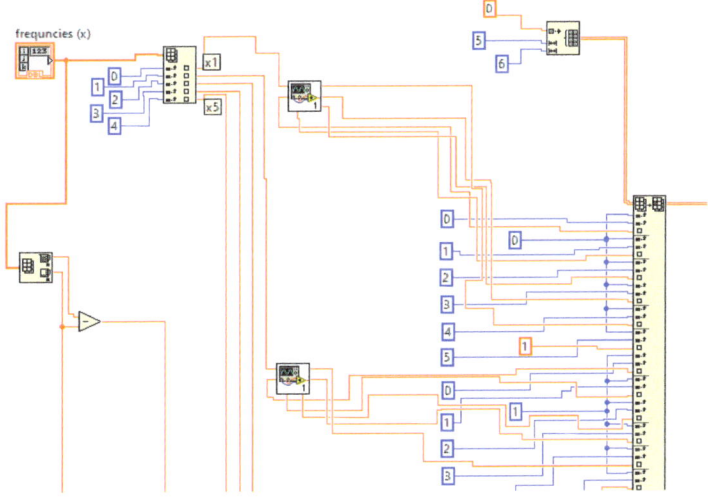

Figure 9. Part of the block schema of the LabView virtual instrumentation to determine the FO^5 order.

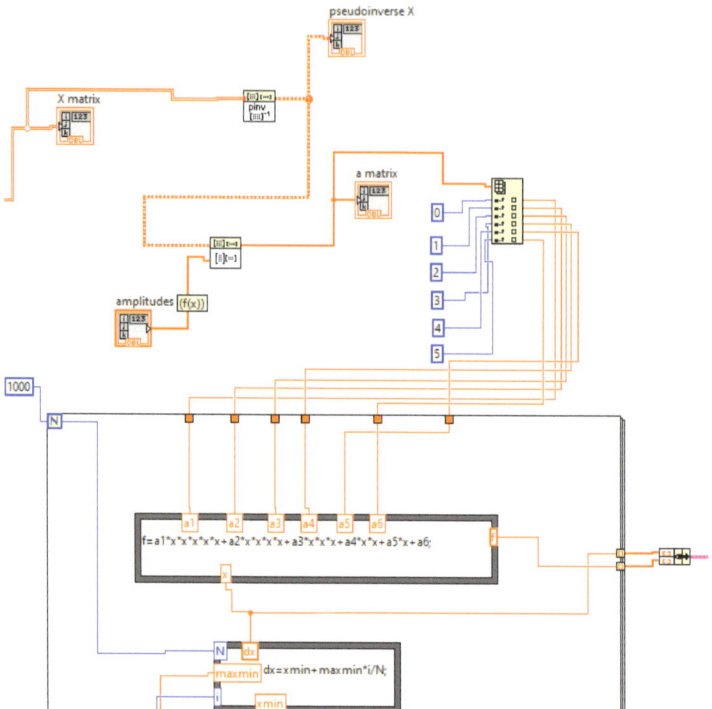

Figure 10. Part of the block diagram represents the *FO (polynomial function of the 5-degree order)* characteristic.

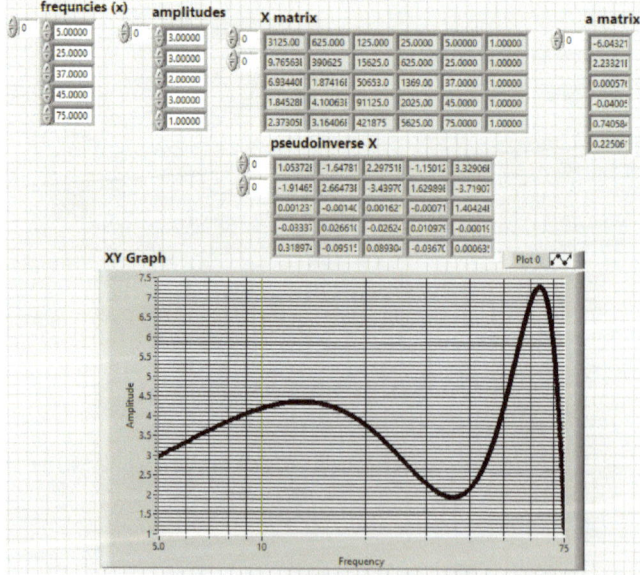

Figure 11. Front panel with the results of the optimization function *FO* for known points from the *DBSVM* (the maximal values from the Fourier spectrum).

2.6. Description of the Used Algorithm

The used algorithm includes the following stages, as depicted in Figures 12 and 13: (i) the acquisition of data at different moments of time for the same parameters of wind, power, and speed; (ii) the application of relation (4) for calculating the distances between points d_i (max. amplitude and frequency of the Fourier spectra acquired); (iii) applying relation (5) to determine the average distance, d; (iv) defining group 1 of the DBSVM after checking the condition $d_i < d$; (v) establishing the boundary curve of DBSVM; (vi) analysis of Fourier spectra from group 1; (vii) defining the 5 maximum points from the Fourier spectra both for the upwind position and for the downwind position of the sensors; (viii) the use of LabView virtual instrumentation to determine the 5th-order objective functions; (ix) plotting multiple objective functions for Fourier spectra acquired during three months of operation, under the same conditions of wind, power, and speed; (x) defining the maximum points of the objectively drawn functions in order to determine the trend; (xi) determining the coefficients of the 5th-order objective functions of the trend for both low and high frequencies, as well as for upwind and downwind of the gearbox sensors positions.

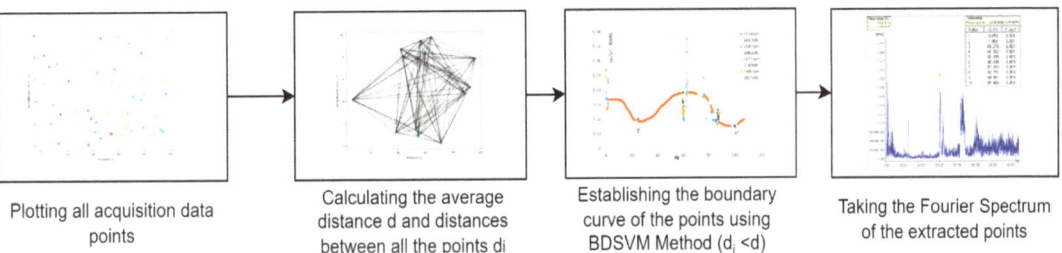

Figure 12. Block schema of the part of the used algorithm to establish the DBSVM of the collected data.

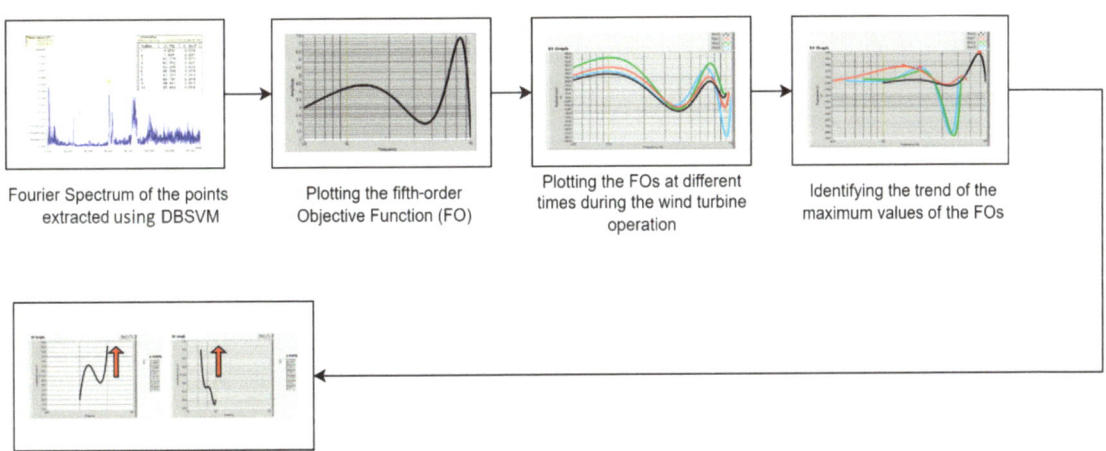

Figure 13. Block schema of the part of the used algorithm to establish the objective functions (*FOs*) using the Fourier spectrum collected from the boundary of the DBSVM.

3. Results and Analysis

3.1. Establishing FO Boundary of Fourier Spectrum

If the operational limit of the turbine is set at a specific *FO*, a defect can be easily detected through control at each frequency. This can be performed by checking if the operational point (frequency, magnitude) is in the normal functioning area or outside of this. In this way, it is possible to determine the maximum permissible magnitude of vibration.

In this case, the equation of the FO will be

$$FO = -6.043x^5 + 2.233x^4 + 0.0005x^3 - 0.04x^2 + 0.74x + 0.225 \qquad (9)$$

For predictive maintenance, the following relation would be applied:

$$FO_i(f_i) < FO_j(x_i) \qquad (10)$$

where x_i is the frequency for the imposed five points $\in DBSVM$, the points from the boundary limits, and f_i represents all the current frequencies that must be checked. If this condition is false, the respective points could be the potential mechanical wear.

Using the Fourier spectra, the objective functions (FO_i) were constructed the objective functions (FO_i) for each of these datasets. All these FOs are shown in Figures 12–15, for upwind and downwind sensors from the wind turbine gearbox. All objective functions, FOs, were determined using the maximal values of magnitude from each of the used Fourier spectra; see the table of each acquisition Fourier spectrum.

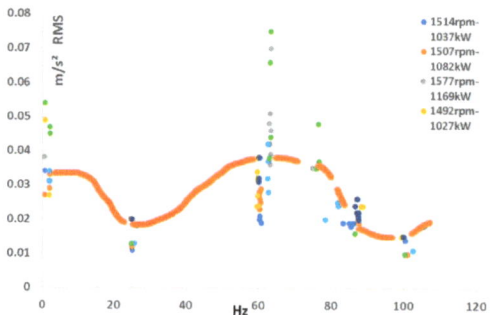

Figure 14. The acquisition data distribution and the establishment of boundary values for group 1, representing meaningful points of DBSVM, occur under similar dynamic conditions of speed and power. This characteristic is constructed by applying the DBSVM algorithm.

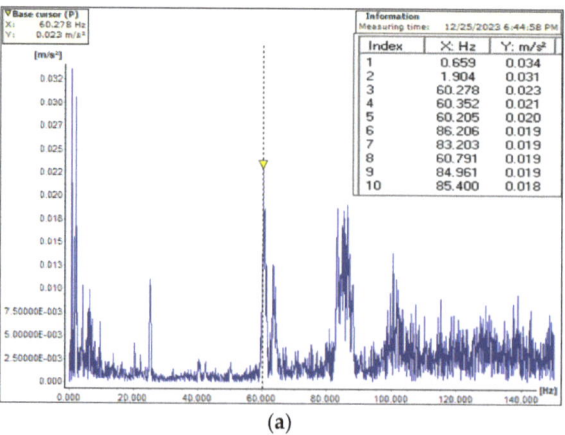

(a)

Figure 15. Cont.

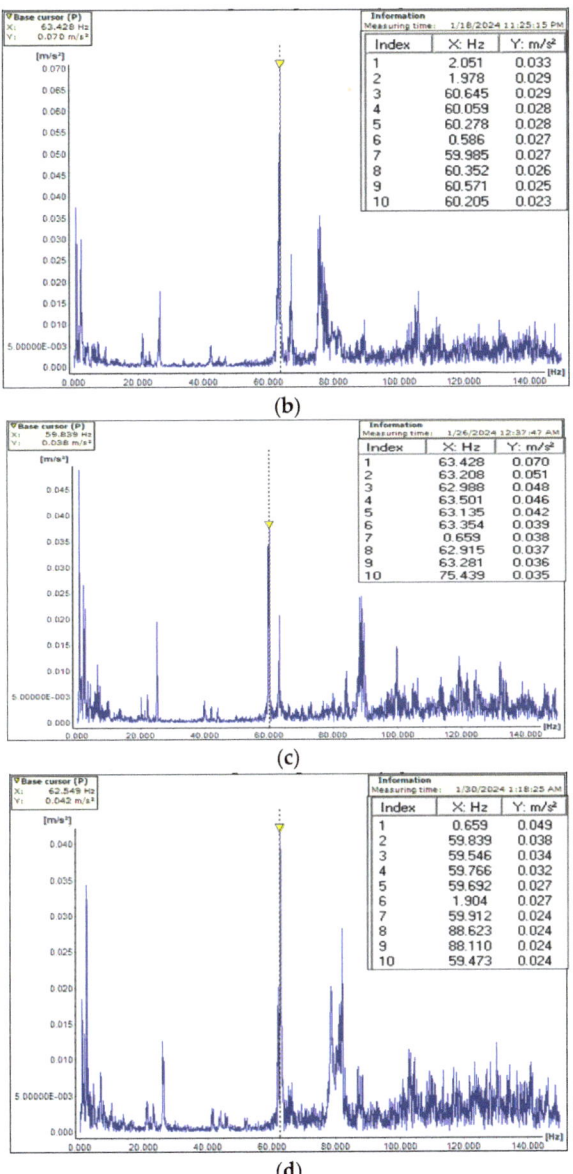

Figure 15. *Cont.*

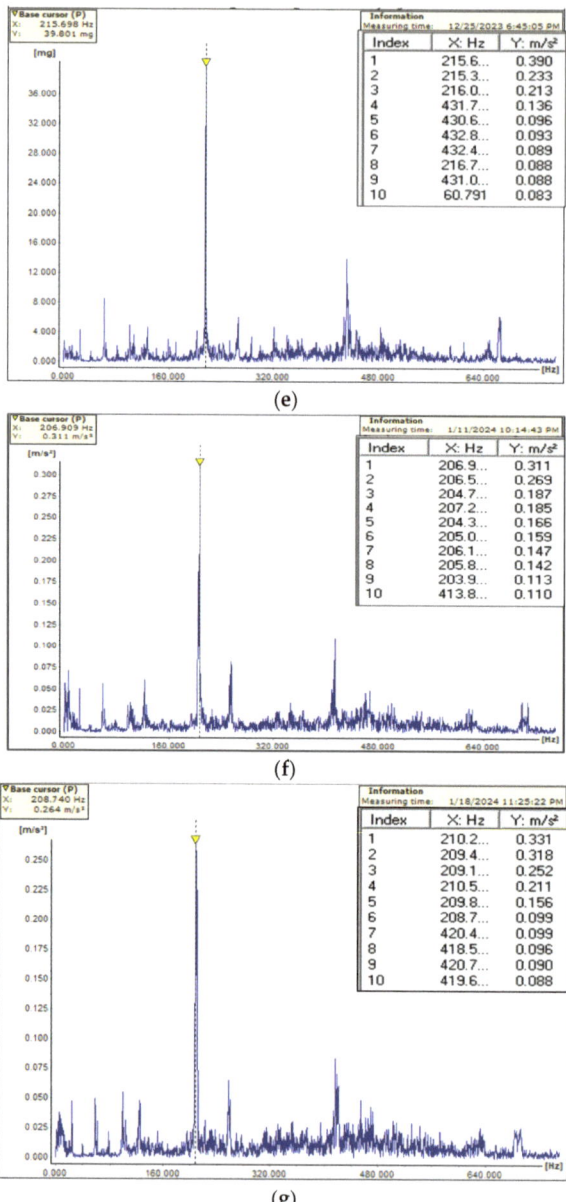

Figure 15. *Cont.*

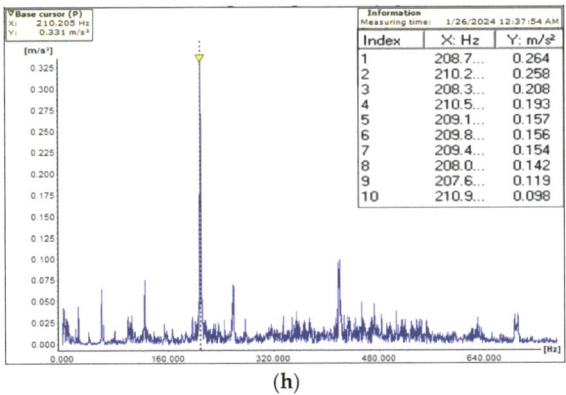

(h)

Figure 15. Fourier spectrum from data acquisition between December 2023 and February 2024, in an upwind and downwind position of the sensors in the gearbox of WTs. (**a**) Fourier spectrum at 1514 RPM and 1037.4 kW on 25 December 2023, in an upwind position. (**b**) Fourier spectrum at 1577 RPM and 1169.4 kW on 18 January 2024, in an upwind position. (**c**) Fourier spectrum at 1492 RPM and 1027.6 kW on 26 January 2024, in an upwind position. (**d**) Fourier spectrum at 1552 RPM and 1158 kW, on 30 January 2024, in an upwind position. (**e**) Fourier spectrum at 1523 RPM and 1054 kW on 25 December 2023, in a downwind position. (**f**) Fourier spectrum at 1455 RPM and 971.4 kW on 11 January 2024, in a downwind position. (**g**) Fourier spectrum at 1471 RPM and 982 kW on 18 January 2024, in a downwind position. (**h**) Fourier spectrum at 1481 RPM and 1006.5 kW on 26 January 2024, in a downwind position.

3.2. Construct the Objective Functions FO for All Selected Fourier Spectra

To construct the FO for the data acquisition and establish the trend of the maximum values of the vibration magnitude vs. frequency, four Fourier spectra were used for the upwind and downwind bearings; see Figure 15. The results of FO_i are shown in Figures 16–19.

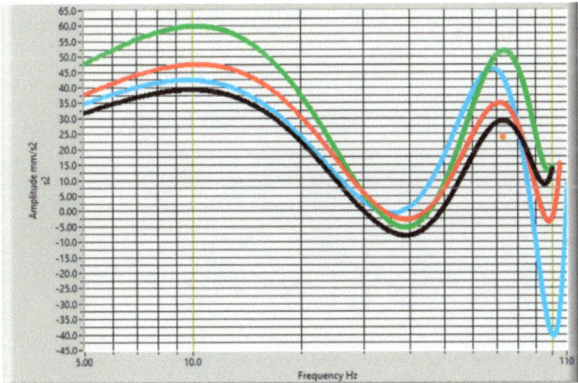

Figure 16. Objective functions (FO_i) for all four selected acquisition data spectra in the upwind sensor position.

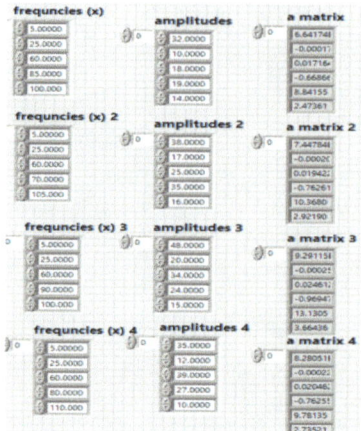

Figure 17. The front panel of the used LabView *VI*-s with input and output data for the upwind position sensor.

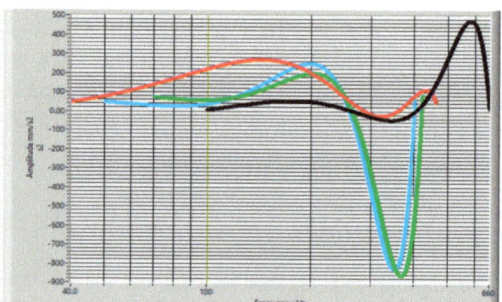

Figure 18. Objective functions (FO_i) for all four selected acquisition data spectra in a downwind sensor position.

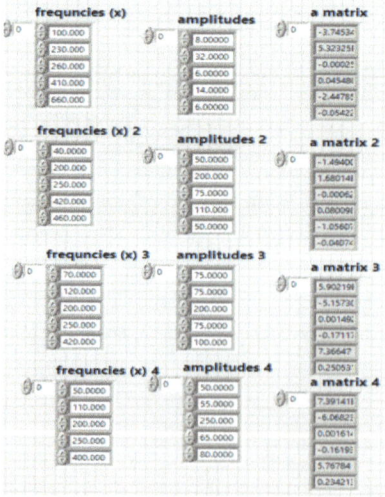

Figure 19. The front panel of the used LabView *VI*-s with input and output data for the downwind position sensor.

To validate the mathematical vibration model proposed (Figure 13), the vibration data are obtained from the CMS of a 2.0 MW industrial WT gearbox, based on the acceleration position and data acquisition shown in Figures 2 and 5. The gearbox is a planetary type with a transmission ratio of 116. This model was applied to synthesize the data acquisition of the wind turbine in the period between December 2023 and February 2024. The conditions that were imposed are the following: (i) the data acquisition was for the same, or very similar, wind turbines; (ii) the data acquisition was carried out from the sensors on the gearbox, B3-LSS and B5-HSS, with upwind LSS bearing radial and similarly downwind HSS bearing radial; (iii) the data acquisition was performed in the similar dynamic conditions of wind intensity, speed, and power; (iv) the acquisition data that was synthesized are the data that fall under the condition to be classified as a meaningful point, $x_i \in$ group 1, $x_i \in DBSVM$; see Figure 14.

Using the data from the column matrices a_i, the fifth-order equation for FO_i will be determined. The FO_i for the upwind position of the sensor is shown in relation (11) and for the downwind position in relation (12).

$$FO = 6.64x^5 - 0.00017x^4 + 0.017x^3 - 0.0668x^2 + 8.84x + 2.473 \tag{11}$$

$$FO = 7.44x^5 - 0.0002x^4 + 0.019x^3 - 0.762x^2 + 10.368x + 2.921$$

$$FO = 9.291x^5 - 0.00025x^4 + 0.00246x^3 - 0.969x^2 + 13.1305x + 3.664$$

$$FO = 8.28x^5 - 0.00022x^4 + 0.02x^3 - 0.762x^2 + 9.781x + 2.735$$

$$FO = -3.745x^5 + 5.323x^4 - 0.00025x^3 + 0.045x^2 - 2.447x - 0.054 \tag{12}$$

$$FO = -1.494x^5 + 1.68x^4 - 0.00062x^3 + 0.08x^2 - 1.056x - 0.04$$

$$FO = 5.902x^5 - 5.157x^4 + 0.00014x^3 - 0.171x^2 + 7.366x + 0.25$$

$$FO = 7.391x^5 - 6.068x^4 + 0.00161x^3 - 0.161x^2 + 5.767x + 0.234$$

All determined FOs represent different stages of the mechanical condition of the turbine gearbox assembly.

3.3. Determine the FO for the Trend

With the help of these functions, the trend of potential defects in the turbine gearbox area can be assessed. The characteristic frequencies of the WT gearbox in the damage case are presented in Figures 20 and 21, corresponding to LSS-upwind and HSS-downwind. The frequency spectrum of acceleration for the LSS-upwind position shows the fundamental frequency of the planet pin (Figure 20) and the frequency spectrum of the HSS-downwind position shows the existence of the gear mesh frequency (GMF) generated by the HSS pin gear and planet pin gear. In the case of a faulty gear, the amplitude is much higher, reaching up to 10 times higher than in the normal condition case.

At any given moment, it is possible to check whether the function is approaching the period close to the appearance of a defect or not [4,37]. Throughout this timeframe, it will be possible to examine whether the points (frequency, magnitude) fall within the first or last FO or between them, providing information on the proximity of a potential defect, as per relations (11) and (12). The trends of these functions are depicted in Figure 22, represented by the maximum of the FO for each of the cases.

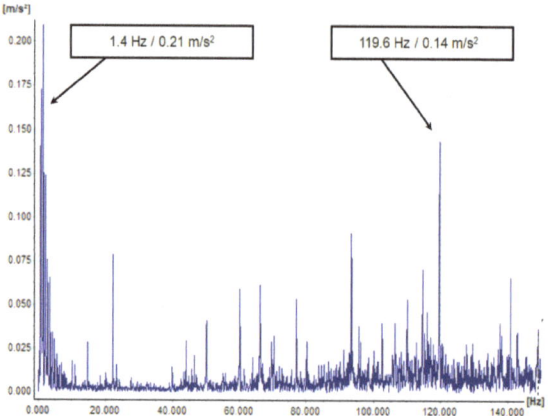

Figure 20. The frequency spectrum at the LSS position in the case of gearbox defect.

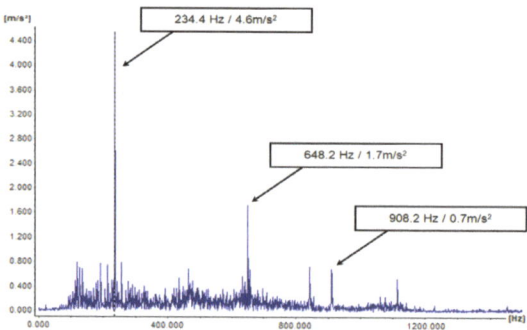

Figure 21. The frequency spectrum at the HSS position in the case of gearbox defect.

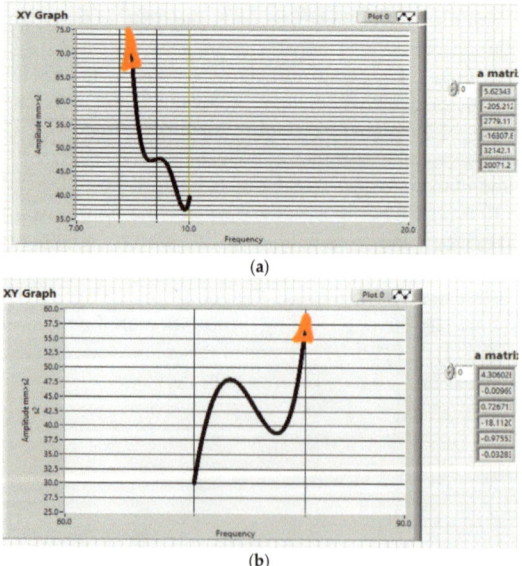

Figure 22. *Cont.*

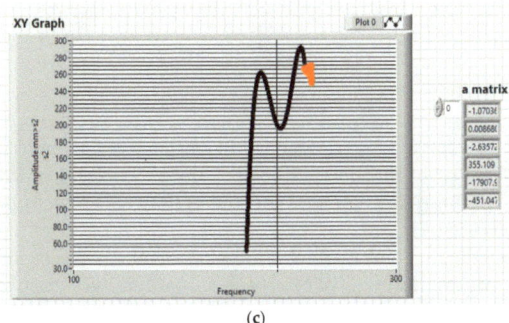

(c)

Figure 22. The trend of the magnitude–frequency points from the FO. (**a**) Trend of the FO in the upwind position of the gearbox sensor in a low frequency. (**b**) Trend of the FO in the upwind position of the gearbox sensor in a high frequency. (**c**) Trend of the FO in the downwind position of the gearbox sensor.

The trend functions are the following:

- for the low frequency in the upwind position,

$$FO = 5.6234x^5 - 205.21x^4 + 2779.11x^3 - 16307.64x^2 + 32142.12x + 20071.2 \qquad (13)$$

- for high frequency in the upwind position,

$$FO = 4.306x^5 - 0.0096x^4 + 0.7267x^3 - 18.112x^2 - 0.9755x - 0.0328$$

- for high frequency in the downwind position,

$$FO = -1.0703x^5 + 0.0086803x^4 - 2.6357x^3 + 355.109x^2 - 17907.9x - 451.047$$

$$(Magnitude, frequency)_{icatastrophic\ wear_{upwind\ or\ downwind}} \in FO_{trend_{upwind\ or\ downwind}} \qquad (14)$$

If the first FO objective function is defined after an intervention when the gearbox is working correctly, and the last function is determined close to the appearance of a defect, the position of any point (frequency, amplitude) can be determined between these limits. Intermediate FOs define the intermediate limits. Using this method, it will be possible to implement preventive maintenance and also monitor the normal operation of the gearbox of the wind turbine. The validation of this developed method can be carried out by checking whether the maximum points (frequency, magnitude) from the Fourier spectrum belonging to a certain trend as identified by the objective functions, correspond to any known instances of gearbox malfunction or failure in wind turbines. This would be performed through a collaboration with a wind turbine expert.

4. Conclusions and Future Work

This paper presents a novel approach to addressing the complexities of vibration monitoring and analysis in wind turbine gearboxes. By leveraging mathematical modeling and AI techniques, we have developed a method for evaluating gearbox conditions during operation that can help make meaningful interpretations from uncategorized vibration data from wind turbines. After analyzing the obtained results, the objective functions, and the trend of the monitoring results, we can make the following remarks: (i) the applied method is general and can be applied to many other dynamic monitoring processes; (ii) the designed LabView instrumentation for the synthetic analysis of the obtained acquisition data opens the way to applying more virtual instrumentation in monitoring the dynamic behavior across various mechanical fields; (iii) using DBSVMs to filter out the meaningful data adds a new front to applying machine learning in monitoring processes; (iv) establishing the trend of the FO for each position of the gearbox sensors ensures the design of an intelligent monitoring system for predictive maintenance; (v) the trend for the low frequency in the upwind sensor position is a decrease in frequency and an increase in magnitude; (vi) conversely, the trend involves an increase in both frequency and magnitude for the high frequency; and (vii) in the downwind sensor position, the trend is characterized by an increase in frequency and a decrease in magnitude.

In future work, we propose the generalization of this method and leveraging of neural networks for the rapid establishment of weight matrices, objective functions, and wear trends in wind turbines across all sensors. This will be integrated into a comprehensive matrix comprising objective functions, alongside a monitoring and trend matrix.

In the next stage of this research, SVM Regression analysis will be implemented to predict the magnitude of vibrations based on various input features (e.g., frequency, time). This information will help obtain a quantitative measure of potential defects. Upon a further assessment of the FFT spectra of vibrations leading up to failures or defects, we also aim to study and explore other features (fluctuations in phase, etc.) that could indicate upcoming defects. This condition-based maintenance strategy can also be further enhanced by incorporating supervised classification. We plan to label the datasets indicating different points (labeled points) in time leading up to the developing fault. This would be conducted through collaboration with industry specialists. The classification algorithm can be employed to identify the definite states of the system (normal operation, potential fault, critical fault). The combination of regression and classification would allow for a more comprehensive predictive maintenance approach.

The proposed method is intended to be applied in other industrial applications in the case of condition monitoring of machine tool spindles.

Author Contributions: Conceptualization, C.B. and A.O.; methodology, C.B. and A.O.; software, C.B. and A.O.; validation, C.B., A.O., S.O. and A.A.; formal analysis A.O., S.O. and A.A.; investigation, C.B., A.O., S.O., A.A., N.M. and H.U.; resources, C.B.; writing—original draft preparation, A.O. and C.B.; writing—review and editing, C.B., A.O., S.O., A.A., N.M. and H.U.; supervision, A.O.; funding acquisition, C.B. and A.O. All authors have read and agreed to the published version of the manuscript.

Funding: This work was supported by a grant from the National Program for Research of the National Association of Technical Universities—GNAC ARUT 2023.

Data Availability Statement: The requested data are currently unavailable as the owner has deemed them confidential for economic reasons.

Conflicts of Interest: The authors declare no conflicts of interest.

References

1. Nie, Y.; Li, F.; Wang, L.; Li, J.; Sun, M.; Wang, M.; Li, J. A mathematical model of vibration signal for multistage wind turbine gearboxes with transmission path effect analysis. *Mech. Mach. Theory* **2021**, *167*, 104428. [CrossRef]
2. Spinato, F.; Tavner, P.; van Bussel, G.; Koutoulakos, E. Reliability of wind turbine subassemblies. *IET Renew. Power Gener.* **2009**, *3*, 387–401. [CrossRef]
3. Dao, C.; Kazemtabrizi, B.; Crabtree, C. Wind turbine reliability data review and impacts on levelised cost of energy. *Wind Energy* **2019**, *22*, 1848–1871. [CrossRef]
4. Ren, B.; Chi, Y.; Zhou, N.; Wang, Q.; Wang, T.; Luo, Y.; Ye, J.; Zhu, X. Machine learning applications in health monitoring of renewable energy systems. *Renew. Sustain. Energy Rev.* **2024**, *189*, 114039. [CrossRef]
5. Sheng, S.; Veers, P. Wind Turbine Drivetrain Condition Monitoring—An Overview. In Proceedings of the Mechanical Failures Prevention Group: Applied Systems Health Management Conference 2011, Virginia Beach, VA, USA, 10–12 May 2011.
6. Zhu, Y.; Xie, B.; Wang, A.; Qian, Z. Fault diagnosis of wind turbine gearbox under limited labeled data through temporal predictive and similarity contrast learning embedded with self-attention mechanism. *Expert Syst. Appl.* **2024**, *245*, 123080. [CrossRef]
7. Teng, W.; Ding, X.; Zhang, X.; Liu, Y.; Ma, Z. Multi-fault detection and failure analysis of wind turbine gearbox using complex wavelet transform. *Renew. Energy* **2016**, *93*, 591–598. [CrossRef]
8. Vamsi, I.; Sabareesh, G.; Penumakala, P. Comparison of condition monitoring techniques in assessing fault severity for a wind turbine gearbox under non-stationary loading. *Mech. Syst. Signal Process.* **2019**, *124*, 1–20. [CrossRef]
9. Teng, W.; Ding, X.; Tang, S.; Xu, J.; Shi, B.; Liu, Y. Vibration Analysis for Fault Detection of Wind Turbine Drivetrains—A Comprehensive Investigation. *Sensors* **2021**, *21*, 1686. [CrossRef] [PubMed]

10. Liu, L.; Liang, X.; Zuo, M.J. Vibration signal modeling of a planetary gear set with transmission path effect analysis. *Measurement* **2016**, *85*, 20–31. [CrossRef]
11. Liang, X.; Zuo, M.J.; Liu, L. A windowing and mapping strategy for gear tooth fault detection of a planetary gearbox. *Mech. Syst. Signal Process.* **2016**, *80*, 445–459. [CrossRef]
12. Feng, Z.; Ma, H.; Zuo, M.J. Vibration signal models for fault diagnosis of planet bearings. *J. Sound Vib.* **2016**, *370*, 372–393. [CrossRef]
13. Wang, T.; Han, Q.; Chu, F.; Feng, Z. Vibration based condition monitoring and fault diagnosis of wind turbine planetary gearbox: A review. *Mech. Syst. Signal Process.* **2019**, *126*, 662–685. [CrossRef]
14. Salameh, J.P.; Cauet, S.; Etien, E.; Sakout, A.; Rambault, L. Gearbox condition monitoring in wind turbines: A review. *Mech. Syst. Signal Process.* **2018**, *111*, 251–264. [CrossRef]
15. Hameed, Z.; Hong, Y.; Cho, Y.; Ahn, S.; Song, C. Condition monitoring and fault detection of wind turbines and related algorithms: A review. *Renew. Sustain. Energy Rev.* **2009**, *13*, 1–39. [CrossRef]
16. Zhang, M.; Wang, K.; Wei, D.; Zuo, M.J. Amplitudes of characteristic frequencies for fault diagnosis of planetary gearbox. *J. Sound Vib.* **2018**, *432*, 119–132. [CrossRef]
17. McNiff, B.; Keller, J.; Fernandez-Sison, A.; Demtroder, J. A Revised International Standard for Gearboxes in Wind Turbine Systems. In Proceedings of the Conference for Wind Power Drives, Aachen, Germany, 21–22 March 2023.
18. Cao, L.; Liu, S. Vibration Suppression of an Input-Constrained Wind Turbine Blade System. *Mathematics* **2023**, *11*, 3946. [CrossRef]
19. Peng, J.; Bian, Y.; Tian, P.; Liu, P.; Gao, Z. Vibration alleviation for wind turbine gearbox with flexible suspensions based on modal interaction. *J. Low Freq. Noise Vib. Act. Control* **2023**, *42*, 1390–1418. [CrossRef]
20. Global Energy Research Council. Global Wind Report. 2009. Available online: http://www.gwec.net/fileadmin/documents/Publications/Global_Wind_2007_report/GWEC_Global_Wind_2009_Report_Lowres_15th.520Apr./pdf (accessed on 1 April 2024).
21. *BS ISO 10816-21:2015*; Mechanical Vibration. Evaluation of Machine Vibration by Measurements on Non-Rotating Parts Horizontal Axis Wind Turbines with Gearbox. International Organization for Standardization: Geneva, Switzerland, 2015.
22. Verbruggen, T.W. Condition Monitoring: Theory and Practice. In Proceedings of the 2009 Wind Turbine Condition Monitoring Workshop, Broomfield, CO, USA, 8–9 October 2009.
23. Germanischer Lloyd. *Guideline for the Certification of Condition Monitoring Systems for Wind Turbines*; Germanischer Lloyd WindEnergie GmbH: Hamburg, Germany, 2007.
24. Gellermann, T.; Walter, G. *Requirements for Condition Monitoring Systems for Wind Turbines*; AZT Report No. 03.01.068; AZT: London, UK, 2003.
25. Wind Stats Newsletter, 2003–2009, Vol. 16, No. 1 to Vol. 22, No. 4. Haymarket Business, Media: London, UK. Available online: https://www.nrel.gov/docs/fy13osti/58774.pdf (accessed on 2 April 2024).
26. Veers, P. Databases for Use in Wind Plant Reliability Improvement. In Proceedings of the 2009 Wind Turbine Condition Monitoring Workshop, Broomfield, CO, USA, 8–9 October 2009.
27. Bisu, C.F.; Zapciu, M.; Cahuc, O.; Gérard, A.; Anica, M. Envelope Dynamic Analysis: A New Approach for Milling Process Monitoring. *Int. J. Adv. Manuf. Technol.* **2011**, *62*, 471–486. [CrossRef]
28. Zhang, M.; Cui, H.; Li, Q.; Liu, J.; Wang, K.; Wang, Y. An improved sideband energy ratio for fault diagnosis of planetary gearboxes. *J. Sound Vib.* **2021**, *491*, 115712. [CrossRef]
29. Pattabiraman, T.R.; Srinivasan, K.; Malarmohan, K. Assessment of sideband energy ratio technique in detection of wind turbine gear defects. *Case Stud. Mech. Syst. Signal Process.* **2015**, *2*, 1–11.
30. Hanna, J.; Hatch, C.; Kalb, M. *Detection of Wind Turbine Gear Tooth Defects Using Sideband Energy Ratio*; GE Enregy: New York, NY, USA, 2012.
31. Verbruggen, T.W. *Wind Turbine Operation and Maintenance Based on Condition Monitoring*; Energy Research Center of the Netherlands: Petten, The Netherlands, 2003. Available online: https://www.ecn.nl/publicaties/PdfFetch.aspx?nr=ECN-C--03-047 (accessed on 20 April 2024).
32. Dodge, Y. *The Concise Encyclopedia of Statistics*; Springer: Berlin/Heidelberg, Germany, 2008.
33. Nazari, Z.; Kang, D.; Endo, H. Density Based Support Vector Machines. In Proceedings of the 29th International Technical Conference on Circuits/Systems, Computers and Communications (ITC-CSCC), Phuket, Thailand, 1–4 July 2014; pp. 1–3.
34. El Moutaouakil, K.; El Ouissari, A.; Olaru, A.; Palade, V.; Ciorei, M. OPT-RNN-DBSVM: OPTimal Recurrent Neural Network and Density-Based Support Vector Machine. *Mathematics* **2023**, *11*, 3555. [CrossRef]
35. Available online: https://www.geeksforgeeks.org/covariance-matrix/ (accessed on 2 April 2024).
36. Nazari, Z.; Kang, D. Density Based Support Vector Machines for Classification. *Int. J. Adv. Res. Artif. Intell.* **2015**, *4*, 040411. [CrossRef]
37. Valikhani, M.; Jahangiri, V.; Ebrahimian, H.; Moaveni, B.; Liberatore, S.; Hines, E. Inverse modeling of wind turbine drivetrain from numerical data using Bayesian inference. *Renew. Sustain. Energy Rev.* **2023**, *171*, 113007. [CrossRef]

Disclaimer/Publisher's Note: The statements, opinions and data contained in all publications are solely those of the individual author(s) and contributor(s) and not of MDPI and/or the editor(s). MDPI and/or the editor(s) disclaim responsibility for any injury to people or property resulting from any ideas, methods, instructions or products referred to in the content.

Article

Real-Time EtherCAT-Based Control Architecture for Electro-Hydraulic Humanoid

Maysoon Ghandour [1,*,†], Subhi Jleilaty [1], Naima Ait Oufroukh [1], Serban Olaru [2] and Samer Alfayad [1]

1 IBISC Laboratory, University of Evry, University of Paris Saclay, 91034 Évry-Courcouronnes, France; jleilatysubhi@gmail.com (S.J.); naima.aitoufroukh@univ-evry.fr (N.A.O.); samer.alfayad@univ-evry.fr (S.A.)
2 Department of Robotics and Production System, National University of Science and Technology Politecnica Bucharest, 060042 Bucharest, Romania; serban1978@yahoo.com
* Correspondence: maysoon.ghandour@gmail.com
† Current address: IBISC Laboratory, 40 rue de Pelvoux, 91080 Évry-Courcouronnes, France.

Abstract: Electro-hydraulic actuators have witnessed significant development over recent years due to their remarkable abilities to perform complex and dynamic movements. Integrating such an actuator in humanoids is highly beneficial, leading to a humanoid capable of performing complex tasks requiring high force. This highlights the importance of safety, especially since high power output and safe interaction seem to be contradictory; the greater the robot's ability to generate high dynamic movements, the more difficult it is to achieve safety, as this requires managing a large amount of motor energy before, during, and after the collision. No matter what technology or algorithm is used to achieve safety, none can be implemented without a stable control system. Hence, one of the main parameters remains the quality and reliability of the robot's control architecture through handling a huge amount of data without system failure. This paper addresses the development of a stable control architecture that ensures, in later stages, that the safety algorithm is implemented correctly. The optimum control architecture to utilize and ensure the maximum benefit of electro-hydraulic actuators in humanoid robots is one of the important subjects in this field. For a stable and safe functioning of the humanoid, the development of the control architecture and the communication between the different components should adhere to some requirements such as stability, robustness, speed, and reduced complexity, ensuring the easy addition of numerous components. This paper presents the developed control architecture for an underdeveloped electro-hydraulic actuated humanoid. The proposed solution has the advantage of being a distributed, real-time, open-source, modular, and adaptable control architecture, enabling simple integration of numerous sensors and actuators to emulate human actions and safely interact with them. The contribution of this paper is an enhancement of the updated rate compared to other humanoids by 20% and by 40 % in the latency of the master. The results demonstrate the potential of using EtherCAT fieldbus and open-source software to develop a stable robot control architecture capable of integrating safety and security algorithms in later stages.

Keywords: humanoid; real-time software; control system architecture-based EtherCAT

MSC: 28-06

Citation: Ghandour, M.; Jleilaty, S.; Ait Oufroukh, N.; Olaru, S.; Alfayad, S. Real-Time EtherCAT-Based Control Architecture for Electro-Hydraulic Humanoid. *Mathematics* 2024, 12, 1405. https://doi.org/10.3390/math12091405

Academic Editor: Junyong Zhai

Received: 31 March 2024
Revised: 25 April 2024
Accepted: 30 April 2024
Published: 3 May 2024

Copyright: © 2024 by the authors. Licensee MDPI, Basel, Switzerland. This article is an open access article distributed under the terms and conditions of the Creative Commons Attribution (CC BY) license (https://creativecommons.org/licenses/by/4.0/).

1. Introduction

The high demand for high-performance robots resulted in the growth of the development of hardware, software, and control architecture, ensuring a safe and stable interaction of the robot with the environment. Several research efforts have been targeted to develop and evolve the hardware and software of robotic systems. Developing a humanoid capable of safely performing human tasks, such as navigating rough terrains and interacting socially with humans, may require numerous degrees of freedom. A robot's control architecture aims to organize and distribute the multiple controllers responsible for controlling the actuators and sensors. This is carried out while considering their communication to ensure that all components work towards the overall objective. Therefore, the development of the control architecture for numerous actuated joints is challenging because of the requirements that the system will impose.

HYDROïD, shown in Figure 1 is a hydraulically actuated full-size humanoid robot with 51 degrees of freedom designed to operate in dangerous environments, assist the elderly, and support human needs in industry [1]. Therefore, developing a safe interaction capability for its joints is necessary. To ensure dynamic motion in HYDROïD, all the mechatronics subsystems in a robot should be highly reactive, starting from the sensors and actuators and going into the software architecture of the robot.

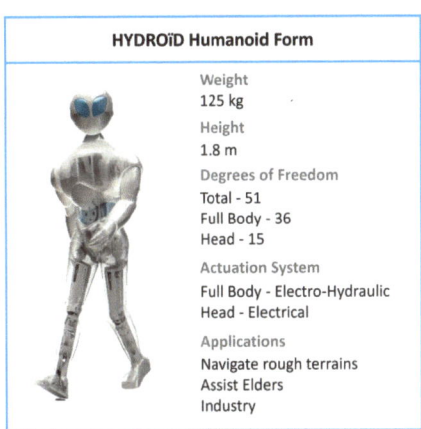

Figure 1. HYDROïD humanoid general description.

The actuation system in humanoids plays a crucial role in determining their performance and the overall robot capability. Although hydraulic actuators have proven better performance than electrical actuators in force, power-to-weight, and power-to-volume ratio, they still suffer from oil leakage, control complexity, and decreased social acceptance. One of the main issues in hydraulic actuators is the losses generated from using only one pump that activates all the robot's joints where many hoses are used to drive the hydraulic power; each hydraulic cylinder requires two hoses and thus four connection points, all of which are susceptible to leakage. This is also a drawback from the energy efficiency point of view. However, these issues are resolved in HYDROïD's latest generation, where the hydraulic power is locally generated at the actuated joint. This decreases the need for hydraulic tubes and gives the joint the required power to enhance energy efficiency. This is achieved by developing a patented actuation system called Servo Electro-Hydraulic Actuator (SEHA) [2]. SEHA is an all-in-one compact actuator that is characterized by increased power-to-weight and power-to-volume ratios, increased safety through a force compensation module integrated into the actuator that is activated in case pressure exceeds a preset value, and enhanced joint movement through a flexible control, making it a suitable actuator for robotics applications and especially those performing heavy-duty tasks.

Moreover, the control architecture is responsible for the communication between the hardware layer represented by the actuator, sensors, and low-level controllers, and the higher-level software managing hardware. The control architecture aims to address the numerous actuators and sensors utilized in the robot and ensure the maximum benefit, which is one of the critical issues in developing the dynamic motion of robots. This raises the question of how to organize and physically distribute the robot's controllers to achieve its dynamic balance and ability to interact safely with the user. The control architecture has to be reliable in that it is able to handle huge amounts of data from sensors and actuators without system failure.

One of the additional challenges encountered in developing humanoid robots is the crucial need for a real-time environment that will adopt the controller and ensure the robot's stability and safety. This requirement is essential when there is a complex robot that will collaborate with humans where the real-time environment will ensure the safety of the robot's interaction. Hence, developing a Real-Time Operating System (RTOS) is required to ensure deterministic behavior by handling the interrupts with predefined time [3]. RTOS is designed for applications requiring immediate critical task processing within a predefined and guaranteed time frame. This is particularly important in the case of humanoids, as they need to respond precisely to interact safely with their environment and perform complex tasks. The real-time concept is required at the low-level joint controller, the high-level master controller, and the communication between these two layers.

In general, the high-level master controller is restricted by the low-level controller as it is built on top of it. The low update rate achieved at the low-level controller will limit the control system's performance. The update rate, which refers to the frequency at which the data are updated by the system, is constrained by the communication field bus utilized. Despite all the improvements in modern communication and the low-level controller, many robots still have relatively low update rates, often restricted by the communication fieldbus. Therefore, the communication protocol used in humanoid should adhere to some requirements, such as being stable, robust, fast, and capable of handling all the data needed to be transmitted.

This paper presents a real-time control architecture based on the EtherCAT fieldbus communication protocol for an electro-hydraulic humanoid robot HYDROïD. The presented work includes the implementation on both the joint and software levels, as well as the communication interface between them. Hence, a distributed, real-time, open-source, modular, and adaptable control architecture is proposed, enabling the simple integration of numerous sensors and actuators to emulate human actions while ensuring safe collaboration between humans and robots in several tasks.

Section 2 presents the previous works of the implemented control architecture of humanoid robots. The mechatronics overview of HYDROïD is presented in Section 3, the modeling and simulation of one joint, and the Inverse Geometric Model and Inverse Kinematic Model of the hybrid ankle mechanism are provided in Section 4, while the proposed real-time control architecture is presented in Section 5. The conducted experiment and the results are shown in Section 6. And finally, we conclude and present the future work in Section 7.

2. Previous Works

The dynamic balance of the robot, as well as its ability to safely interact with the user, depend mainly on the developed control architecture and its distribution in the robot. Hence, a centralized approach was first suggested. This approach involves a single central unit that controls all the robot joints. NAO [4], HRP-2 [5], and PETMAN [6] adopted this approach. It is simple but limited due to the high computations required, especially in the case of many degrees of freedom. Moreover, the failure of this controller will lead to the failure of the whole system. Hence, the decentralized approach was introduced, where multiple processors are distributed for each joint or multiple joints, each of which operates with a degree of authority. This approach was used in Valkryie [7] and LOLA [8] humanoids. Compared to the centralized approach, this approach will allow more efficient control of the robot joints with the processing of the sensor data without the high computational power required in the centralized approach. However, an issue of coordination emerged mainly when the robot was performing complex tasks since some controllers may have conflicting goals affecting the overall behavior of the robot. Also, the system will introduce some delays due to the required communication between the multiple controllers. Hence, the researcher's efforts led to the development of the distributed control architecture. In this architecture, multiple controllers are distributed among the robot joints. Each controller operates independently and communicates with each other in a peer-to-peer manner, with the ability to make its own decisions. Each controller can join or leave the system without affecting the overall performance. However, coordination and synchronization among distributed agents can be complex and require robust communication mechanisms. It is worth mentioning that in HRP-2, the centralized control system was initially adopted. However, some problems occurred in the electrical system, leading to the disconnection between the interface boards mounted in the main computer and the sensors/motors, and the HRP-2 became out of control. Hence, they shifted to distributed control architecture in HRP-3p and HRP-5P [9]. Overall, the control architecture topology for humanoid robots has progressed with several available solutions; each has its own advantages, and deciding which approach should be taken is highly important.

One of the additional challenges encountered when designing the robot's software is the crucial need for a real-time environment for implementing the controller and ensuring the robot's stability. This requirement is essential when there is a complex robot that will collaborate with humans, and the real-time environment is indispensable for ensuring the safety of the robot's interaction. Hence, developing a Real-Time Operating System (RTOS) is required. The development of the RTOS is made on either the high-level master controller or the joint controller. The choice of the operating system at the high level or the main PC is highly important. One of the common practices is adopting open-source operating systems such as Ubuntu, as it is considered cost-effective and flexible. However, Ubuntu is not a real-time OS, but several approaches can be taken to support Linux in real time. This includes the RT Patch, GPL, RTAI, and Xenomia. These approaches are based either on the double-kernel method, where a real-time kernel is installed into the standard operating system kernel, or on the enhancement of the Linux scheduler. RT Patch can fulfill real-time requirements for periodic tasks with a minimal jitter for long-term measurements, increasing confidence in the stability of this approach [10].

The communication protocol handles the data transmission between components and elements like sensors, actuators, and computational units. The Controller Area Network (CAN) is one of the most popular protocols. The CAN bus network provides a cost-effective networking solution for low-bandwidth applications where the bandwidth is 1 Mbps. It is a serial communications protocol that supports distributed real-time. CAN Bus protocol was used in different robots like HUBO [11], KHR-2 [12], PETMAN, and HRP-3. It provides robust communication; however, it is limited in bandwidth. Moreover, the Powerlink protocol [13] is one of the well-known protocols for real-time motion control. It is an open-source library that can be integrated into developing a real-time system. The performance of this protocol depends on the topology used, and it needs high computational power. Finally, the Ethernet for Control Automation Technology (EtherCAT) protocol [14] from Beckhoff Automation is a well-supported protocol that meets the demands of real-time and high-bandwidth applications. It can achieve a bandwidth of up to 1 Gbps or even 10 Gbps. This communication protocol was used in the Atlas [15], TALOS [16], Hydra [17], TOCABI [18], and WalkMan [19] robots.

The main control frequency of the HUBO robot using CANbus is 100 Hz, while the control frequency of the joint motor is 1 kHz. The control loops in iCub with CAN communication protocol operate at 100 Hz. Escher humanoid uses CANopen, and the configuration values and joint space set points are transmitted at 500 Hz rate [20]. Petman from Boston Dynamics utilized a modified CANbus, and the update rate was 1 kHz. In general, the limitations in the bandwidth of CANbus and that it does not allow handling multiple point-to-point connections were major drawbacks [21].

Several efforts were made to enhance the performance of this protocol using multiple parallel CAN networks, but the update rate was relatively low. Also, Sercos-II, with a bandwidth of 16 Mbit/s, is used for the robot TORO, and the achieved update rate is 1 kHz. LOLA robot chose the SERCOS-III protocol to resolve the issues of CANbus, and they achieved a 1 kHz update rate. EtherCAT bus is used in RoboSimian [22], Hydra, and ARMAR-6 [23], and the update rate achieved in these robots is 1 kHz. Some robots switched to EtherCAT to take advantage of this protocol. For example, Boston Dynamics switched from CANBus in Petman into EtherCAT in Atlas, and the LOLA robot switched to EtherCAT, achieving a 2 kHz update rate [8]. Higher update rates were achieved using dual channel EtherCAT in the TOCABI robot, where a 4 kHz update rate was achieved.

3. HYDROïD's Mechatronics Overview

HYDROïD is a humanoid robot that comprises 51 DoFs emulating human joints. 36 DoF of which are hydraulically actuated, composing the body, and 15 electrically actuated, composing the head. In this paper, our primary focus is on controlling the hydraulically actuated joints. The initial version of HYDROïD is based on an electro-hydraulic actuation system with a double-stage servo valve. The mechanisms in HYDROïD and the DoFs [24] are shown in Table 1. The robot and its kinematic structure are shown in Figure 2.

Table 1. Representation of HYDROïD's DoF.

Mechanism	DoF/Mechanism	Quantity	Total DoF
Toe	1	2	2
Ankle	3	2	6
Knee	1	2	2
Hip	3	2	6
Torso	4	1	4
Shoulder	4	2	8
Elbow	1	2	2
Wrist	3	2	6
			36

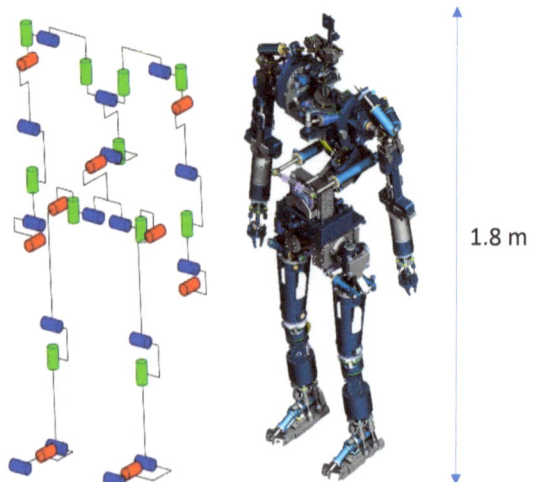

Figure 2. HYDROïD's kinematic structure.

These mechanisms were developed upon studying human morphology and appearance, leading to compact, lightweight, and modular mechanisms that can achieve the required range of motion, torque, and speed.

Hydroid consists mainly of two hybrid mechanisms, each with a rotating hydraulic actuator carrying a parallel structure. The first mechanism is dedicated to the hip, shoulder, and torso, while the second was chosen for the ankle and the wrist.

In the hybrid mechanism of the hip, shoulder, and torso, the requirement is a wide range of motion in the pitch axis, while for the yaw and roll axes, nearly the same range of motion is required. Hence, the choice of the rotary actuator for these mechanisms is for achieving a wide range of motion in the pitch axes, and two other linear actuators are used for the yaw and roll rotation [25].

On the other hand, for the ankle and wrist mechanisms, the small space allocated is a challenge. Also, the center of gravity for these mechanisms is preferred to be closer to the knee/elbow to reduce energy consumption, but the range of motion on the three axes is nearly the same. Hence, the rotary actuator for these mechanisms is chosen for the roll rotation, and it is placed near the knee/elbow, while four other linear actuators are implemented in parallel structure and responsible for transmitting the motion to the ankle/wrist in the roll and pitch rotations. The choice of the rotary actuator leads to a compact design that can be easily integrated into the wrist and the ankle while achieving the needed torque [26].

4. Control Architecture Development Methodology

Understanding the mechanisms and the actuator utilized in the robot is an essential step before designing the control architecture. It is important to comprehend the complexity imposed by the control, which is essential to develop an optimized and efficient control architecture. Rough calculations of the computational cost that can be added to the system are essential for determining the distribution of the controllers in later stages.

Considering the mechanical design of HYDROïD, there are two existing layers: the joint layer, represented by the servo valve and the hydraulic actuator, and the mechanisms, such as the mechanism of the wrist, ankle, hip, and shoulder. Figure 3 shows an example of HYDROïD's leg; at the joint level, we have a servo valve that takes current i as input and gives the flow Q as output, leading to a motion of the whole mechanism. The complexity of the control will impact the decisions made in later stages regarding the control architecture. For this purpose, a study is conducted in this section regarding (i) the inverse kinematics and inverse geometry at the ankle mechanism and (ii) the simulation and modeling at the joint level containing a servo valve and rotary hydraulic actuator. This study is beneficial for understanding the complexity and calculating the computational cost for developing the control architecture.

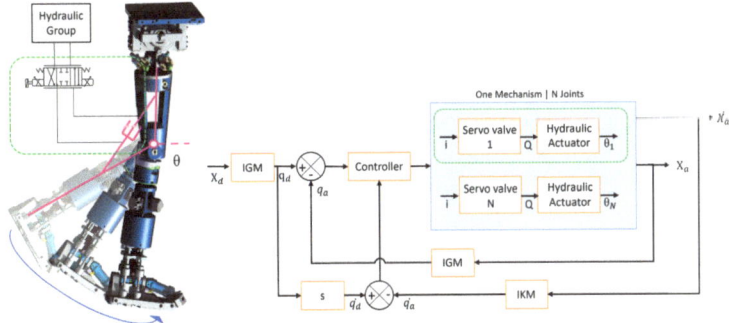

Figure 3. Control of one mechanism in HYDROïD.

4.1. Kinematic and Inverse Geometric Model of Hybrid Mechanism

The ankle mechanism of HYDROïD is a 3 DoF hybrid mechanism achieved with serial and parallel substructures. The merge of the serial and parallel mechanisms leads to an ankle design that respects the size constraints that do not overload the actuator, producing 3 DoF motion with easy control models. This mechanism is chosen as the case study for two main reasons: (i) this mechanism is considered a complicated mechanism, and (ii) calculating the computation costs of such mechanism is beneficial in designing the control architecture. This section represents the Inverse Geometric Model(IGM) and the Inverse Kinematics Model (IKM) for the ankle mechanism.

The kinematic structure of the ankle mechanism is shown in Figure 4. In the ankle mechanism, there are four closed loops named Ch_j, and each is composed of the links: Co_1, Co_2, Co_3, Co_1^j, Co_2^j, Co_3^j, Co_4^j, Co_5^j, and Co_6^j with $j = 1, 2, 3, 4$. There are four linear actuators grouped in couples: the couple (r_1^1, r_1^3) allows the actuation of the joint q_s and the couple (r_1^2, r_1^4) allows the actuation of the joint q_f. The orientation of the end effector with respect to the base is identified with three angles, roll, yaw, and pitch, and grouped in vector $X_a = (\theta_r, \theta_p, \theta_y)$. The inputs of the hybrid mechanism are the length of the linear actuator r_{ij} and the roll rotation θ_r, while the outputs are the angles q_v, q_s, and q_f.

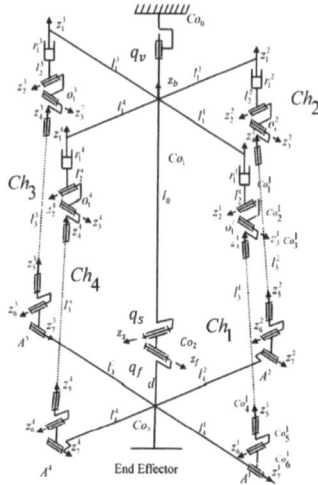

Figure 4. Kinematic structure of HYDROïD's ankle mechanism.

4.1.1. Inverse Geometric Model

The IGM is used to determine the required stroke of the linear actuators. So r_{1j} will be calculated for a given posture of the end effector through the IGM. To obtain the IGM of the ankle mechanism, the roll rotation is straightforward since it belongs to the serial part where $q_v = \theta_r$; however, for the yaw and pitch, which belong to the parallel mechanism, more calculations are needed, which be presented in this section. Upon the calculations of the IGM, the following notations are adopted:

- The jth closed kinematic chain is designated as a chain $Loop_j$ for j = 1, 2, 3, 4. The mechanism outputs are grouped into a vector q = (q_s, q_f, q_v).
- The linear joint positions are the mechanism inputs and are named r_1^j for j = 1, 2, 3, 4.
- The rotation of the ith joint in the jth closed loop is represented by θ_i^j.
- All the joints are passive joints except for r_1^j and q_v are the active joint variables.

To carry out the IGM, we consider the open mechanism maintained by breaking the z_5^j joints as shown in Figure 4. Hence, the IGM can be presented as follows: For $Loop_1$ and $Loop_3$, The rotation of the second and third joints are represented in Equations (1) and (2).

$$\theta_3^j = \arcsin(\frac{d \cdot S_{qv}}{l_3^j}) \qquad (1)$$

$$\theta_2^j = \arcsin(\frac{(d \cdot S_{qs} \cdot C_{qf}) + (l_4^j \cdot C_{qs}) - l_1^j}{l_3^j \cdot C_3^j}) \qquad (2)$$

Hence, the active joint variables are written as in Equation (3),

$$r_1^j = l_0 - l_2^j - l_4^j \cdot S_{qs} + d \cdot C_{qs} C_{qf} - l_3^j C_2^j C_3^j \tag{3}$$

Similarly, the same relations are established for $Loop_2$ and $Loop_4$. The rotations are shown in Equations (4) and (5).

$$\theta_3^j = \arcsin\left(\frac{d \cdot S_{qv} + l_4^j \cdot C_{qf} - l_1^j}{l_3^j}\right) \tag{4}$$

$$\theta_2^j = \arcsin\left(\frac{(d \cdot S_{qs} \cdot C_{qf}) - (l_4^j \cdot S_{qs} \cdot S_{qf})}{l_3^j \cdot C_3^j}\right) \tag{5}$$

The active joint variable in these loops is, therefore, presented in Equation (6).

$$r_1^j = l_0 - l_2^j - l_4^j \cdot C_{qs} \cdot S_{qf} + d \cdot C_{qs} C_{qf} - l_3^j C_2^j C_3^j \tag{6}$$

4.1.2. Inverse Kinematic Model

Upon calculating the IGM, the calculation of IKM is essential. These models are useful for the simulation of the robot, for the control, and for the motion planning of the robot. Hence, these models will be integrated into the developed control architecture. The IKM is used to determine the relation between the angular velocities around the roll, pitch, and yaw axes grouped in the vector $\dot{X}_a = (\dot{\theta}_r, \dot{\theta}_p, \dot{\theta}_r)^t$ and the active joint velocities grouped in the vector $\dot{r}_a = (\dot{r}_1^1, \dot{r}_1^2, \dot{r}_1^3, \dot{r}_1^4, \dot{r}_1^5, \dot{q}_1)^t$.

The kinematic model will be expressed as in Equation (7) for the four closed loops. Detailed calculations of the IKM can be found in the Appendix A.

$$\begin{bmatrix} L_4^1 W^1 & L_4^1 S_{qs} V^1 & 0 \\ 0 & L_4^2(C_{qs} W^2 - S_{qs} U^2) & 0 \\ -L_4^3 W^3 & -L_4^3 S_{qs} V^3 & 0 \\ 0 & L_4^4(C_{qs} W^4 - S_{qs} U^4) & 0 \\ 0 & 0 & 1 \end{bmatrix} \cdot \begin{bmatrix} \dot{q}_s \\ \dot{q}_f \\ \dot{q}_v \end{bmatrix} = \begin{bmatrix} W^1 & & 0 \\ & W^2 & \\ & W^3 & \\ & W^3 & \\ 0 & & 1 \end{bmatrix} \cdot \begin{bmatrix} \dot{r}_1^1 \\ \dot{r}_1^2 \\ \dot{r}_1^3 \\ \dot{r}_1^4 \\ \dot{q}_v \end{bmatrix} \tag{7}$$

Equation (7) has the classical matrix form shown in Equation (8).

$$A \cdot \dot{q} = B \cdot \dot{r} \tag{8}$$

Moreover, the global kinematic variable $\dot{X}$ can be written as shown in Equation (9).

$$\dot{X} = \dot{q}_s Z_b + \dot{q}_v Z_v + \dot{q}_f Z_f = \begin{bmatrix} \dot{q}_s S_{qv} + \dot{q}_f C_{qv} C_{qs} \\ -\dot{q}_s C_{qv} + \dot{q}_f S_{qv} C_{qs} \\ \dot{q}_v + \dot{q}_f S_{qs} \end{bmatrix} = \begin{bmatrix} 0 & C_{qv} C_{qs} & S_{qv} \\ 0 & S_{qv} C_{qs} & -C_{qv} \\ 1 & S_{qs} & 0 \end{bmatrix} \begin{bmatrix} \dot{q}_v \\ \dot{q}_f \\ \dot{q}_s \end{bmatrix} \tag{9}$$

Equation (9) can be reformulated and written as in 10.

$$[\dot{X} = D \cdot \dot{q}] \tag{10}$$

Finally, substituting Equation (10) in Equation (8) results in the final kinematic model of the proposed ankle mechanism, and this is presented in (11).

$$A \cdot D^{-1} \cdot \dot{X} = B \cdot \dot{r} \tag{11}$$

4.2. Modeling and Simulation of Electro-Hydraulic Actuator

The joints of HYDROïD are designed using linear hydraulic actuators [27] or rotary hydraulic actuators with Flapper-Nozzle double-stage servo valves. As rotary actuators are more complex, we will consider them for our study. A study of the servo valve's dynamic behavior is also considered to improve the dynamic performance and reduce the complexity of the implemented control. An analysis stage is an important stage in determining the influence of the parameters affecting the dynamic performance of the servo valve. This study is indispensable for ensuring the maximum benefit of the servo valve and, therefore, ensuring the best performance of the system, reducing the control algorithm complexity that will be implemented in the control architecture.

This section presents the modeling, simulation, and validation of the mathematical model; the parameters that have the greatest influence on the dynamic performance of the servo valve are also determined. An interpretation of the servo valve's dynamic behavior is based on determining the stationary error, the damping factor, the natural frequency, the proper frequency, the overshoot, the response time, the transient time, etc.

The joint level comprises a double-stage servo valve and a rotating hydraulic motor, as shown in Figure 5. Table 2 presents the notations used throughout this section.

Table 2. Terms, Notation, and Units.

Terms	Notation	Units
Geometric capacity of engine	q_m	cm^3
Gradient of flow losses proportional to the pressure in the engine	a_m	$cm^5/daNs$
Reduced moment of inertia at the shaft of rotating hydraulic motor	J_r	$daNcms^2/rad$
Moment of inertia at the motor	J_M	$daNcms^2/rad$
Moment of inertia at the stator	J_s	$daNcms^2/rad$
Resisting moment	M_r	$daNcm$
Moment of friction	M_f	$daNcms$
Gradient of moment losses proportional to angular velocity	b_m	$daNcms/rad$
Dry friction coefficient	c_{f_u}	-
Angular speed of the rotating hydraulic motor shaft	ω	rad/s
Modulus of elasticity of oil	E	daN/cm^2
Damping factor	ζ	-
Natural pulsation	ω_n	rad/s
Current output flow from servo valve	$Q_{current}$	cm^3/s
Angular acceleration	ϵ	rad/s^2
Angular space	θ	rad
Active moment	M_a	$daNcm$
Active power	N	W
Active pressure	p_M	daN/cm^2
Pressure on discharge path	p_0	daN/cm^2

The analysis of the parameters and performance of the dynamic behavior was carried out based on the analysis of the indicial characteristics of speed, acceleration, angular space, moment, power, flow, and pressure. The indicial characteristics were determined based on the indicial functions, whose relationships were established depending on the damping factor for each component, respectively, the servo valve and rotary hydraulic motor. The angular velocity indicial functions were determined taking into account the expression of the transfer function of the rotary hydraulic motor, assimilated with a PT2-type function as in Equation (12). The input is the flow from the servo valve, Q, and the output is the angular velocity of the rotating hydraulic shaft ω.

$$H(s) = \frac{\omega(s)}{Q(s)} = \frac{a_0}{b_2 s^2 + b_1 s + b_0} \quad (12)$$

where the coefficients of the transfer function of the rotary hydraulic motor are represented in Equations (13)–(16).

$$a_0 = \frac{q_m(1 - c_{f_u})}{2\pi} \quad (13)$$

$$b_0 = \left(\frac{q_m}{2\pi}\right)^2 (1 - c_{f_u}) + a_m b_m \quad (14)$$

$$b_1 = (J_m + J_s) a_m + \frac{q_m}{2E} b_m \quad (15)$$

$$b_2 = (J_m + J_s) \frac{q_m}{2E} \quad (16)$$

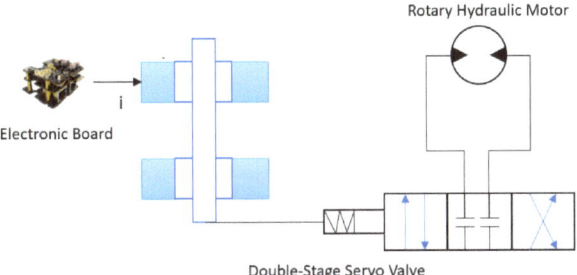

Figure 5. Simplified schematic of the robot's joint level composed of a double-stage servo valve, hydraulic rotating motor, and an electronic board.

The parameters of the transfer function, the damping factor, and the natural pulsation are shown in Equation (17).

$$\begin{aligned} \xi &= \frac{b_1}{2\sqrt{b_0 \cdot b_2}} \\ \omega_n &= \sqrt{\frac{b_0}{b_2}} \end{aligned} \quad (17)$$

For each operating point, the damping factor ξ is determined and compared with the value 1 to establish the slope of the indicial characteristic of the angular velocity ω.

If the damping factor is greater than one, then the response is over-damped. The expression for the angular velocity of the rotary hydraulic motor servo system is shown in Equation (18).

$$\omega_1 = \frac{a_0}{b_0} \cdot Q_{current} \cdot \left[1 + \frac{b_2 \cdot e^{-b_1 t}}{b_1 - b_2} - \frac{b_1 \cdot e^{-b_2 t}}{b_1 - b_2}\right] - \frac{M_f \cdot a_m \cdot \left[1 - e^{-\frac{b_0 t}{a_m \cdot (J_M + J_S)}}\right]}{b_0} \quad (18)$$

If the damping factor is unity, then the response is critically damped, and the expression for the angular velocity of the rotary hydraulic servo system is represented in Equation (19).

$$\omega_2 = \frac{a_0}{b_0} \cdot Q_{current} \cdot \left[1 - e^{-\omega_n t} \cdot (1 - \omega_n t)\right] - \frac{M_f \cdot a_m \cdot \left[1 - e^{-\frac{b_0 t}{a_m \cdot (J_M + J_S)}}\right]}{b_0} \quad (19)$$

Finally, if the damping factor is less than one, then the response is underdamped, and the expression for the angular velocity of the rotary hydraulic motor is in Equation (20).

$$\omega_3 = \frac{a_0}{b_0\sqrt{1-\xi^2}} \cdot Q_{current} \cdot \left[1 - e^{-\omega_n \xi t} \cdot \sin\left(\omega_n\sqrt{1-\xi^2} \cdot t + \arctan\frac{\sqrt{1-\xi^2}}{\xi}\right)\right]$$
$$- \frac{M_f \cdot a_m \cdot \left[1 - e^{-\frac{b_0 t}{a_m \cdot (J_M + J_S)}}\right]}{b_0} \quad (20)$$

Based on the angular velocity calculations, the angular acceleration's current value is determined in Equation (21).

$$\epsilon_{current} = \frac{\omega_{current}}{\Delta t} \quad (21)$$

Hence, the current value of the angular space is shown in Equation (22).

$$\theta_{current} = \frac{\omega_{current} \cdot \Delta t}{2} \quad (22)$$

Based on the current values of active pressure and the active flow, the current value of active moment and active power are calculated in Equation (23) and in Equation (24):

$$M_{a_{current}} = \frac{q_M}{2\pi} \cdot (p_{M_{current}} - p_0) \quad (23)$$

$$N_{current} = p_{M_{current}} \cdot Q_{current} \cdot 0.1 \quad (24)$$

The analysis of the dynamic behavior based on the transfer functions and the inverse Laplace transform was conducted through a program in MatLab—Simulink. Using the mathematical model and after applying the inverse Laplace transform, a proportional type transfer function with second-order inertia (PT2) is obtained for the rotary hydraulic motor and the servo valve.

To enhance the performance of the servo valve, a correction was applied, an electronic RC correction with the transfer function of first-order shown Equation (25) with T representing the time constant with a constant value of 0.006 s. The results of this correction are shown in Figure 6. The figure on the left shows the results without correction of the servo system, and the figure on the right shows the results with the corrections included.

$$H(s) = \frac{1}{Ts + 1} \quad (25)$$

The anticipation correction is also applied using the transfer function in Equation (26). This transfer function is considered a correction of the anticipation in case T_d is greater than T_i, and otherwise, it is a correction on the inertial. Both corrections were applied with T_d equal to 0.01 s and T_i equal to 0.001s in case of anticipation correction and with T_d equal to 0.0001 s and T_i equal to 0.001 s for the inertial correction. The results of these corrections are shown in Figure 7, with the left graph showing the anticipation correction results and the right one showing the graph on the left showing the inertial corrections.

$$H(s) = \frac{T_d s + 1}{T_i s + 1} \quad (26)$$

Also, the first-order inertial and the second-order inertial corrections are applied and shown in Figure 8. The left graph shows the results of the first-order correction, and the right graph shows the results of the second-order corrections. The transfer function for the first order is shown in Equation (27) with T_i equals 0.02 s, and that of the second order is shown in Equation (28) with T_{i1} equals 0.0005 s and T_{i2} equals 0.02 s.

$$H(s) = \frac{1}{T_i s + 1} \quad (27)$$

$$H(s) = \frac{1}{T_{i2}s^2 + T_{i1}s + 1} \tag{28}$$

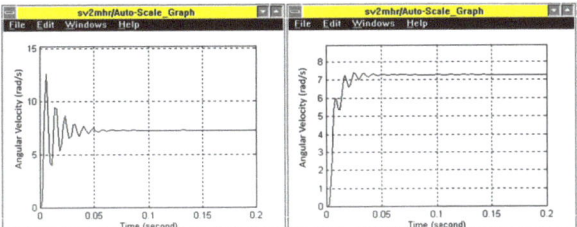

Figure 6. Rotary hydraulic actuator with a servo valve simulation results, the graph on the left is without the servo valve correction there is an overshoot, the graph on the right is with the servo valve correction included with T = 0.006 s, the overshoot is reduced and the response time is increased.

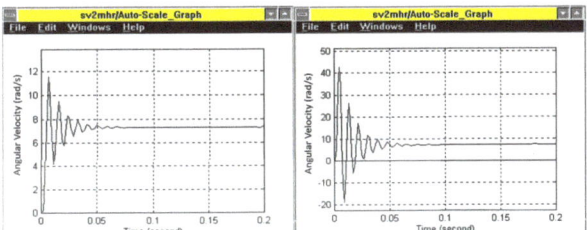

Figure 7. Rotary hydraulic actuator with a servo valve simulation results with anticipation and inertial correction of the servo valve, the graph on the left is with inertial correction with $T_d = 0.0001$ and $T_i = 0.001$, the overshoot increased and the response time decreased, and the graph on the right is with anticipation correction with $T_d = 0.01$ and $T_i = 0.001$ the overshoot increased, the response time decreased and the transient time increased.

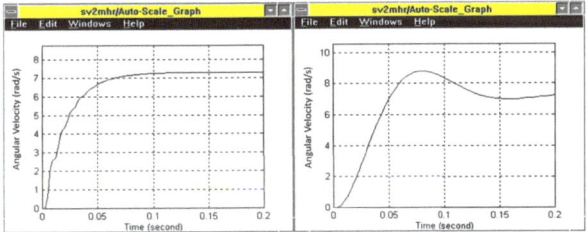

Figure 8. Rotary hydraulic actuator with a servo valve simulation results with inertial correction of the servo valve, the graph on the left is with first-order inertial correction included with $T_i = 0.02$ sec the overshoot is reduced, but the response time is increased, and the graph on the right is with second-order inertial correction with $T_{i2} = 0.0005$ s and $T_{i1} = 0.02$ s, a large response time is introduced and the transient time is increased but operation is stable.

The following can be observed from the simulation results: (i) first-order inertial correction reduces the overshoot and the transient time but increases the response time, so the operation of the system will be slower; (ii) the anticipation correction increases the overshoot; reduces the response time and increases the transient time, ensuring good promptness so that the system will be fast, accurate, but at the limit of stability; (iii) the second-order inertial correction causes a large delay in the response, introducing a low-frequency vibratory component, it increases the transient time, the operation of the system is transferred to low frequencies so that the response will be prolonged, with oscillations, but the operation is stable.

4.3. Computational Cost Estimations

A control loop based on the IGM and the KM could be implemented to control the ankle mechanism. PID controller would also be integrated, as shown in Figure 9. We aim to approximate the computational cost for such a control algorithm to develop an efficient control architecture. For this purpose, we implemented the code for each block on Matlab, estimated the execution time, and calculated the number of executions for each algorithm. The results are shown in Table 3. The execution time is the average time for 100 runs of the algorithm and the number of operations presents the operations presented in the algorithm, such as multiplication, subtraction, addition, and trigonometric functions.

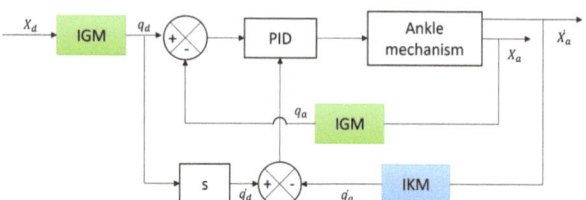

Figure 9. Control loop using the IGM and the KM model.

Table 3. Estimation of the control loop computation costs.

Algorithm	Number of Operations	Execution Time (ms)
IGM	52	5.3
IKM	150	15.4
PID	15	4.6

Table 3 shows a rough estimation of the required computations for integrating a control loop in the control architecture. This study is essential for deciding the characteristics of the MCU at the joint level, as the MCU should be capable of handling all these computations in addition to all the computations required for the sensor readings, communication tasks, and any other control task. The utilized MCU at the joint level has a CPU speed of 480 MHz, and it can achieve 1027 DMIPS, meaning that up to 480 million cycles are performed per second and 1027 million instructions can be executed per second. Hence, the MCU can handle the required computations for implementing the control loop for one mechanism, which can be integrated directly into the joint level.

5. Proposed Real-Time Control Architecture

The development of the control architecture of HYDROïD follows a set of specific requirements: (i) achieving high bandwidth in hardware and software, (ii) achieving the deterministic property and having real-time software, (iii) implementing a cost-effective control architecture, (iv) robust and stable distribution of controllers that respects the human-like appearance avoiding the bulky cables, (v) ease of integrating the numerous sensors and actuators, and (vi) respecting the safety factors for the protection of the robot and the human. The proposed control architecture is a distributed, open-source, modular, and real-time architecture to fulfill these requirements. Three main layers represent HYDROïD's hierarchical architecture: (i) the top layer, the central control unit, (ii) the hardware abstraction layer, and (iii) the lowest layer represented by the firmware of the embedded electronic board on each actuator. This section covers all the necessary selections to develop the control architecture that meets all the requirements.

5.1. Joint Controller

A customized in-house electronic board, shown in Figure 10, is developed on the joint level, represented by SEHA, responsible for controlling the actuator. The compactness of this electronic board is beneficial for the robot's human-like appearance. The board comprises three different boards. The first is mainly responsible for the power generation of all the actuator sensors besides the sensor's connectors. The second board consists of (i) the microprocessor STM32H7 that belongs to the cortex M7, which is the highest performance of the member; (ii) the LAN9252, which is the EtherCAT slave with a bandwidth of 100 Mbits/s; (iii) the conditioning circuits for the sensors integrated into the actuator including position, force, pressure, and temperature sensors, (vi) and the driving circuit of the servo valve. The third board contains the EtherCAT ports, ensuring the EtherCAT communication with the main PC.

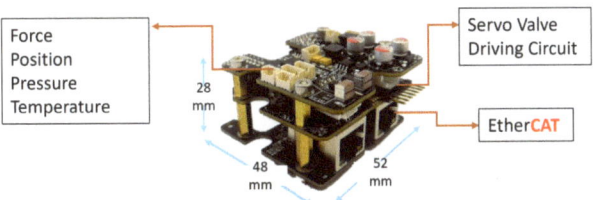

Figure 10. Developed electronic board at the joint level.

5.2. Joint Controller Distribution

A distributed control architecture is adopted for the control of HYDROïD. The centralized architecture is inefficient in controlling 36 DoF, requiring many computations. Hence, the distributed approach is chosen. Although the distributed approach is adopted, the joint controller will not be placed on each joint. Having a controller on each joint has the following drawbacks: (i) adding a joint controller on each joint might make the robot bulky without respecting the human appearance; (ii) the hybrid and parallel mechanisms in HYDROïD need synchronization between them to avoid mechanical problems and possible errors. For this purpose, the controllers will be distributed based on the mechanism, with each controller responsible for controlling a specific mechanism. This will result in a total of 15 controllers for the robot. This distribution is presented in Figure 11, highlighting the mechanisms one joint controller will control.

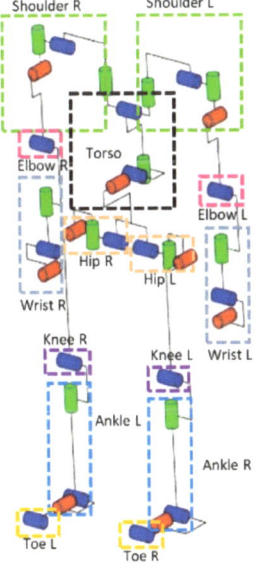

Figure 11. Distribution of the joint controllers of HYDROïD upon mechanism.

5.3. Communication Protocol

Because a distributed control architecture is utilized, communication between the joint controllers and the central control unit is needed, and the communication protocol selection is essential. The communication protocol ensures the achievement of the bandwidth requirements and the deterministic property. Comparing the different communication protocols, EtherCAT was selected as the communication protocol for data transmission because it shows the best performance besides enabling both hard and soft real-time communication with high bandwidth, leading to an advanced control architecture. This communication protocol is based on the Master-Slave topology. The central control unit layer represents one master device, and the joint controllers represent multiple slave devices. A special EtherCAT configuration is implemented on both levels to identify the master and the slave.

5.4. Real-Time EtherCAT Joint Controller

FreeRTOS, an open-source, real-time, and portable firmware for embedded systems, is selected as the firmware for the joint controller. The advantage of FreeRTOS is that it is suited for applications that require soft and hard real-time requirements, and it is cost-effective as it is open-source. It is a real-time kernel on top of which embedded applications can be built to meet the hard real-time requirements. The application will then be organized as a collection of independent threads of execution. We have developed different threads or tasks for the board responsible for controlling the servo valve of SEHA. These are mainly: (i) Thd-SensorMeasurementProcess, which is the thread responsible for all the sensor measurements; (ii) Thd-ControlValveProcess, which is the thread responsible for the current control of the servo valve; (iii) Thd-ControlPositionProcess, which is the process for controlling the position of the servo valve; (iv) lastly, there are the threads responsible for the EtherCAT processing that are Thd-EtherCAT-Sync1-Handle which handles the interrupts of the EtherCAT slave from the SYNC pin of the LAN9252, Thd-EtherCAT-IRQ-Event-Handle which handles the interrupts of the IRQ pin of the LAN9252, and finally the Thd-EtherCAT-Timer-Handle that performs the EtherCAT check operation triggered by a timer. The developed tasks are shown in the Figure 12. The highest priority is given to the communication task to ensure that any interruption from this side is handled directly. Then, the second highest priority is given to the control task, followed by the priority of the sensor readings.

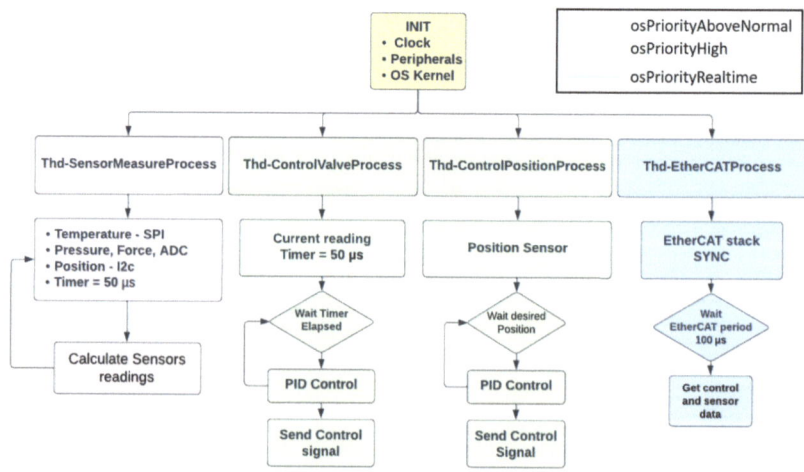

Figure 12. Developed tasks for the joint controller.

The Slave Stack Code (SSC) tool is utilized to develop the EtherCAT frame structure on the board. The EtherCAT slave device can be configured by the master device using an EtherCAT SubDevice Information (ESI) configuration file, an XML document containing all the information needed to set up a slave device properly for communication. The ESI configuration file for the developed joint controller board is generated using the Slave Stack Code (SSC) tool. The generated file has a special element hierarchy, which describes the slave's physical properties and the details of the communication protocol. This includes RxPDO and TxPDO elements, representing a single Process Data Object (PDO). PDOs represent the process data exchanged between the master and slave devices of EtherCAT and are updated cyclically. A unique index must define each PDO. The RxPdo elements describe the data transmitted from the slave device to the master device, while the TxPdo elements describe data transmitted from the master device to the slave device. The configured RxPDO and TxPDO for the joint controller of SEHA are shown in Table 4.

Table 4. PDO mapping of the EtherCAT frame.

PDO	Object	Data Type	Index
RxPDO	Index	Uint32	0x1600
	Data1	Uint32	0x1600
	Data2	Uint32	0x1600
	Total bits	96 bits	
TxPDO	Valve Current	Uint32	0x1A00
	Temperature	Uint32	0x1A00
	Position	Uint32	0x1A00
	Force	Uint32	0x1A00
	System State	Uint16	0x1A00
	Total bits	144 bits	

288 Bus Variables for communicating the 36 DoF are used. The total size of the TxPDO is 144 bits, and that of the RxPDO is 96. Hence, the total size of the input/output frame is 240 bits for each DoF. The total frame size for controlling the 36 DoF of HYDROïD is 8640 bits or 1080 bytes. Hence, with a link speed of 100 Mbits/s, the simplified theoretical transmission delay is calculated in Equation (29).

$$\tau = \frac{8*x}{C} \quad (29)$$

This leads to a transmission delay τ equal to 86.4 µs. To obtain the minimum cycle time, the propagation delays and the latencies within the slave must be added to the transmission delay. According to [28], the minimum achievable cyclic time is calculated as in Equation (30). The notations are shown in Table 5.

$$\Gamma = (2n-1)\ell + 2n\delta + \tau \quad (30)$$

Table 5. Terms, Notation, and Units.

Terms	Notation	Units
Minimum cycle time	Γ	s
Transmission delay	τ	s
Network device latency	ℓ	s
Propagation delay	δ	s
Link capacity	C	bits/s
Payload	x	bytes
Number of network devices (slaves)	n	

The typical propagation delay is 0.3 μs/slave. In EtherCAT communication, a frame is sent by the master, and slaves can read and write data on the fly. The duration of reading or writing operations corresponds only to the network device latency (ℓ), independent of frame size, and the same for all slaves. In addition, if we assumed that the network device latency = 0.3 μs/slave, the minimum cycle time will be around 130 μs. Hence, this configuration allows the transmission and receipt of data, ideally at approximately 7 kHz. The decision for the proposed control architecture is to operate on 5 kHz, making sure not to exceed the 7 kHz that is allowed and ensuring that if the DoF increases, no losses will occur. These calculations were repeated to further address the adaptability of the developed control architecture, considering that 50 DoF are connected. This resulted in a minimum cycle time of 164.7 μs, and hence, the transmission and reception of data is allowed at 6 kHz.

5.5. Real-Time Software

On the central control unit, represented by the master PC controlling the robot, a Real-Time Operating System is developed. The selected operating system is Ubuntu, a cost-effective OS that guarantees the software's real-time performance. Therefore, an RT-preemptible kernel is used to ensure the deterministic property.

Due to the increased complexity of robotic applications, robotics middleware was developed to reduce complexity, improve the software application, and simplify the software design. When using middleware, the development cost will be reduced as the developer will build components representing different parts of the robot and easily integrate these components with other existing components [29]. Among the different existing middleware, OROCOS, a real-time middleware, is used to obtain the hardware abstraction layer with the environment. OROCOS is an open-source middleware that supports four C++ libraries: the Real-Time Toolkit (RTT), the Kinematics and Dynamics Library, the Bayesian Filtering Library, and the OROCOS Component Library [30]. The RTT provides the infrastructure and the functionalities to build robotics applications in real time. It is a component-based tool, and components are connected via defined ports. The port holds a certain message type that could be transmitted using a defined frequency with a big advantage: respecting real-time constraints. This advantage allows the creation of precise control for robotic systems. OROCOS manages the hardware interface between the EtherCAT master and slaves and sends/receives data within specified time boundaries. This hardware interface is developed using the Simple Open Master EtherCAT (SOEM) library, which manages the data transmission using the rtt—soem package. In OROCOS, each EtherCAT hardware requires a specific driver to be developed. It is worth noting that some of the drivers for Beckhoff technology already exist. However, in our case, the driver and its corresponding messages were developed according to the EtherCAT frame of the SEHA controller board. The high-level control is implemented on OROCOS with a period of up to 5 KHz. To take advantage of the large packages available on Robot Operating System (ROS) [31], we designed a mixed hybrid architecture that integrates ROS, enabling hydraulic actuator control through ROS. OROCOS has an integrated interface with the ROS system, which makes it easy to exchange data safely with the ROS system without perturbing real-time performance inside OROCOS. ROS is an open-source middleware that includes many libraries and tools for developing robotic applications. One of the additional advantages of ROS is that it facilitates the integration of our control architecture with other existing robotics systems by developing a bridge and taking advantage of the libraries and the plugins available.

Moreover, to facilitate the usage of the developed control architecture, a Graphical User Interface (GUI) is developed on ROS that integrates the different modes of control, giving the user the ability to switch between these modes easily. Figure 13 shows the implemented software architecture to control HYDROïD.

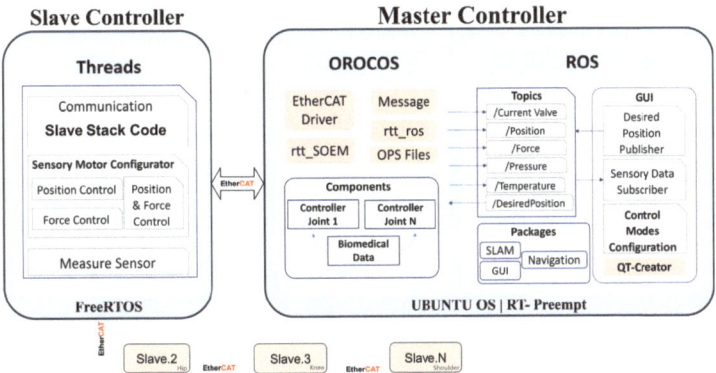

Figure 13. Developed control architecture for HYDROïD's control.

6. Experimental Validation

The performance of the developed real-time control architecture will be evaluated in three different ways. First, the performance of the joint controller will be evaluated by testing the maximum update rate that the developed board can handle. Second, the performance of the EtherCAT Bus communication is evaluated by testing the network latency. Lastly, the performance of the patched Real-Time Operating System is evaluated by measuring its latency.

6.1. Joint Controller Performance

To evaluate the joint controller from the communication perspective, a test was made to check the maximum frequency that could be sent to the EtherCAT slave from the master without losing any frame. Hence, the aim is to check the maximum frequency that could be handled by the EtherCAT slave joint controller without any communication error. For this purpose, a counter was implemented in the EtherCAT slave. A frame is then sent from the master within a specified time, and the counter is updated upon receiving a new frame each time. In case the slave receives a new frame, the counter will be updated. Otherwise, in the case of communication problems where the board cannot receive a new frame within the requested time, the counter will not update, and the same value as the previous counter will be shown. Hence, several frequencies were set from the master, starting from 1 kHz frequency and reaching 20 kHz. For the update rate ranging from 1 kHz to 9 kHz, the results are shown in Figure 14. The counter and the error are plotted, but at these update rates, no errors were detected. However, for the update rates ranging from 10 kHz to 20 kHz, the results are shown in Figure 15; the error is highlighted in red. The board normally operated without communication issues for all the frequencies below 10 kHz. The error percentages are shown in Figure 16; below 10 kHz, the error is 0%, which increases to 0.2% at 10 kHz and to 44.37% at 20 kHz.

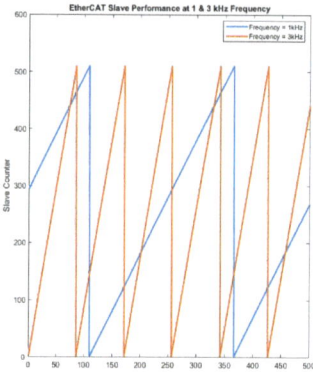 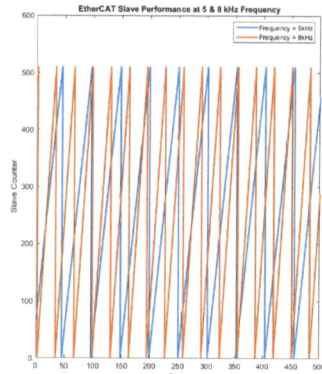

Figure 14. EtherCAT Slave performance at update rate ranging from 1 kHz to 9 kHz, the slave is operating normally, and all the frames are received with no errors.

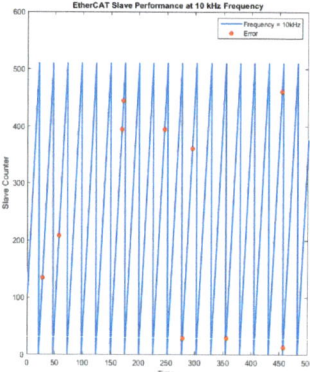 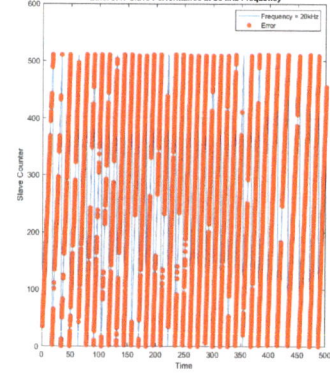

Figure 15. EtherCAT Slave performance from 10 kHz to 20 kHz, errors, highlighted in red, start to occur and increase with increasing the update rate.

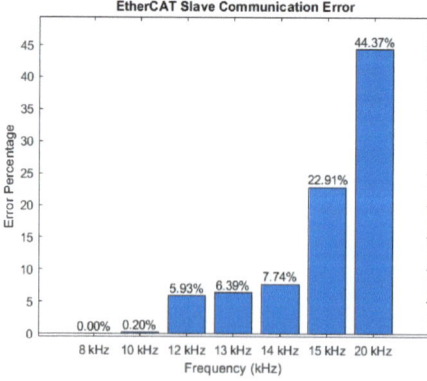

Figure 16. Error percentage at EtherCAT slave among different update rates.

6.2. EtherCAT Bus Performance

A test was conducted to evaluate the latency in the slave by operating eight slave boards equivalent to controlling the leg of HYDROïD. The network latency represents the delay between the starting execution time of two tasks on the first and last slave in the network. The boards are connected to the master controller. The first board is taken as the reference slave. From the master side, an OROCOS component was developed to toggle the servo valve output of the first and last board. The frequency of the toggle is 5 kHz. An oscilloscope is connected to the first and the last board to measure the output delay. The test was conducted for 65 h. Figure 17 shows the schematics of the conducted experiment. The resulting data shows that the delay did not exceed 167 µs as shown in Figure 18, which does not exceed the 200 µs or the 5 kHz.

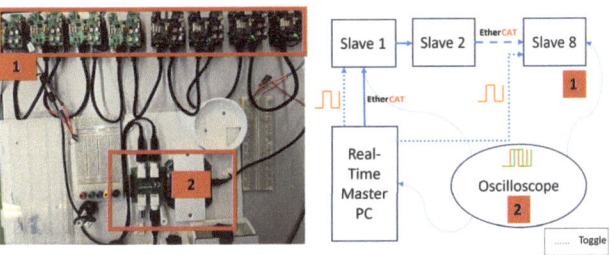

Figure 17. Conducted test for evaluating network latency.

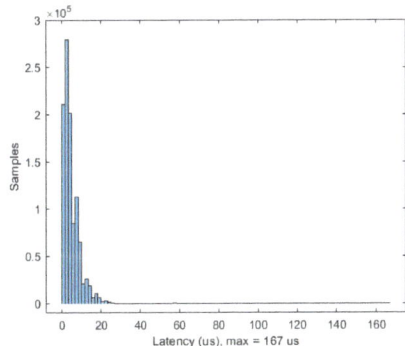

Figure 18. Latency on the slave joint controller-max latency = 167 µs.

6.3. Software Performance

To evaluate the real-time capabilities of the developed real-time software, a cyclic test was performed for 17 h with the entire functioning of the system, besides launching some threads to overcharge the PC, like CPU stress utility, network flood, and graphical stress, to ensure the reliability of the system. The cyclic test is a test that aims to measure the latency of the patched kernel. The latency is the delay before executing the task. This test was conducted twice, one on a standard kernel of Ubuntu without any modifications and the other on the modified kernel. The test on the standard kernel was launched for approximately 3 h, and the latency was 99,577 µs, as shown in Figure 19. However, the latency for the test conducted on the real-time kernel was 58 µs for the operation of 17 h, as shown in Figure 20. The achieved latency in the real-time software assures that our system will not exceed the desired cycle time, 200 µs equivalent to 5 kHz. Hence, in our proposed solution, the concept of real-time in terms of achieving tasks within a pre-specified time is respected.

To evaluate the overall performance of the control architecture, a preliminary experiment was conducted directly on HYDROïD by operating four mechanisms: the left and right knee and the right and left hip. The aim was to test the control architecture and the ease of implementing any control algorithm via an experiment to achieve the gate walking cycle through the control of the left and right hip and the left and right knee.

The high-level master controller transmits the desired position to achieve the movement based on biomedical data of a healthy person with a 63 cm step length at 100 cm/s velocity, while the low-level joint controller controls the current sent to the servo valve based on the desired position from the high-level controller. Figure 21 shows a snapshot of the robot in operation, and human-like motion was achieved.

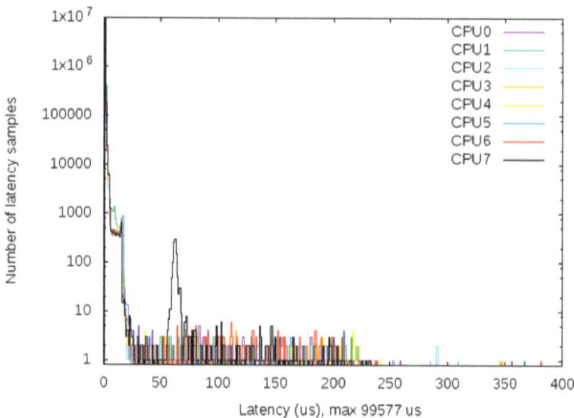

Figure 19. Latency of standard Ubuntu kernel.

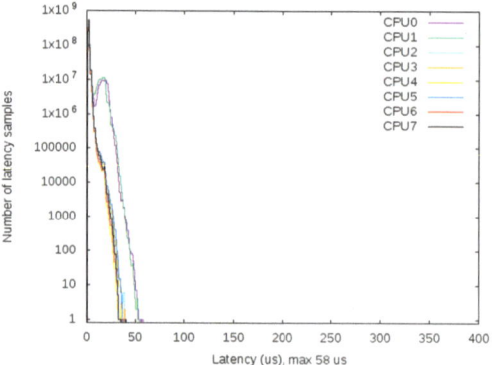

Figure 20. Latency of real-time Ubuntu kernel.

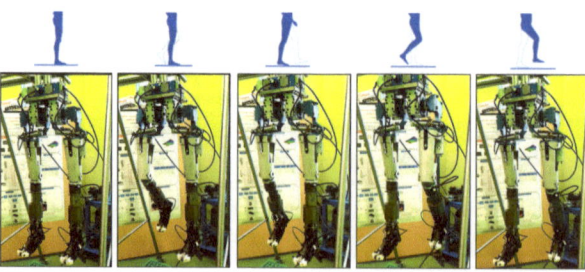

Figure 21. HYDROïD in operation applying the proposed control architecture.

7. Conclusions

Hydraulic actuators have proven to be highly effective due to their force capabilities; on the other hand, electrical actuators are useful from the cost perspective and the ease of control. SEHA is a hybrid technology that combines the advantages of the electrical and hydraulic actuators and avoids the disadvantages, leading to a high-performance actuator.

This paper proposes a real-time control architecture for HYDROïD humanoid. The modeling and simulation of one joint of HYDROïD composed of a rotary hydraulic actuator and servo valve were presented and analyzed with the aim of enhancing the performance for developing an efficient control algorithm with reduced complexity. The IGM and the IKM were also calculated and presented, and a rough estimation of the computational cost for implementing a control loop was also conducted to develop an efficient control algorithm. The requirement upon developing the control architecture is to have adaptable software from the perspective that adding sensors and actuators should not be complex. Moreover, the software should ensure the deterministic properties, and hence, there is real-time software for the safe interaction of the robot with its environment. Moreover, high bandwidth is required on both the low-level joint controller and the high-level master controller. Hence, this paper presents a distributed, real-time, and modular control architecture. The proposed architecture is based on EtherCAT communication protocol; the operating system is Ubuntu with a preemptible kernel ensuring the deterministic behavior and a hybrid middleware approach that integrates OROCOS and ROS. The choice of EtherCAT and real-time development is essential for ensuring the safety and stability of the robot, and the hybrid middleware will reduce the complexity of integrating new sensors and actuators, making it an adaptable one.

The results show that (i) the developed in-house board can handle an update rate up to 10 kHz, and communication errors start at 10 kHz. (ii) the network latency while connecting seven boards and operating at 5 kHz did not exceed 167 μs. (iii) the software latency does not exceed 58 μs, and the deterministic behavior is achieved. Our results show an improvement of 20% for the update rate over those reported by [18], where they achieved a 4 kHz update rate, which is the highest rate according to our knowledge. Also, the results show a reduction by approximately 40% in the control task latency compared to those achieved in [32].

As a future step to achieve the real-time capabilities of the OS, Ubuntu 22.04 will be used. This is the latest version of Ubuntu that offers a real-time kernel. Also, ROS2 [33], the 2nd version of ROS, will be integrated into the system. Moreover, the whole robot will be operated, and the robot's motion planning will be executed to operate the robot in different environmental setups. Integrating safety algorithms is one of the future works that will be implemented. To further address the safety concerns, the developed control architecture, due to its distributed structure, enables the addition of safety algorithms either on the low-level controller or on the high-level controller on OROCOS or ROS, depending on the algorithm's requirements.

Author Contributions: Conceptualization, M.G. and S.A.; methodology, M.G. and S.J.; software, M.G. and S.J.; validation, M.G.; formal analysis, M.G. and S.J.; investigation, M.G.; resources, M.G. and N.A.O.; data curation, M.G.; writing—original draft preparation, M.G.; writing—review and editing, S.A., N.A.O. and S.O.; visualization, M.G. and S.O.; supervision, S.A. and N.A.O.; project administration, S.A.; funding acquisition, S.A. All authors have read and agreed to the published version of the manuscript.

Funding: This research was funded by KALYSTA Actuation Company and the industrial excellence chair between the University of Evry and Kalysta Actuation.

Data Availability Statement: The raw data supporting the conclusions of this article will be made available by the authors on request.

Conflicts of Interest: The authors declare no conflicts of interest.

Appendix A. Detailed Calculation of IKM

Using the kinematic composition formula, the kinematic of the jth closed loop can be described as in Equation (A1). Where $T_{c_{S_p/S_b}}$ is the kinematic wrench of the foot relative to the base of the parallel mechanism, $T^j_{c_{S_1}/S_b}$ is the kinematic wrench of the foot relative to the base of the parallel mechanism, $T^j_{c_{S_2}/S_1}$ is the kinematic wrench of the jth cable relative to the jth linear actuator, and $T^j_{c_{Sp}/S_2}$ is the kinematic wrench of the foot relative to the jth cable.

$$T_{c_{Sp}/S_b} = T^j_{c_{S_1}/S_b} + T^j_{c_{S_2}/S_1} + T^j_{c_{Sp}/S_2} \tag{A1}$$

This relation is expressed with the individual screws as shown in Equation (A2). With $\dot{\theta}^j_i$ representing the derivative of θ^j_i and $\$^j_k$ is the kinematic screw of the kth joint in the jth closed chain.

$$T_{c_{Sp}/S_b} = \dot{r}^j_1\$^j_1 + \dot{\theta}^j_2\$^j_2 + \dot{\theta}^j_3\$^j_3 + \dot{\theta}^j_4\$^j_4 + \dot{\theta}^j_5\$^j_5 + \dot{\theta}^j_6\$^j_6 + \dot{\theta}^j_7\$^j_7 \tag{A2}$$

Since r^j_1 is the active variable in the closed chain, we can define the reciprocal screw of this variable named $\$^{R_j}_1$ that satisfies the condition in Equation (A3)

$$\$^j_i \$^{R_j}_1 = 0, i = 2, 3, 4, 5, 6, 7 \tag{A3}$$

The frame R^j_1 placed on O^j_1 and parallel to R_b is chosen as the working reference frame for each closed chain. The jth cable projected in this frame can be written with the coordinates as in Equation (A4).

$$A^j O^j_1 = \begin{bmatrix} U_j & V_j & W_j \end{bmatrix}_{R^j_1} \tag{A4}$$

Solving Equation (A3) will result in the desired screw shown in Equation (A5).

$$\$^{R_j}_1 = \frac{1}{\sqrt{U^2_j + V^2_j + W^2_j}} [U_j\ V_j\ W_j\ 0\ 0\ 0] \tag{A5}$$

Multiplying Equation (A5) by Equation (A2) and choosing A^j, shown in Equation (A1), as the working point for the jth closed chain, the determined relation is shown in Equation (A6).

$$\$^{R_j}_1 T_{c_{Sp}/S_b}(A^j) = \dot{r}^j_1 \$^j_1 \$^{R_j}_1 \tag{A6}$$

The kinematic wrench of the foot relative to the base S_b project on R_b is written in Equation (A7).

$$T_{c_{Sp}/S_b}(A_0) = \dot{q}_s z_s + \dot{q}_f z_f = \begin{bmatrix} \dot{q}_f C_{qs} & -\dot{q}_s & \dot{q}_f S_{qs} \end{bmatrix} \tag{A7}$$

Because of the hybrid mechanism, the first rotation joint is q_v independent of the two other joints q_s and q_f. Therefore, replacing Equations (A5) and (A7) in Equation (A6) for the four closed loops, the kinematic model will be expressed as in Equation (A8).

$$\begin{bmatrix} L^1_4 W^1 & L^1_4 S_{qs} V^1 & 0 \\ 0 & L^2_4(C_{qs}W^2 - S_{qs}U^2) & 0 \\ -L^3_4 W^3 & -L^3_4 S_{qs} V^3 & 0 \\ 0 & L^4_4(C_{qs}W^4 - S_{qs}U^4) & 0 \\ 0 & 0 & 1 \end{bmatrix} \cdot \begin{bmatrix} \dot{q}_s \\ \dot{q}_f \\ \dot{q}_v \end{bmatrix} = \begin{bmatrix} W^1 & & 0 \\ & W^2 & \\ & W^3 & \\ & W^3 & \\ 0 & & 1 \end{bmatrix} \cdot \begin{bmatrix} \dot{r}^1_1 \\ \dot{r}^2_1 \\ \dot{r}^3_1 \\ \dot{r}^4_1 \\ \dot{q}_v \end{bmatrix} \tag{A8}$$

References

1. Ibrahim, A.A.H.; Ammounah, A.; Alfayad, S.; Tliba, S.; Ouezdou, F.B.; Delaplace, S. Hydraulic Robotic Leg for HYDROïD Robot: Modeling and Control. *J. Robot. Mechatron.* **2022**, *34*, 576–587. [CrossRef]
2. Alfayad, S.; Kardofaki, M.; Sleiman, M. Hydraulic Actuator with Overpressure Compensation. WO Patent WO2020173933A1, 3 September 2020.

3. Fischmeister, S.; Lam, P. Time-Aware Instrumentation of Embedded Software. *IEEE Trans. Ind. Inform.* **2010**, *6*, 652–663. [CrossRef]
4. Gouaillier, D.; Hugel, V.; Blazevic, P.; Kilner, C.; Monceaux, J.; Lafourcade, P.; Marnier, B.; Serre, J.; Maisonnier, B. Mechatronic design of NAO humanoid. In Proceedings of the 2009 IEEE International Conference on Robotics and Automation, Kobe, Japan, 12–17 May 2009; pp. 769–774. [CrossRef]
5. Kaneko, K.; Kaminaga, H.; Sakaguchi, T.; Kajita, S.; Morisawa, M.; Kumagai, I.; Kanehiro, F. Humanoid Robot HRP-5P: An Electrically Actuated Humanoid Robot with High-Power and Wide-Range Joints. *IEEE Robot. Autom. Lett.* **2019**, *4*, 1431–1438. [CrossRef]
6. Nelson, G.; Saunders, A.; Neville, N.; Swilling, B.; Bondaryk, J.; Billings, D.; Lee, C.; Playter, R.; Raibert, M. PETMAN: A Humanoid Robot for Testing Chemical Protective Clothing. *J. Robot. Soc. Jpn.* **2012**, *30*, 372–377. [CrossRef]
7. Radford, N.A.; Strawser, P.; Hambuchen, K.; Mehling, J.S.; Verdeyen, W.K.; Donnan, A.S.; Holley, J.; Sanchez, J.; Nguyen, V.; Bridgwater, L.; et al. Valkyrie: NASA's First Bipedal Humanoid Robot. *J. Field Robot.* **2015**, *32*, 397–419. [CrossRef]
8. Sygulla, F.; Wittmann, R.; Seiwald, P.; Berninger, T.; Hildebrandt, A.; Wahrmann, D.; Rixen, D. An EtherCAT-Based Real-Time Control System Architecture for Humanoid Robots. In Proceedings of the 2018 IEEE 14th International Conference on Automation Science and Engineering (CASE), Munich, Germany, 20–24 August 2018; pp. 483–490. [CrossRef]
9. Akachi, K.; Kaneko, K.; Kanehira, N.; Ota, S.; Miyamori, G.; Hirata, M.; Kajita, S.; Kanehiro, F. Development of humanoid robot HRP-3P. In Proceedings of the 5th IEEE-RAS International Conference on Humanoid Robots, San Diego, CA, USA, 5–7 December 2005; pp. 50–55. [CrossRef]
10. Cereia, M.; Bertolotti, I.C.; Scanzio, S. Performance of a Real-Time EtherCAT Master Under Linux. *IEEE Trans. Ind. Inform.* **2011**, *7*, 679–687. [CrossRef]
11. Park, I.W.; Kim, J.Y.; Lee, J.; Oh, J.H. Mechanical design of humanoid robot platform KHR-3 (KAIST Humanoid Robot 3: HUBO). In Proceedings of the 5th IEEE-RAS International Conference on Humanoid Robots, Tsukuba, Japan, 5 December 2005; pp. 321–326. [CrossRef]
12. Park, I.W.; Kim, J.Y.; Park, S.W.; Oh, J.H. Development of humanoid robot platform KHR-2 (KAIST humanoid robot-2). In Proceedings of the 4th IEEE/RAS International Conference on Humanoid Robots, Santa Monica, CA, USA, 10–12 November 2004; Volume 1, pp. 292–310. [CrossRef]
13. Cena, G.; Seno, L.; Valenzano, A.; Vitturi, S. Performance analysis of Ethernet Powerlink networks for distributed control and automation systems. *Comput. Standards Interfaces* **2009**, *31*, 566–572. [CrossRef]
14. Jansen, D.; Buttner, H. Real-time Ethernet: The EtherCAT solution. *Comput. Control Eng.* **2004**, *15*, 16–21. [CrossRef]
15. Nelson, G.; Saunders, A.; Playter, R. The PETMAN and Atlas Robots at Boston Dynamics. In *Humanoid Robotics: A Reference*; Springer: Dordrecht, The Netherlands, 2019; pp. 169–186. [CrossRef]
16. Stasse, O.; Flayols, T.; Budhiraja, R.; Giraud-Esclasse, K.; Carpentier, J.; Mirabel, J.; Del Prete, A.; Souères, P.; Mansard, N.; Lamiraux, F.; et al. TALOS: A new humanoid research platform targeted for industrial applications. In Proceedings of the 2017 IEEE-RAS 17th International Conference on Humanoid Robotics (Humanoids), Birmingham, UK, 15–17 November 2017; pp. 689–695. [CrossRef]
17. Kaminaga, H.; Ko, T.; Masumura, R.; Komagata, M.; Sato, S.; Yorita, S.; Nakamura, Y. Mechanism and Control of Whole-Body Electro-Hydrostatic Actuator Driven Humanoid Robot Hydra. In *Proceedings of the 2016 International Symposium on Experimental Robotics, Nagasaki, Japan, 3–8 October 2016*; Kulić, D., Nakamura, Y., Khatib, O., Venture, G., Eds.; Springer: Cham, Switzerland, 2017; pp. 656–665. [CrossRef]
18. Ahn, J.; Park, S.; Sim, J.; Park, J. Dual-Channel EtherCAT Control System for 33-DOF Humanoid Robot TOCABI. *IEEE Access* **2023**, *11*, 44278–44286. [CrossRef]
19. Ferrati, M.; Settimi, A.; Muratore, L.; Cardellino, A.; Rocchi, A.; Mingo Hoffman, E.; Pavan, C.; Kanoulas, D.; Tsagarakis, N.G.; Natale, L.; et al. The Walk-Man Robot Software Architecture. *Front. Robot. AI* **2016**, *3*, 25. [CrossRef]
20. Nori, F.; Traversaro, S.; Eljaik, J.; Romano, F.; Del Prete, A.; Pucci, D. iCub Whole-Body Control through Force Regulation on Rigid Non-Coplanar Contacts. *Front. Robot. AI* **2015**, *2*, 6. [CrossRef]
21. Stasse, O.; Flayols, T. An Overview of Humanoid Robots Technologies. In *Biomechanics of Anthropomorphic Systems*; Venture, G., Laumond, J.P., Watier, B., Eds.; Springer Tracts in Advanced Robotics; Springer International Publishing: Cham, Switzerland, 2019; Volume 124, pp. 281–310. [CrossRef]
22. Karumanchi, S.; Edelberg, K.; Baldwin, I.; Nash, J.; Reid, J.; Bergh, C.; Leichty, J.; Carpenter, K.; Shekels, M.; Gildner, M.; et al. Team RoboSimian: Semi-autonomous Mobile Manipulation at the 2015 DARPA Robotics Challenge Finals. *J. Field Robot.* **2017**, *34*, 305–332. [CrossRef]
23. Asfour, T.; Waechter, M.; Kaul, L.; Rader, S.; Weiner, P.; Ottenhaus, S.; Grimm, R.; Zhou, Y.; Grotz, M.; Paus, F. ARMAR-6: A High-Performance Humanoid for Human-Robot Collaboration in Real-World Scenarios. *IEEE Robot. Autom. Mag.* **2019**, *26*, 108–121. [CrossRef]
24. Ammounah, A. Architecture de Contrôle pour un robot Humanoïde à Actionnement Hydraulique. Ph.D. Thesis, Université Paris-Saclay, Paris, France, 2021.
25. Alfayad, S.; Tayba, A.M.; Ouezdou, F.B.; Namoun, F. Kinematic Synthesis and Modeling of a Three Degrees-of-Freedom Hybrid Mechanism for Shoulder and Hip Modules of Humanoid Robots. *J. Mech. Robot.* **2016**, *8*, 041017. [CrossRef]
26. Abdellatif Hamed Ibrahim, A.; Alfayad, S.; Hildebrandt, A.C.; Ouezdou, F.; Mechbal, N.; Zweiri, Y. Development of a New Hydraulic Ankle for HYDROïD Humanoid Robot. *J. Intell. Robot. Syst.* **2018**, *92*, 293–308. [CrossRef]

27. Alfayad, S.; Kardofaki, M.; Sleiman, M.; Arlot, R. Verin a Capteur de Position Integre. WO Patent WO2023088972A1, 25 May 2023.
28. Robert, J.; Georges, J.P.; Rondeau, E.; Divoux, T. Minimum Cycle Time Analysis of Ethernet-Based Real-Time Protocols. *Int. J. Comput. Commun. Control* **2012**, *7*, 743–757. [CrossRef]
29. Elkady, A.; Sobh, T. Robotics Middleware: A Comprehensive Literature Survey and Attribute-Based Bibliography. *J. Robot.* **2012**, *2012*, 959013. [CrossRef]
30. Bruyninckx, H. Open robot control software: The OROCOS project. In Proceedings of the 2001 ICRA—IEEE International Conference on Robotics and Automation (Cat. No.01CH37164), Seoul, Republic of Korea, 21–26 May 2001; Volume 3, pp. 2523–2528. [CrossRef]
31. Quigley, M.; Gerkey, B.; Conley, K.; Faust, J.; Foote, T.; Leibs, J.; Berger, E.; Wheeler, R.; Ng, A. ROS: An open-source Robot Operating System. In Proceedings of the ICRA Workshop on Open Source Software, Kobe, Japan, 12–17 May 2009.
32. Puck, L.; Keller, P.; Schnell, T.; Plasberg, C.; Tanev, A.; Heppner, G.; Roennau, A.; Dillmann, R. Performance Evaluation of Real-Time ROS2 Robotic Control in a Time-Synchronized Distributed Network. In Proceedings of the 2021 IEEE 17th International Conference on Automation Science and Engineering (CASE), Lyon, France, 23–27 August 2021; pp. 1670–1676. [CrossRef]
33. Macenski, S.; Foote, T.; Gerkey, B.; Lalancette, C.; Woodall, W. Robot Operating System 2: Design, Architecture, and Uses in the Wild. *Sci. Robot.* **2022**, *7*, eabm6074. [CrossRef] [PubMed]

Disclaimer/Publisher's Note: The statements, opinions and data contained in all publications are solely those of the individual author(s) and contributor(s) and not of MDPI and/or the editor(s). MDPI and/or the editor(s) disclaim responsibility for any injury to people or property resulting from any ideas, methods, instructions or products referred to in the content.

Article

Robust Control Based on Adaptative Fuzzy Control of Double-Star Permanent Synchronous Motor Supplied by PWM Inverters for Electric Propulsion of Ships

Djamel Ziane [1], Samir Zeghlache [2], Mohamed Fouad Benkhoris [1,*] and Ali Djerioui [2]

1. IREENA Laboratory, Nantes University, 44600 Saint Nazaire, France; djamel.ziane@univ-nantes.fr
2. Laboratoire d'Analyse des Signaux et Systèmes, Departement of Electrical Engineering, Faculty of Technology, University of M'sila, BP 166 Ichbilia, M'sila 28000, Algeria; samir.zeghlache@univ-msila.dz (S.Z.); ali.djerioui@univ-msila.dz (A.D.)
* Correspondence: mohamed-fouad.benkhoris@univ-nantes.fr

Abstract: This study presents the development of an adaptive fuzzy control strategy for double-star PMSM-PWM inverters used in ship electrical propulsion. The approach addresses the current and speed tracking challenges of double-star permanent magnet synchronous motors (DSPMSMs) in the presence of parametric uncertainties. Initially, a modeling technique employing a matrix transformation method is introduced, generating decoupled and independent star windings to eliminate inductive couplings, while maintaining model consistency and torque control. The precise DSPMSM model serves as the foundation for an unknown nonlinear backstepping controller, approximated directly using an adaptive fuzzy controller. Through the Lyapunov direct method, system stability is demonstrated. All signals in the closed-loop system are ensured to be uniformly ultimately bounded (UUB). The proposed control system aims for low tracking errors, while also mitigating the impact of parametric uncertainties. The effectiveness of the adaptive fuzzy nonlinear control system is validated through tests conducted in hardware-in-the-loop (HIL) simulations, utilizing the OPAL-RT platform, OP4510.

Keywords: adaptive fuzzy control; double-star permanent magnet synchronous motor (DSPMSM); OPAL-RT (OP4510); model transformation

MSC: 93C42

1. Introduction

In the current context of energy transition and the search for sustainable solutions for maritime transport, the use of electric propulsion systems is emerging as a promising alternative. Double-star permanent magnet synchronous motors (DSPMSMs) supplied with pulse width modulation (PWM) inverters constitute a popular configuration for the electric propulsion of ships due to their high performance and increased energy efficiency [1]. In these large-scale drives, multi-phase machines offer crucial advantages [2], such as power distribution over multiple branches, a reduction in torque ripple amplitude, a decrease in current harmonics, and fault tolerance due to the high number of phases.

However, in marine environments, electric propulsion systems face significant variations in essential internal machine parameters, such as resistance, inductance, inertia, and friction. These internal variations, induced by dynamic operational conditions such as changes in load and speed, can substantially influence the performance of the propulsion system, thereby affecting power distribution and dynamic response. These internal variations are complemented by external variations, such as changing weather conditions and interactions with water. These external factors introduce disturbances, thus affecting the performance of the propulsion system and requiring dynamic adaptation of control strategies.

The DSPMSM has been the subject of several scientific articles, both in terms of modeling and control. The initial work emerged in the 1990s and 2000s, with proposed modeling and control approaches powered by voltage inverters applied for railway and ship propulsion [3,4].

Modeling work on the machine is diverse and conducted with various approaches. The difference lies in considering the machine as a six-phase machine with a connected neutral or considering it as equivalent to two three-phase machines with the two neutrals of the two machines separated. In the consideration as two three-phase motors, the modeling approach relies on the use of coordinate transformation based on synchronous rotating coordinate transformation (dq) [5,6].

Another modeling method exists which involves vector space decomposition (VSD), which is a machine modeling technique. The machine is divided into orthogonal subspaces using this method: one subspace producing a single flux/torque (α-β) and numerous subspaces not producing flux/torque (x-y) [7].

The two methods mentioned earlier have been used as modeling tools in several works, including the double-star machine. In this article, we employ another approach to establish the model based on the general approach dynamical modeling of multi/three-phase machines developed in [8]. The elaborate modeling approach utilizes a novel decoupling transformation to eliminate couplings of multi-phase permanent magnet synchronous machines in a generic modular configuration.

In the literature, various approaches to synchronous machine control can be found, with most of them focusing on field-oriented control, classic direct torque control (DTC), adaptive fuzzy DTC, and neural DTC. If we analyze the different objectives targeted in these papers, we can summarize them into two components: optimal torque control and speed control. However, in these studies, the system under investigation is always considered to be time-invariant and without disturbance elements, thus not reflecting the reality of the system in real cases. This is because energy conversion systems undergo parametric variations related to heating, aging, magnetic circuit saturation, and other external constraints.

In Reference [9], a new method for direct torque control of permanent magnet synchronous machines (PMSMs) was presented. The simulation results confirmed the advantages of this approach compared to the conventional Direct Torque Control (DTC) approach. The proposed method offers a constant inverter switching frequency, reduces torque ripples, and exhibits good robustness to variations in stator resistance. However, it is observed that the only parametric variations considered are those related to stator resistance.

Vector control of rotating machines is recognized for its efficiency due to its simplicity of design and implementation, as well as its natural decoupling between flux and currents. This type of control is typically achieved using proportional–integral (PI) controllers, whose parameters are calculated directly from the machine characteristics using conventional analytical methods. However, this approach requires careful calculation and a good understanding of all machine parameters.

Historically, Fuzzy Logic Systems (FLSs) have a stellar reputation as effective approximators [10]. Their universal approximation qualities have led to their considerable usage in modeling and regulating unpredictable nonlinear systems. For diverse types of nonlinear systems, many adaptive fuzzy control methods have emerged in recent years [11–20]. The adaptive fuzzy control techniques for uncertain nonlinear systems were developed in [21–23], using a backstepping methodology. The stability of the closed-loop systems was achieved using the famous Lyapunov direct method. In this work, we suggest using this robust approximation method to address uncertainties and unknown dynamics inside the DSPMSM.

Motivated by the previous discussion, in this paper, the problem of currents and speed control is investigated for DSPMSM subject to parametric uncertainties via fuzzy approximation-based adaptive control. The FLS is used with the assistance of adaptive estimators in order to approximate unknown nonlinear dynamics. Additionally, a robust adaptive compensation is utilized in order to mitigate the impact of parametric uncertainties and to correct approximation errors.

Our objective is to replace the PI controllers with adaptive controllers based on fuzzy logic in order to achieve a more robust control. We take into account the specific constraints encountered by maritime propulsion systems. Our proposal involves using an adaptive fuzzy control technique as an effective solution to mitigate the effects of internal parametric variations, such as stator resistance, machine inductances, inertia of the machine-load system, or viscous friction due to aging. This paper's key contributions are as follows: (i) the suggestion of an adaptive fuzzy control algorithm for DSPMSM that is resilient against uncertainty and can dampen the external disturbances; (ii) by integrating FLS, there is no reliance on the mathematical model; and (iii) the global closed-loop system is demonstrated to exhibit UUB stability.

The present study focuses on creating an adaptive fuzzy control method for DSPMSM systems that are exposed to external disturbances and uncertainty. The system dynamics are presumed to be unknown, and the controller settings are adjusted in response to the emergence of uncertainty. Making use of the fact that the system dynamics will be transformed into a strict-feedback form, if we include the models of uncertainty, external disturbances may affect the system model. A new fuzzy adaptive control approach is combined with a nonlinear control method to address this class of nonlinear systems. There are two control terms in the suggested adaptive control law. An adaptive fuzzy control rule is used as the initial control term to adjust the parameters online in order to deal with the uncertain system dynamics. To address the issue of fuzzy approximation errors, uncertainties, and external disturbances, the second term serves as a robust control by using the tangent hyperbolic function. The Lyapunov technique is used to examine the stability of the closed-loop system and guarantee the tracking error's convergence to zero.

The main contributions of this paper are the introduction of a new adaptive control strategy based on Takagi–Sugeno fuzzy inference systems. This strategy is designed to handle all types of uncertainties and external disturbances that may arise in the system dynamics. This work is compared to existing works in the same area, as referenced in [24–39]. The suggested research aims to address complex non-linear control issues with fewer assumptions compared to the existing literature. The following points encapsulate the contributions made by this work:

- The suggested fuzzy adaptive controller for uncertain systems reduces the amount of online learning parameters, making it easier to tune and suited for real-time implementations. Furthermore, regardless of the order of the nonlinear system, the suggested technique requires just basic fuzzy inference systems, while in [24–27], the number of updating parameters is still determined by the system's order.
- The control techniques proposed in [24–27] are based on backstepping, which is known to have the drawback of complexity growing. However, in the proposed method, the controllers have simpler structures and fewer design parameters, as the causes for the complexity growing problem were completely eliminated.
- The suggested adaptive control techniques may accomplish an a priori intended transient and steady-state performance in addition to ensuring the stability of the whole control system by adding prescribed performance. As a consequence, the suggested methods guarantee that the tracking error always converges to a predetermined, arbitrarily tiny residual set, which is not possible with the prior findings in the literature [24–26].
- By using the adaptive fuzzy control approach developed, the singularity and explosion of complexity concerns are effectively avoided in comparison to the backstepping control algorithms presented in [28,29]. In order to improve the tracking performance, the robust adaptive compensation techniques are also made to adjust for approximation errors and lessen the impact of parametric uncertainties.
- In contrast to the references mentioned in [30–34], the proposed controller is more flexible, as it does not require any knowledge of the mathematical model. On the other

hand, the suggested controller takes a systematic approach to handling unknown uncertainties and external disturbances.

- Restrictive assumptions regarding external perturbation were made by the authors in Refs. [35–37]. The external disturbance is modeled in Ref. [36] using time-varying free-models with derivable bounds, whereas in Ref. [35], it is described as an exogenous neutral stable system. In Ref. [37], it is divided into two parts, one of which represents an estimated portion and the other of which is generated by an exogenous system. The proposed work, on the other hand, assumes external perturbations under only the boundedness mild condition without considering any additional information.
- The developed controllers in Refs. [38–40] are intended for systems where the control gain must be a simple constant, which is a limiting constraint. The latter constraint is lifted in the suggested method to include a broader category of dynamical systems. To encompass a wide range of dynamical systems, such as inverted pendulums, induction motor drives, single-link robot arms, mass–spring–damper systems, flexible spacecraft, quadrotors, and many more, we actually presume that the system dynamics are unknown, with the control gain as an unknown nonlinear function.

Prior to presenting the obtained results, we delineate a modeling approach in the first section, followed by the mathematical development of the adaptive fuzzy control technique in the second section. Finally, the results of our tests, accompanied by a detailed analysis, are presented. These tests were conducted using the real-time simulator OP4510 from OPAL-RT.

2. Description of the Studied System

Our study focuses on a complex system, consisting of a permanent magnet synchronous machine with two stator windings. Each of these windings is powered by a three-phase inverter. To ensure the precise control of these two power electronic structures, we implemented adaptive fuzzy controllers, enabling the flexible and efficient management of the system in the presence of disturbances. The depicted system is illustrated in Figure 1.

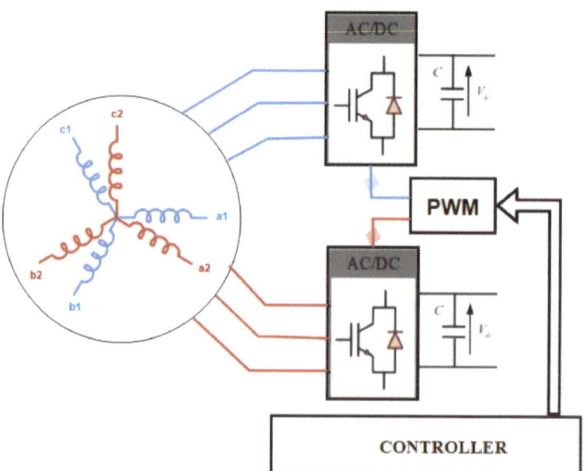

Figure 1. Structure of the considered system.

The PMSM is in the smoothed pole machine. The study is based on the following assumptions:

(a) The multi-phase winding consists of 2×3 identical phases.
(b) Variable reluctance effects and saturation phenomena are neglected.
(c) Only the first space harmonic is taken into account.
(d) The temperature effects are neglected.

(e) The capacitive effect between the windings is neglected.
(f) The semiconductor components constituting the inverters are supposed to be perfect.

3. Mathematical Modeling of DSPMSM

As shown in Figure 2, the double-star permanent magnet synchronous motors (DSPMSM) considered in our study are composed of two three-phase windings phase-shifted by an angle, γ.

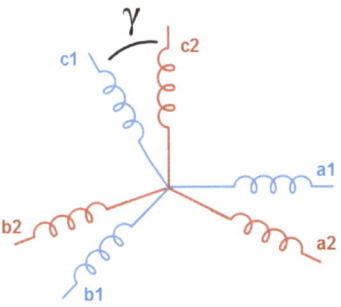

Figure 2. Double-star machine winding.

The establishment of the decoupled dynamic model of the DSPMSM is performed in three steps, which are described in the following subsections.

3.1. Electrical Model in the (a1 b1 c1 a2 b2 c2) Frame

The general electrical equation of the DSPSM in this natural basis can be written as follows:

$$[V_S] = R_S[I_S] + [L_S]\frac{d}{dt}[I_S] + [E_S] \quad (1)$$

where $[V_s]$ represents the supply voltage vector of the stator windings. It is defined as follows:

$$[V_s] = \left[[V_{S1}]^t \ [V_{S2}]^t\right]^t \quad (2)$$

where

$$[V_{si}] = \begin{bmatrix} v_{ai} \\ v_{bi} \\ v_{ci} \end{bmatrix} \ i = 1, 2 \quad (3)$$

where $[I_s]$ represents the stator's currents vector. It is defined as follows:

$$[I_s] = \left[[I_{S1}]^t \ [I_{S2}]^t\right]^t \quad (4)$$

where

$$[I_{si}] = \begin{bmatrix} i_{ai} \\ i_{bi} \\ i_{ci} \end{bmatrix} \ i = 1, 2 \quad (5)$$

where $[E_s]$ represents the EMF voltage vector. It is defined as follows:

$$[E_s] = \left[[E_{S1}]^t \ [E_{S2}]^t\right]^t \quad (6)$$

where

$$[E_{si}] = \begin{bmatrix} E_{ai} \\ E_{bi} \\ E_{ci} \end{bmatrix} = -\sqrt{2}\omega\varphi_f \begin{bmatrix} \sin(\theta - (i-1)\gamma) \\ \sin\left(\theta - \frac{2\pi}{3} - (i-1)\gamma\right) \\ \sin\left(\theta + \frac{2\pi}{3} - (i-1)\gamma\right) \end{bmatrix} \ i = 1, 2 \quad (7)$$

where R_s is the resistance of each winding, and $[L_s]$ is the stator's inductance matrix. $[L_s]$ is defined as follows:

$$[L_s] = \begin{bmatrix} [L_{s1}] & [M_{s12}] \\ [M_{s12}]^t & [L_{s2}] \end{bmatrix} \quad (8)$$

where $[L_{si}]$ (i = 1, 2) represents the matrix inductance of each star and is defined as follows:

$$[L_{s1}] = [L_{s2}] = \begin{bmatrix} l_{fs} + M_{ss} & M_{ss}\cos(\frac{2\pi}{3}) & M_{ss}\cos(\frac{4\pi}{3}) \\ M_{ss}\cos(\frac{4\pi}{3}) & l_{fs} + M_{ss} & M_{ss}\cos(\frac{2\pi}{3}) \\ M_{ss}\cos(\frac{2\pi}{3}) & M_{ss}\cos(\frac{4\pi}{3}) & l_{fs} + M_{ss} \end{bmatrix} \quad (9)$$

where l_{fs} is the leakage inductance, and M_{ss} is the maximal mutual inductance between two windings.

$[M_{s12}]$ is the mutual inductance matrix between the two three windings. It is given by the following relation:

$$[M_{s12}] = \begin{bmatrix} M_{ss}(\gamma) & M_{ss}(\gamma + \frac{2\pi}{3}) & M_{ss}(\gamma + \frac{4\pi}{3}) \\ M_{ss}(\gamma + \frac{4\pi}{3}) & M_{ss}(\gamma) & M_{ss}(\gamma + \frac{2\pi}{3}) \\ M_{ss}(\gamma + \frac{2\pi}{3}) & M_{ss}(\gamma + \frac{4\pi}{3}) & M_{ss}(\gamma) \end{bmatrix} \quad (10)$$

From the inductance matrix, it can be easily shown that the matrix is fully coupled, so that the control the motor's currents in this frame are complicated.

3.2. Electrical Dynamical Model in the ($\alpha_1\ \beta_1\ \alpha_2\ \beta_2$)

To write the electrical equations of the DSPSM in this reference frame, first we apply the Concordia transformation to each star (Figure 3A). Second, a rotation of an angle, γ, is applied to the second star (Figure 3B).

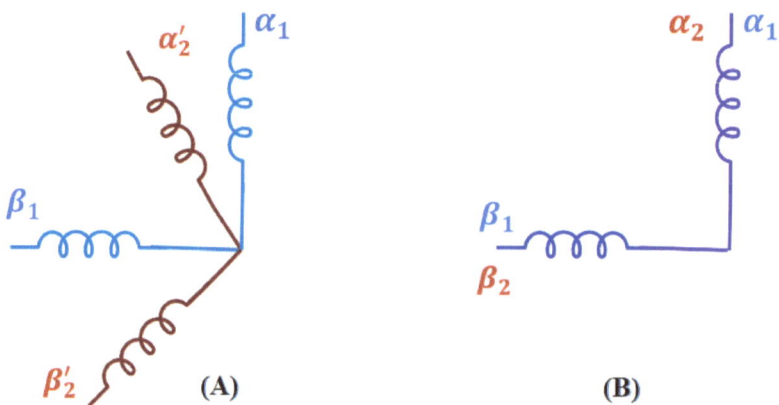

Figure 3. Equivalent windings in ($\alpha_1\beta_1o_1$ and $\alpha_2\beta_2o_2$)

This transformation from the reference frame (a1 b1 c1 a2 b2 c2) to the reference frame ($\alpha 1\ \beta 1\ o 1\ \alpha 2\ \beta 2\ o 2$) is defined as follows:

$$[T_6]^t = \begin{bmatrix} \begin{bmatrix} [[T_{32}]]^t \\ [T_{31}]^t \end{bmatrix} & [0]_{3\times 3} \\ [0]_{3\times 3} & \begin{bmatrix} [[T_{32}]\cdot P(-\gamma)]^t \\ [T_{31}]^t \end{bmatrix} \end{bmatrix} \quad (11)$$

where

$$[T_{32}] = \sqrt{\frac{2}{3}} \begin{bmatrix} 1 & -\frac{1}{2} & -\frac{\sqrt{3}}{2} \\ 0 & -\frac{1}{2} & \frac{\sqrt{3}}{2} \end{bmatrix}^t \quad [T_{31}] = \begin{bmatrix} \frac{1}{\sqrt{2}} & \frac{1}{\sqrt{2}} & \frac{1}{\sqrt{2}} \end{bmatrix}^t \quad (12)$$

$$P(\gamma) = \begin{bmatrix} cos(\gamma) & sin(\gamma) \\ -sin(\gamma) & cos(\gamma) \end{bmatrix} \quad (13)$$

The electrical equation of the DSPMSM in this frame is as follows:

$$[T_6]^t[V_S] = R_S[T_6]^t[I_S] + [T_6]^t[L_S]\frac{d}{dt}[I_S] + [T_6]^t[E_S] \quad (14)$$

By applying this transformation, the stator's inductance matrix [L$_s$], the inductance matrix in this frame can be deduced:

$$[L_s'] = [T_6]^t[L_s][T_6] \quad (15)$$

Thus, after the development of the calculation, we obtain the following:

$$[L_s'] = \begin{bmatrix} \begin{bmatrix} l_{fs} + \frac{3}{2}M_{ss} & 0 & 0 \\ 0 & l_{fs} + \frac{3}{2}M_{ss} & 0 \\ 0 & 0 & l_{fs} \end{bmatrix} & \begin{bmatrix} \frac{3}{2}M_{ss} & 0 & 0 \\ 0 & \frac{3}{2}M_{ss} & 0 \\ 0 & 0 & 0 \end{bmatrix} \\ \begin{bmatrix} \frac{3}{2}M_{ss} & 0 & 0 \\ 0 & \frac{3}{2}M_{ss} & 0 \\ 0 & 0 & 0 \end{bmatrix} & \begin{bmatrix} l_{fs} + \frac{3}{2}M_{ss} & 0 & 0 \\ 0 & l_{fs} + \frac{3}{2}M_{ss} & 0 \\ 0 & 0 & l_{fs} \end{bmatrix} \end{bmatrix} \quad (16)$$

The inductance matrix [L$_s$'] is not completely diagonal. In this new reference frame, windings following the same axis, α or β, are coupled. Therefore, their mutual inductance is non-zero and equal to 3M$_{ss}$/2. However, since the axes α and β are orthogonal, the mutual inductances between the windings following these axes are zero.

The electromagnetic torque in this frame is defined as follows:

$$\Gamma = \frac{e_{\alpha 1}i_{\alpha 1} + e_{\beta 1}i_{\beta 1} + e_{\alpha 2}i_{\alpha 2} + e_{\beta 2}i_{\beta 2}}{\Omega} \quad (17)$$

It is important to notice that the transformation P(γ) applied to the second star and defined above only introduces a rotation of the EMF vector and does not modify its module. We then have the following:

$$\begin{aligned} e_{\alpha 1} = e_{\alpha 2} = e_\alpha = -\sqrt{3}\omega\varphi_f sin(\theta) \\ e_{\beta 1} = e_{\beta 2} = e_\beta = \sqrt{3}\omega\varphi_f cos(\theta) \end{aligned} \quad (18)$$

Then, the torque expression can be simplified and becomes as follows:

$$\Gamma = \frac{e_\alpha(i_{\alpha 1} + i_{\alpha 2}) + e_\beta(i_{\beta 1} + i_{\beta 2})}{\Omega} \quad (19)$$

3.3. Electrical Dynamical Model in the (α β z$_1$ z$_2$ z$_3$ z$_4$)

Based in the next expression of the torque, a new change of variable based on the sum of the currents is introduced. And in order to preserve the order of the system (6), we introduce the difference in currents which will also have the advantage of eliminating the coupling terms present in the inductance matrix. These current differences have no effect on the torque, but this ensures the bijectivity of the transformation matrix from one frame of reference to another. Thus, in order to write the electrical equations in the new frame, where the inductance matrix is diagonal, we also apply it to the voltage, current, and EMF vectors.

From this, to transform the quantities of the (α_1 β_1 o_1 α_2 β_2 o_2) reference to this new reference, called (α β z_1 z_2 z_3 z_4), the following normalized matrix is defined:

$$\begin{bmatrix} x_\alpha \\ x_\beta \\ x_{z1} \\ x_{z2} \\ x_{z3} \\ x_{z'} \end{bmatrix} = \frac{1}{\sqrt{2}} \begin{bmatrix} \begin{bmatrix} 1 & 0 & 0 \\ 0 & 1 & 0 \\ 0 & 0 & 1 \end{bmatrix} & \begin{bmatrix} 1 & 0 & 0 \\ 0 & 1 & 0 \\ 0 & 0 & 1 \end{bmatrix} \\ \begin{bmatrix} 1 & 0 & 0 \\ 0 & 1 & 0 \\ 0 & 0 & 1 \end{bmatrix} & -\begin{bmatrix} 1 & 0 & 0 \\ 0 & 1 & 0 \\ 0 & 0 & 1 \end{bmatrix} \end{bmatrix} \begin{bmatrix} x_{\alpha 1} \\ x_{\beta 1} \\ x_{01} \\ x_{\alpha 2} \\ x_{\beta 1} \\ x_{02} \end{bmatrix} \quad (20)$$

where x = v, i, and e.

By applying this transformation to the electrical equation in (α_1 β_1 o_1 α_2 β_2 o_2), the new electrical equation is obtained.

Now, the electrical equation in this ($\alpha\beta z1z2\ z3z4$) frame can be easily deduced:

$$\begin{cases} v_\alpha = R_s i a + L_c M_{ss} \frac{d}{dt}(ia) - \sqrt{6}\omega \varphi_f \sin(\theta) \\ v_\beta = R_s i_\beta + L_c \frac{d}{dt}(i_\beta) + \sqrt{6}\omega \varphi_f \cos(\theta) \\ v_{zj} = R_s i_{zj} + l_{fs} \frac{d}{dt}(i_{zj})\ j = 1,\ 4 \end{cases} \quad (21)$$

where

$$L_c = l_{fs} + 3M_{ss} \quad (22)$$

Finally, the transformation matrix from the initial the (a1 b1 c1 a2 b2 c2) coordinate system to the final (α β z_1 z_2 z_3 z_4) coordinate system is as follows:

$$[T]^t = \frac{1}{\sqrt{2}} \begin{bmatrix} \begin{bmatrix} 1 & 0 & 0 \\ 0 & 1 & 0 \\ 0 & 0 & 1 \end{bmatrix} & \begin{bmatrix} 1 & 0 & 0 \\ 0 & 1 & 0 \\ 0 & 0 & 1 \end{bmatrix} \\ \begin{bmatrix} 1 & 0 & 0 \\ 0 & 1 & 0 \\ 0 & 0 & 1 \end{bmatrix} & -\begin{bmatrix} 1 & 0 & 0 \\ 0 & 1 & 0 \\ 0 & 0 & 1 \end{bmatrix} \end{bmatrix} \begin{bmatrix} \begin{bmatrix} [T_{32}]^t \\ [T_{31}]^t \end{bmatrix} & [0]_{3\times 3} \\ [0]_{3\times 3} & \begin{bmatrix} [[T_{32}] \cdot P(-\gamma)]^t \\ [T_{31}]^t \end{bmatrix} \end{bmatrix} \quad (23)$$

In the case where the angular offset is $\gamma = \frac{\pi}{6}$, this leads to the matrix T, as presented in Equation (24):

$$[T] = \begin{bmatrix} \sqrt{\frac{1}{3}} & -\sqrt{\frac{1}{11}} & -\sqrt{\frac{1}{11}} & \frac{1}{2} & -\frac{1}{2} & 0 \\ 0 & \frac{-1}{2} & \frac{1}{2} & -\sqrt{\frac{1}{11}} & -\sqrt{\frac{1}{11}} & \sqrt{\frac{1}{3}} \\ \sqrt{\frac{1}{6}} & \sqrt{\frac{1}{6}} & \sqrt{\frac{1}{6}} & \sqrt{\frac{1}{6}} & \sqrt{\frac{1}{6}} & \sqrt{\frac{1}{6}} \\ \sqrt{\frac{1}{3}} & -\sqrt{\frac{1}{11}} & -\sqrt{\frac{1}{11}} & -\frac{1}{2} & \frac{1}{2} & 0 \\ 0 & -\frac{1}{2} & \frac{1}{2} & \sqrt{\frac{1}{11}} & \sqrt{\frac{1}{11}} & \sqrt{\frac{1}{11}} \\ \sqrt{\frac{1}{6}} & \sqrt{\frac{1}{6}} & \sqrt{\frac{1}{6}} & -\sqrt{\frac{1}{6}} & -\sqrt{\frac{1}{6}} & -\sqrt{\frac{1}{6}} \end{bmatrix} \quad (24)$$

3.4. Electrical Equation in Park's Frame

By applying the classical Park transformation only to the ab component, the dynamical electrical model in the $(dqzj)_{j=1,4}$ frame can be established:

$$\begin{cases} V_d = R i_d + L_c \frac{d}{dt} i_d - \omega L_c i_q \\ V_q = R i_q + L_c \frac{d}{dt} i_q + \omega L_c i_d + \sqrt{6}\omega \varphi_f \\ V_{zj} = R i_{zj} + l_{fs} \frac{d}{dt} i_{zj}\ j = 1,\ 4 \end{cases} \quad (25)$$

The electromagnetic torque equation is as follows:

$$\Gamma = \frac{e_d i_d + e_q i_q}{\Omega} = \frac{e_q i_q}{\Omega} = p\frac{e_q i_q}{\omega} = \sqrt{6} p \varphi_f i_q \qquad (26)$$

3.5. Mechanical Equation

The mechanical equation is classical, and it is given by the following relationship:

$$J\frac{d\Omega}{dt} = \Gamma - \Gamma_l - f_v \Omega \qquad (27)$$

where J is the motor inertia, Γ_l is the load torque, and f_v is the viscous friction.

3.6. Modeling Approach for Control Strategies

By combining the electrical Equations (25) and (26) with the mechanical Equation (27), we obtain the model of the DSPMSM used to develop the control strategy, as depicted in Equation (28):

$$\begin{cases} \frac{d\Omega}{dt} = -\frac{\Gamma_l}{J} - \frac{f_v}{J}\Omega + \frac{\sqrt{6}p\varphi_f}{J}i_q \\ \frac{di_q}{dt} = -p\Omega i_d - \frac{\sqrt{6}\varphi_f}{L_6}p\Omega - \frac{R}{L_c}i_q + \frac{1}{L_c}V_q \\ \frac{di_d}{dt} = p\Omega i_q - \frac{R}{L_c}i_d + \frac{1}{L_c}V_d \\ \frac{di_{zj}}{dt} = \frac{R}{l_{fs}}i_{zj} + \frac{1}{l_{fs}}V_{zj} \quad for\ j = 1, \ldots, 4 \end{cases} \qquad (28)$$

The DSPMSM model in (28) may be reorganized in the following manner:

$$\begin{cases} \frac{d\Omega}{dt} = f_1 + g_1 i_q \\ \frac{di_q}{dt} = f_2 + g_2 V_q \\ \frac{di_d}{dt} = f_3 + g_3 V_d \\ \frac{di_{zj}}{dt} = f_4 i_{zj} + g_4 V_{zj} \quad for\ j = 1, \ldots, 4 \end{cases} \qquad (29)$$

where $f_1, \ldots, f_4$ and $g_1, \ldots, g_4$ are unknown continuous nonlinear functions.

$$\begin{cases} f_1 = -\frac{\Gamma_l}{J} - \frac{f_v}{J}\Omega, \quad and \quad g_1 = \frac{\sqrt{6}p\varphi_f}{J} \\ f_2 = -p\Omega i_d - \frac{\sqrt{6}\varphi_f}{L_c}p\Omega - \frac{R}{L_c}i_q, \; g_2 = g_3 = \frac{1}{L_c} \\ f_3 = p\Omega i_q - \frac{R}{L_c}i_d \\ f_4 = \frac{R}{l_f}, \; g_4 = \frac{1}{l_f} \end{cases} \qquad (30)$$

4. Nonlinear Control Design-Based Model for DSPMSM

In this section, a nonlinear control design-based model for DSPMSM is synthesized in order to obtain good tracking performances for speed and torque; to achieve this goal, some realistic assumptions are introduced.

Assumption 1. *The reference signals* Ω^*, i_q^*, i_d^*, *and* i_{zj}^*, *as well as their first derivatives, exhibit boundedness and continuity.*

Assumption 2. *The rotor speed, and stator current are measurable greatness.*

For the reference signals Ω^*, i_q^*, i_d^*, and i_{zj}^*, we may define the tracking errors and their corresponding filtered errors as follows:

$$Z_\Omega = \Omega^* - \Omega;\, S_\Omega = Z_\Omega + \lambda_\Omega \int_0^t Z_\Omega(\tau)d\tau;\, with\, Z_\Omega(0) = 0 \qquad (31)$$

$$Z_{i_q} = i_q^* - i_q; S_{i_q} = Z_{i_q} + \lambda_{i_q} \int_0^t Z_{i_q}(\tau) d\tau; \text{ with } Z_{i_q}(0) = 0 \tag{32}$$

$$Z_{i_d} = i_d^* - i_d; S_{i_d} = Z_{i_d} + \lambda_{i_d} \int_0^t Z_{i_d}(\tau) d\tau; \text{ with } Z_{i_d}(0) = 0 \tag{33}$$

$$\ldots Z_{i_{zj}} = i_{zj}^* - i_{zj}; S_{zj} = Z_{i_{zj}} + \lambda_{zj} \int_0^t Z_{zj}(\tau) d\tau; \text{ with } Z_{zj}(0) = 0 \; j = 1, \ldots, 4 \tag{34}$$

where $\lambda_\Omega, \lambda_{i_q}, \lambda_{i_d}$ and λ_{zj} are positive design parameters.

The control objectives are $i_q^* = i_d = 0$, $i_q = i_q^*$ and $i_{zj} = i_{zj}^* = 0$ for $j = 1, \ldots, 4$.

Step 1. Speed control.

Using Equation (28), the first filtered error dynamic of (31) is provided by the following:

$$\dot{S}_\Omega = \dot{\Omega}^* - f_1 - g_1 i_q^* + \lambda_\Omega Z_\Omega \tag{35}$$

Let us select the Lyapunov function candidate as $V_{1\Omega} = \frac{1}{2} S_\Omega^2$, and its time derivative is as follows:

$$\dot{V}_{1\Omega} = S_\Omega \dot{S}_\Omega = S_\Omega \left(\dot{\Omega}^* - f_1 - g_1 i_q^* + \lambda_\Omega Z_\Omega \right) \tag{36}$$

The control law, i_q^*, is formulated as follows:

$$i_q^* = \frac{1}{g_1} \left[\dot{\Omega}^* - f_1 + \lambda_\Omega Z_\Omega \right] + c_\Omega S_\Omega \tag{37}$$

where c_Ω is the positive design parameter.

It is simply verifiable, using (9), that

$$\dot{V}_{1\Omega} = -c_\Omega S_\Omega^2 < 0 \tag{38}$$

Step 2. Currents control.

Select the candidate Lyapunov function with augmentation as follows:

$$\begin{cases} V_{2i} = \frac{1}{2} S_{i_q}^2 + \frac{1}{2} S_{i_d}^2 \\ V_{2j} = \frac{1}{2} S_{zj}^2 \text{ for } j = 1, \ldots, 4 \end{cases} \tag{39}$$

The filtered error dynamics of (4) to (6) are given by the following:

$$\dot{S}_{i_q} = \frac{di_q^*}{dt} - f_2 - g_2 V_q + \lambda_{i_q} Z_{i_q} \tag{40}$$

$$\dot{S}_{i_d} = \frac{di_d^*}{dt} - f_3 - g_3 V_d + \lambda_{i_d} Z_{i_d} \tag{41}$$

$$\dot{S}_{zj} = \frac{di_{zj}^*}{dt} - f_4 - g_4 V_{zj} + \lambda_{zj} Z_{zj} \text{ for } j = 1, \ldots, 4 \tag{42}$$

After that, the time derivative of (39) is written as follows:

$$\begin{cases} \dot{V}_{2i} = S_{i_q} \left(\frac{di_q^*}{dt} - f_2 - g_2 V_q + \lambda_{i_q} Z_{i_q} \right) + S_{i_d} \left(\frac{di_d^*}{dt} - f_3 - g_3 V_d + \lambda_{i_d} Z_{i_d} \right) \\ \dot{V}_{2j} = S_{zj} \left(\frac{di_{zj}^*}{dt} - f_4 - g_4 V_{zj} + \lambda_{zj} Z_{zj} \right) \quad \text{for } j = 1, \ldots, 4 \end{cases} \tag{43}$$

The control laws V_q, V_d, and V_{zj} for $j = 1, \ldots, 4$ are designed as follows:

$$V_q = \frac{1}{g_2} \left[\frac{di_q^*}{dt} - f_2 + \lambda_{i_q} Z_{i_q} \right] + c_{i_q} S_{i_q} \tag{44}$$

$$V_d = \frac{1}{g_3}\left[\frac{di_d^*}{dt} - f_3 + \lambda_{id}Z_{i_d}\right] + c_{i_d}S_{i_d} \qquad (45)$$

$$V_{zj} = \frac{1}{g_4}\left[\frac{di_{zj}^*}{dt} - f_4 + \lambda_{zj}Z_{zj}\right] + c_{zj}S_{zj} \text{ for } j = 1, \ldots, 4 \qquad (46)$$

where c_{i_q}, c_{i_d}, and c_{zj} for $j = 1, \ldots, 4$ are positive design parameters.

Using (44), (45), and (46), it is simple to demonstrate that

$$\begin{cases} \dot{V}_{2i} = -c_{i_q}S_{i_q}^2 - c_{i_d}S_{i_d}^2 < 0 \\ \dot{V}_{zj} = -c_{zj}S_{zj}^2 < 0 \end{cases} \qquad (47)$$

The control laws i_q^*, V_q, V_d, and V_{zj} for $j = 1, \ldots, 4$ can be expressed as follows [25]:

$$i_q^* = I_q^* + c_\Omega S_\Omega \qquad (48)$$

$$V_q = U_q + c_{i_q}S_{i_q} \qquad (49)$$

$$V_d = U_d + c_{i_d}S_{i_d} \qquad (50)$$

$$V_{zj} = U_{zj} + c_{zj}S_{zj} \text{ for } j = 1, \ldots, 4 \qquad (51)$$

where the ideal controls I_q^*, U_q, U_d, and U_{zj} for $j = 1, \ldots, 4$ are given by the following:

$$I_q^* = \frac{1}{g_1}\left[\dot{\Omega}^* - f_1 + \lambda_\Omega Z_\Omega\right] \qquad (52)$$

$$U_q = \frac{1}{g_2}\left[\frac{di_q^*}{dt} - f_2 + \lambda_{iq}Z_{i_q}\right] \qquad (53)$$

$$U_d = \frac{1}{g_3}\left[\frac{di_d^*}{dt} - f_3 + \lambda_{id}Z_{i_d}\right] \qquad (54)$$

$$U_{zj} = \frac{1}{g_4}\left[\frac{di_{zj}^*}{dt} - f_4 + \lambda_{zj}Z_{zj}\right] \text{ for } j = 1, \ldots, 4 \qquad (55)$$

Given that $\dot{V}_{1\Omega}(t)$, $\dot{V}_{2i}(t)$, and $\dot{V}_{zj}(t)$ for $j = 1, \ldots, 4$ are negative semi-definite, it follows that $V_{1\Omega}(t) \leq V_{1\Omega}(0)$, $V_{2i}(t) \leq V_{2i}(0)$ and $V_{zj}(t) \leq V_{zj}(0)$ for $j = 1, \ldots, 4$.

Consequently, S_Ω, S_{i_q}, S_{i_d}, and S_{zj} for $j = 1, \ldots, 4$ exhibit uniform boundedness. This indicates that the closed-loop signals S_Ω, S_{i_q}, S_{i_d}, and S_{zj} for $j = 1, \ldots, 4$, i_q^*, V_q, V_d, and V_{zj} for $j = 1, \ldots, 4$ are constrained within certain limits.

Given that $V_{1\Omega}(0)$, $V_{2i}(0)$, and $V_{zj}(0)$ for $j = 1, \ldots, 4$ are limited, and $V_{1\Omega}$, V_{2i}, and V_{zj} for $j = 1, \ldots, 4$ are non-increasing and limited from below, it can be concluded that the $\lim_{t\to\infty} V_{1\Omega}(t)$, $\lim_{t\to\infty} V_{2i}(t)$, and $\lim_{t\to\infty} V_{zj}(t)$ for $j = 1, \ldots, 4$ exist. By using Barbalat's Lemma [41], it can be deduced that $(S_\Omega, S_{i_q}, S_{i_d}, S_{zj} \text{ for } j = 1, \ldots, 4) \to 0$ as $t \to \infty$, indicating the asymptotic convergence of filtered errors to zero.

The control laws I_q^*, U_q, U_d, and U_{zj} for $j = 1, \ldots, 4$ given in (52) to (55), and they may be readily derived if the nonlinear functions $f_1, \ldots, f_4$ and $g_1, \ldots, g_4$ are known; nevertheless, the specific forms of these nonlinear functions remain unidentified. Therefore, seven adaptive fuzzy logic systems are used to directly approach these control laws.

5. Overview of the Fuzzy Logic System

A fuzzy logic system is composed of many components: a fuzzifier, a set of fuzzy if–then rules, a fuzzy inference engine, and a defuzzifier. These components are shown in Figure 4.

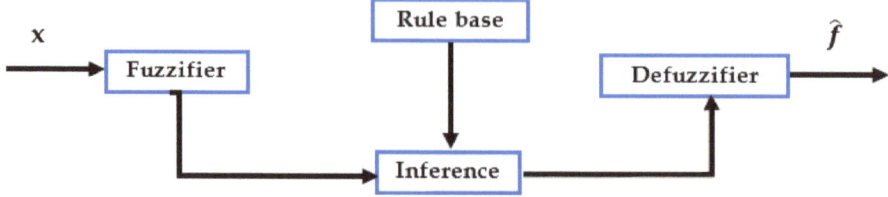

Figure 4. Fuzzy logic system configuration.

The fuzzy inference engine uses fuzzy if–then rules to convert an input vector, $x^T = [x_1, x_2, \ldots, x_n] \in \mathfrak{R}^n$, to an output, $\hat{f} \in \mathfrak{R}^n$. The i-th fuzzy rule is expressed as follows:

Rule (i): if x_1 is B_1^i and ... and x_n is B_n^i then $\hat{f}$ is θ_i (56)

where $B_1^i, B_2^i, \ldots, B_n^i$ are fuzzy sets, and y_i is the fuzzy output singleton in the i_{th} rule. The Singleton fuzzifier, product inference, and center-average defuzzifier produce the fuzzy system's output and may be written as:

$$\hat{f}(x) = \frac{\sum_{i=1}^m y_i \left(\prod_{l=1}^n \mu_{B_l^i}(x_l)\right)}{\sum_{i=1}^m \left(\prod_{l=1}^n \mu_{B_l^i}(x_l)\right)} = \Theta^T \psi(x) \quad (57)$$

The degree of membership of x_l to B_l^i is denoted as $\mu_{B_l^i}(x_l)$. The number of fuzzy rules is represented by m. The adjustable parameter vector denoted by $\Theta^T = [\theta_1, \theta_2, \ldots \theta_m]$ is formed by consequent parameters, and the vector $\psi^T(x) = [\psi_1, \psi_2, \ldots, \psi_m]$ with the following:

$$\psi_i(x) = \frac{\left(\prod_{l=1}^n \mu_{B_l^i}(x_l)\right)}{\sum_{i=1}^m \left(\prod_{l=1}^n \mu_{B_l^i}(x_l)\right)} \quad (58)$$

Referring to the fuzzy basis function (FBF), the assumption that the FBFs are chosen in such a manner that there is always at least one active rule is made throughout the whole of the work [10]; that is to say, $\sum_{i=1}^m \left(\prod_{l=1}^n \mu_{B_l^i}(x_l)\right) > 0$.

The fuzzy system (57) is often used in control systems. Based on the universal approximation findings [42,43], the fuzzy system (16) has the capability to estimate any nonlinear smooth function, $f(x)$, inside a limited working region with a high level of accuracy.

It is crucial to specify the structure of the fuzzy system, including the relevant inputs, the number of membership functions for each input, and the number of rules. Additionally, it is important to accurately define the parameters of the membership functions in advance. The subsequent parameters, Θ, are subsequently calculated by suitable adaption methods.

6. Model Free Control on Adaptive Fuzzy Control Design for DSPMSM

The goal is to develop an appropriate adaptive fuzzy control system for an uncertain DSPMSM model in order to achieve the precise tracking of torque and speed. Fuzzy logic systems are used to approximate the ideal controls, U_{zj}, for $j = 1, \ldots, 4$,, U_q, and U_d.

Lemma 1 ([10]). *For each real continuous function, $f(x)$, defined on a compact subset, $\Phi_f \subset \mathfrak{R}^n$, and for any random $\varepsilon > 0$, there exists a fuzzy logic system, such that we have the following:*

$$\sup_{x \in \Phi_f} \left| f(x) - \Theta^T \psi(x) \right| \leq \varepsilon \quad (59)$$

By utilizing Lemma 1 and referring to the demonstration provided in [10], it can be concluded that fuzzy logic systems possess the ability to universally approximate any smooth function inside a compact set. Given the approximation capacity of fuzzy logic systems, it is reasonable to suppose that the control laws, U_{zj}, for $j = 1, \ldots, 4$, I_q^*, U_q, and U_d may be estimated as follows:

$$U_{zj}(x_j|\Theta_j) = \Theta_j^T \psi_j(x_j) \text{ for } j = 1, \ldots, 4 \tag{60}$$

$$I_q^*(x_5|\Theta_5) = \Theta_5^T \psi_5(x_5) \tag{61}$$

$$U_q(x_6|\Theta_6) = \Theta_6^T \psi_6(x_6) \tag{62}$$

$$U_d(x_7|\Theta_7) = \Theta_7^T \psi_7(x_7) \tag{63}$$

As stated in [29], the optimal parameter vectors, Θ_j^*, for $j = 1, \ldots, 4$, Θ_5^*, Θ_6^*, and Θ_7^* are determined as follows:

$$\Theta_j^* = \arg \min_{\Theta_j \in \Phi_{\theta_j}} \left\{ \sup_{x_j \in \Phi_{x_j}} \left| \hat{U}_{zj}(x_j|\Theta_j) - U_{zj}(t) \right| \right\} \text{ for } j = 1, \ldots, 4 \tag{64}$$

$$\Theta_5^* = \arg \min_{\Theta_5 \in \Phi_{\theta_5}} \left\{ \sup_{x_5 \in \Phi_{x_5}} \left| \hat{I}_q^*(x_5|\Theta_5) - I_q^*(t) \right| \right\} \tag{65}$$

$$\Theta_6^* = \arg \min_{\Theta_6 \in \Phi_{\theta_6}} \left\{ \sup_{x_6 \in \Phi_{x_6}} \left| \hat{U}_q(x_6|\Theta_6) - U_q(t) \right| \right\} \tag{66}$$

$$\Theta_7^* = \arg \min_{\Theta_7 \in \Phi_{\theta_7}} \left\{ \sup_{x_7 \in \Phi_{x_7}} \left| \hat{U}_d(x_7|\Theta_7) - U_d(t) \right| \right\} \tag{67}$$

where Φ_{x_j}, Φ_{x_5}, Φ_{x_6}, and Φ_{x_7} are compact set for x_j, x_5, x_6, and x_7. On the other hand, Φ_{θ_j}, Φ_{θ_5}, Φ_{θ_6}, and Φ_{θ_7} are compact set for θ_j, θ_5, θ_6, and θ_7.

Furthermore, the minimal fuzzy approximation errors ε_j, ε_5, ε_6, and ε_7 are precisely specified as follows:

$$\varepsilon_j = U_{zj}(t) - \hat{U}_{zj}\left(x_j|\Theta_j^*\right) \text{ for } j = 1, \ldots, 4 \tag{68}$$

$$\varepsilon_5 = I_q^*(t) - \hat{I}_q^*\left(x_5|\Theta_5^*\right) \tag{69}$$

$$\varepsilon_6 = U_q(t) - \hat{U}_q\left(x_6|\Theta_6^*\right) \tag{70}$$

$$\varepsilon_7 = U_d(t) - \hat{U}_d\left(x_7|\Theta_7^*\right) \tag{71}$$

The control laws, U_{zj}, for $j = 1, \ldots, 4$, I_q^*, U_q, and U_d may be reformulated as follows:

$$\begin{aligned} U_{zj}(t) &= \hat{U}_{zj}\left(x_j|\Theta_j^*\right) + \varepsilon_j \\ &= \Theta_j^{*T} \psi_j(x_j) + \varepsilon_j \text{ for } j = 1, \ldots, 4 \end{aligned} \tag{72}$$

$$\begin{aligned} I_q^*(t) &= \hat{I}_q^*(x_5|\Theta_5^*) + \varepsilon_5 \\ &= \Theta_5^{*T} \psi_5(x_5) + \varepsilon_5 \end{aligned} \tag{73}$$

$$\begin{aligned} U_q(t) &= \hat{U}_q(x_6|\Theta_6^*) + \varepsilon_6 \\ &= \Theta_6^{*T} \psi_6(x_6) + \varepsilon_6 \end{aligned} \tag{74}$$

$$U_d(t) = \hat{U}_d(x_7|\Theta_7^*) + \varepsilon_7$$
$$= \Theta_7^{*T}\psi_7(x_7) + \varepsilon_7 \tag{75}$$

Let us suppose that the minimal fuzzy approximation errors ε_j, ε_5, ε_6, and ε_7 are, respectively, bounded above by $\bar{\varepsilon}_j > 0$, $\bar{\varepsilon}_5 > 0$, $\bar{\varepsilon}_6 > 0$, and $\bar{\varepsilon}_7 > 0$, meaning that $|\bar{\varepsilon}_j| \leq \bar{\varepsilon}_j$, $|\bar{\varepsilon}_5| \leq \bar{\varepsilon}_5$, $|\bar{\varepsilon}_6| \leq \bar{\varepsilon}_6$, and $|\bar{\varepsilon}_7| \leq \bar{\varepsilon}_7$.

7. Controller Design

The suggested controller in this section uses fuzzy adaptive backstepping and parameter adaptive laws to ensure that all internal signals of the closed-loop system are uniformly ultimately bounded and to minimize filtered errors given in (31) to (34).

In order to estimate the nonlinear control laws, $U_{zj}(t)$, $I_q^*(t)$, $U_q(t)$, and $U_d(t)$, presented in (52) to (55), the fuzzy adaptive control terms in (76) to (79) are defined as $\hat{U}_{zj}(t)$, $\hat{I}_q^*(t)$, $\hat{U}_q(t)$, and $\hat{U}_d(t)$:

$$\hat{U}_{zj}(t) = \Theta_j^T \psi_j(x_j) \tag{76}$$

$$\hat{I}_q^*(t) = \Theta_5^T \psi_5(x_5) \tag{77}$$

$$\hat{U}_q(t) = \Theta_6^T \psi_6(x_6) \tag{78}$$

$$\hat{U}_d(t) = \Theta_7^T \psi_7(x_7) \tag{79}$$

The selected input vectors for the used fuzzy systems are determined as follows:

$$x_j = \begin{bmatrix} i_{zj}, Z_{i_{zj}} \end{bmatrix}^T \text{ for } j = 1, \ldots, 4, \ x_5 = [\Omega, i_q]^T, \ x_6 = \begin{bmatrix} \Omega, i_q, i_q^*, Z_\Omega \end{bmatrix}^T, \ x_6 = [i_d, i_q]^T$$

The adaptive control laws that guarantee the stability of the closed-loop system may be written as follows [28]:

$$V_{zj}(t) = \hat{U}_{zj}(t) + \hat{\varepsilon}_j \tanh\left(\frac{S_{zj}}{\chi_j}\right) + c_{zj}S_{zj} \text{ for } j = 1, \ldots, 4 \tag{80}$$

$$i_q^* = \hat{I}_q^*(t) + \hat{\varepsilon}_5 \tanh\left(\frac{S_\Omega}{\chi_5}\right) + c_\Omega S_\Omega \tag{81}$$

$$V_q(t) = \hat{U}_q(t) + \hat{\varepsilon}_6 \tanh\left(\frac{S_{i_q}}{\chi_6}\right) + c_{i_q} S_{i_q} \tag{82}$$

$$V_d(t) = \hat{U}_d(t) + \hat{\varepsilon}_7 \tanh\left(\frac{S_{i_d}}{\chi_7}\right) + c_{i_d} S_{i_d} \tag{83}$$

where χ_j, χ_5, χ_6, and χ_7 are designed positive constants.

Then, $\hat{\varepsilon}_j, \hat{\varepsilon}_5, \hat{\varepsilon}_6$, and $\hat{\varepsilon}_7$ are adjusted as follows:

$$\dot{\hat{\varepsilon}}_j = \eta_j S_{zj} \tanh\left(\frac{S_{zj}}{\chi_j}\right) - \alpha_j \hat{\varepsilon}_j \tag{84}$$

$$\dot{\hat{\varepsilon}}_5 = \eta_5 S_\Omega \tanh\left(\frac{S_\Omega}{\chi_5}\right) - \alpha_5 \hat{\varepsilon}_5 \tag{85}$$

$$\dot{\hat{\varepsilon}}_6 = \eta_6 S_\Omega \tanh\left(\frac{S_{i_q}}{\chi_6}\right) - \alpha_6 \hat{\varepsilon}_6 \tag{86}$$

$$\dot{\hat{\varepsilon}}_7 = \eta_7 S_\Omega \tanh\left(\frac{S_{i_d}}{\chi_7}\right) - \alpha_7 \hat{\varepsilon}_7 \tag{87}$$

where $\eta_j, \eta_5, \eta_6, \eta_7, \alpha_j, \alpha_5, \alpha_6$, and α_7 are designed positive constants.

The vectors $\Theta_j, \Theta_5, \Theta_6,$ and Θ_7 represent the adaptable parameters of the fuzzy logic system and are defined as follows:

$$\dot{\Theta}_j = \gamma_j S_{zj} \psi_j(x_j) - \sigma_j \dot{\Theta}_j \tag{88}$$

$$\dot{\Theta}_5 = \gamma_5 S_\Omega \psi_5(x_5) - \sigma_5 \dot{\Theta}_5 \tag{89}$$

$$\dot{\Theta}_6 = \gamma_6 S_{i_q} \psi_6(x_6) - \sigma_6 \dot{\Theta}_6 \tag{90}$$

$$\dot{\Theta}_7 = \gamma_7 S_{i_d} \psi_7(x_7) - \sigma_7 \dot{\Theta}_7 \tag{91}$$

By replacing (76) in (80), (77) in (81), (78) in (82), and (79) in (83), we yield the following:

$$V_{zj}(t) = \Theta_j^T \psi_j(x_j) + \hat{\varepsilon}_j \tanh\left(\frac{S_{zj}}{\chi_j}\right) + c_{zj} S_{zj} \text{ for } j = 1, \ldots, 4 \tag{92}$$

$$i_q^* = \Theta_5^T \psi_5(x_5) + \hat{\varepsilon}_5 \tanh\left(\frac{S_\Omega}{\chi_5}\right) + c_\Omega S_\Omega \tag{93}$$

$$U_q(t) = \Theta_6^T \psi_6(x_6) + \hat{\varepsilon}_6 \tanh\left(\frac{S_{i_q}}{\chi_6}\right) + c_{i_q} S_{i_q} \tag{94}$$

$$U_d(t) = \Theta_7^T \psi_7(x_7) + \hat{\varepsilon}_7 \tanh\left(\frac{S_{i_d}}{\chi_7}\right) + c_{i_d} S_{i_d} \tag{95}$$

8. Stability Demonstration Using Lyapunov Theory

Given the following candidate Lyapunov function,

$$V = \frac{1}{2} \sum_{j=1}^{4} \left(S_{zj}^2 + \frac{1}{\gamma_j} \tilde{\Theta}_j^T \tilde{\Theta}_j + \frac{1}{\eta_j} \tilde{\varepsilon}_j^T \tilde{\varepsilon}_j \right) + \frac{1}{2} \left(S_\Omega^2 + \frac{1}{\gamma_5} \tilde{\Theta}_5^T \tilde{\Theta}_5 + \frac{1}{\eta_5} \tilde{\varepsilon}_5^T \tilde{\varepsilon}_5 \right) + \\ \frac{1}{2} \left(S_{i_q}^2 + \frac{1}{\gamma_6} \tilde{\Theta}_6^T \tilde{\Theta}_6 + \frac{1}{\eta_6} \tilde{\varepsilon}_6^T \tilde{\varepsilon}_6 \right) + \frac{1}{2} \left(S_{i_d}^2 + \frac{1}{\gamma_7} \tilde{\Theta}_7^T \tilde{\Theta}_7 + \frac{1}{\eta_7} \tilde{\varepsilon}_7^T \tilde{\varepsilon}_7 \right) \tag{96}$$

where $\tilde{\Theta}_j, \tilde{\Theta}_5, \tilde{\Theta}_6,$ and $\tilde{\Theta}_7$ are the approximation errors, which are given as follows:

$$\tilde{\Theta}_j^T = \Theta_j^{T^*} - \Theta_j^T \tag{97}$$

$$\tilde{\Theta}_5^T = \Theta_5^{T^*} - \Theta_5^T \tag{98}$$

$$\tilde{\Theta}_6^T = \Theta_6^{T^*} - \Theta_6^T \tag{99}$$

$$\tilde{\Theta}_7^T = \Theta_7^{T^*} - \Theta_7^T \tag{100}$$

$\tilde{\varepsilon}_j, \tilde{\varepsilon}_5, \tilde{\varepsilon}_6,$ and $\tilde{\varepsilon}_7$ are the approximation errors expressed in (101) to (104), with $\varepsilon_j^*, \varepsilon_5^*, \varepsilon_6^*,$ and ε_7^* serving as the optimal parameters; and $\hat{\varepsilon}_j, \hat{\varepsilon}_5, \hat{\varepsilon}_6,$ and $\hat{\varepsilon}_7$ are the estimate of $\varepsilon_j^*, \varepsilon_5^*, \varepsilon_6^*,$ and ε_7^*, respectively.

$$\tilde{\varepsilon}_j = \varepsilon_j^* - \hat{\varepsilon}_j \tag{101}$$

$$\tilde{\varepsilon}_5 = \varepsilon_5^* - \hat{\varepsilon}_5 \tag{102}$$

$$\tilde{\varepsilon}_6 = \varepsilon_6^* - \hat{\varepsilon}_6 \tag{103}$$

$$\tilde{\varepsilon}_7 = \varepsilon_7^* - \hat{\varepsilon}_7 \tag{104}$$

The temporal derivative of V is computed as follows:

$$\dot{V} = \sum_{j=1}^{4}\left(S_{zj}\{U_{zj} - V_{zj}\} + \frac{1}{\gamma_j}\tilde{\Theta}_j^T\dot{\tilde{\Theta}}_j + \frac{1}{\eta_j}\tilde{\varepsilon}_j^T\dot{\tilde{\varepsilon}}_j\right) + \left(S_{\Omega}\{I_q^* - i_q^*\} + \frac{1}{\gamma_5}\tilde{\Theta}_5^T\dot{\tilde{\Theta}}_5 + \frac{1}{\eta_5}\tilde{\varepsilon}_5^T\dot{\tilde{\varepsilon}}_5\right) + \\ \left(S_{i_q}\{U_q - V_q\} + \frac{1}{\gamma_6}\tilde{\Theta}_6^T\dot{\tilde{\Theta}}_6 + \frac{1}{\eta_6}\tilde{\varepsilon}_6^T\dot{\tilde{\varepsilon}}_6\right) + \left(S_{i_d}\{U_d - V_d\} + \frac{1}{\gamma_7}\tilde{\Theta}_7^T\dot{\tilde{\Theta}}_7 + \frac{1}{\eta_7}\tilde{\varepsilon}_7^T\dot{\tilde{\varepsilon}}_7\right)$$
(105)

By introducing (72) through (75) and (92) through (95) in (105), we can yield the following:

$$\dot{V} = \sum_{j=1}^{4}\left(S_{zj}\left\{\Theta_j^{T*}\psi_j(x_j) + \varepsilon_j - \Theta_j^T\psi_j(x_j) - \hat{\varepsilon}_j tanh\left(\frac{S_{zj}}{\chi_j}\right) - c_{zj}S_{zj}\right\} + \frac{1}{\gamma_j}\tilde{\Theta}_j^T\dot{\tilde{\Theta}}_j + \frac{1}{\eta_j}\tilde{\varepsilon}_j^T\dot{\tilde{\varepsilon}}_j\right) + \\ \left(S_{\Omega}\left\{\Theta_5^{T*}\psi_5(x_5) + \varepsilon_5 - \Theta_5^T\psi_5(x_5) - \hat{\varepsilon}_5 tanh\left(\frac{S_{\Omega}}{\chi_5}\right) - c_{\Omega}S_{\Omega}\right\} + \frac{1}{\gamma_5}\tilde{\Theta}_5^T\dot{\tilde{\Theta}}_5 + \frac{1}{\eta_5}\tilde{\varepsilon}_5^T\dot{\tilde{\varepsilon}}_5\right) + \\ \left(S_{i_q}\left\{\Theta_6^{T*}\psi_6(x_6) + \varepsilon_6 - \Theta_6^T\psi_6(x_6) - \hat{\varepsilon}_6 tanh\left(\frac{S_{i_q}}{\chi_6}\right) - c_{i_q}S_{i_q}\right\} + \frac{1}{\gamma_6}\tilde{\Theta}_6^T\dot{\tilde{\Theta}}_6 + \frac{1}{\eta_6}\tilde{\varepsilon}_6^T\dot{\tilde{\varepsilon}}_6\right) + \\ \left(S_{i_d}\left\{\Theta_7^{T*}\psi_7(x_7) + \varepsilon_7 - \Theta_7^T\psi_7(x_7) - \hat{\varepsilon}_7 tanh\left(\frac{S_{i_d}}{\chi_7}\right) - c_{i_d}S_{i_d}\right\} + \frac{1}{\gamma_7}\tilde{\Theta}_7^T\dot{\tilde{\Theta}}_7 + \frac{1}{\eta_7}\tilde{\varepsilon}_7^T\dot{\tilde{\varepsilon}}_7\right)$$
(106)

Given that the optimal parameters $\Theta_j^{T*}, \Theta_5^{T*}, \Theta_6^{T*}, \Theta_7^{T*}, \varepsilon_j^*, \varepsilon_5^*, \varepsilon_6^*$, and ε_7^* vary slowly over time, $\left(\dot{\Theta}_j^{T*} = \dot{\Theta}_5^{T*} = \dot{\Theta}_6^{T*} = \dot{\Theta}_7^{T*} = 0\right)$ and $\left(\dot{\varepsilon}_m^* = \dot{\varepsilon}_5^* = \dot{\varepsilon}_6^* = \dot{\varepsilon}_7^* = 0\right)$, the temporal derivative of the approximation errors may be expressed as follows:

$$\begin{cases} \dot{\tilde{\Theta}}_j^T = -\dot{\Theta}_j^T \\ \dot{\tilde{\Theta}}_5^T = -\dot{\Theta}_5^T \\ \dot{\tilde{\Theta}}_6^T = -\dot{\Theta}_6^T \\ \dot{\tilde{\Theta}}_7^T = -\dot{\Theta}_7^T \end{cases}$$
(107)

$$\begin{cases} \dot{\tilde{\varepsilon}}_j = -\dot{\hat{\varepsilon}}_j \\ \dot{\tilde{\varepsilon}}_5 = -\dot{\hat{\varepsilon}}_5 \\ \dot{\tilde{\varepsilon}}_6 = -\dot{\hat{\varepsilon}}_6 \\ \dot{\tilde{\varepsilon}}_6 = -\dot{\hat{\varepsilon}}_6 \end{cases}$$
(108)

By substituting (107) and (108) into (106), we obtain the following:

$$\dot{V} = \sum_{j=1}^{4}\left(-c_{zj}S_{zj}^2 + S_{zj}\tilde{\Theta}_j^T\psi_j(x_j) + S_{zj}\left\{\varepsilon_j - \hat{\varepsilon}_j tanh\left(\frac{S_{zj}}{\chi_j}\right)\right\} - \frac{1}{\gamma_j}\tilde{\Theta}_j^T\dot{\Theta}_j + \frac{1}{\eta_j}\tilde{\varepsilon}_j^T\dot{\hat{\varepsilon}}_j\right) + \\ \left(-c_{\Omega}S_{\Omega}^2 + S_{\Omega}\tilde{\Theta}_5^T\psi_5(x_5) + S_{\Omega}\left\{\varepsilon_5 - \hat{\varepsilon}_5 tanh\left(\frac{S_{\Omega}}{\chi_5}\right)\right\} - \frac{1}{\gamma_5}\tilde{\Theta}_5^T\dot{\Theta}_5 + \frac{1}{\eta_5}\tilde{\varepsilon}_5^T\dot{\hat{\varepsilon}}_5\right) + \left(-c_{i_q}S_{i_q}^2 + S_{i_q}\tilde{\Theta}_6^T\psi_6(x_6) + \\ S_{i_q}\left\{\varepsilon_6 - \hat{\varepsilon}_6 tanh\left(\frac{S_{i_q}}{\chi_6}\right)\right\} - \frac{1}{\gamma_6}\tilde{\Theta}_6^T\dot{\Theta}_6 + \frac{1}{\eta_6}\tilde{\varepsilon}_6^T\dot{\hat{\varepsilon}}_6\right) + \left(\begin{array}{c}S_{i_d}\left\{\Theta_7^{T*}\psi_7(x_7) + \varepsilon_7 - \Theta_7^T\psi_7(x_7) - \hat{\varepsilon}_7 tanh\left(\frac{S_{i_d}}{\chi_7}\right) - c_{i_d}S_{i_d}\right\} \\ + \frac{1}{\gamma_7}\tilde{\Theta}_7^T\dot{\tilde{\Theta}}_7 + \frac{1}{\eta_7}\tilde{\varepsilon}_7^T\dot{\tilde{\varepsilon}}_7\end{array}\right)$$
(109)

$$\dot{V} \leq \sum_{j=1}^{4}\left(-c_{zj}S_{zj}^2 + \frac{1}{\gamma_j}\widetilde{\Theta}_j^T\left\{\gamma_j S_{zj}\psi_j(x_j) - \dot{\Theta}_j\right\} + |S_{zj}|\varepsilon_j^* - S_{zj}\hat{\varepsilon}_j tanh\left(\frac{S_{zj}}{\chi_j}\right) + \frac{1}{\eta_j}\widetilde{\varepsilon}_j\left\{\eta_j S_{zj}tanh\left(\frac{S_{zj}}{\chi_j}\right) - \dot{\hat{\varepsilon}}_j\right\}\right.$$
$$-\varepsilon_j^* S_{zj}tanh\left(\frac{S_{zj}}{\chi_j}\right) + \hat{\varepsilon}_j S_{zj}tanh\left(\frac{S_{zj}}{\chi_j}\right)\right) + \left(-c_\Omega S_\Omega^2 + \frac{1}{\gamma_5}\widetilde{\Theta}_5^T\left\{\gamma_5 S_\Omega \psi_5(x_5) - \dot{\Theta}_5\right\} + |S_\Omega|\varepsilon_5^*\right.$$
$$-S_\Omega \hat{\varepsilon}_5 tanh\left(\frac{S_\Omega}{\chi_5}\right) + \frac{1}{\eta_5}\widetilde{\varepsilon}_5\left\{\eta_5 S_\Omega tanh\left(\frac{S_\Omega}{\chi_5}\right) - \dot{\hat{\varepsilon}}_5\right\} - \varepsilon_5^* S_\Omega tanh\left(\frac{S_\Omega}{\chi_5}\right) + \hat{\varepsilon}_5 S_\Omega tanh\left(\frac{S_\Omega}{\chi_5}\right)\right)$$
$$+\left(-c_{i_q}S_{i_q}^2 + \frac{1}{\gamma_6}\widetilde{\Theta}_6^T\left\{\gamma_6 S_{i_q}\psi_6(x_6) - \dot{\Theta}_6\right\} + |S_{i_q}|\varepsilon_6^* - S_{i_q}\hat{\varepsilon}_6 tanh\left(\frac{S_{i_q}}{\chi_6}\right)\right.$$
$$+\frac{1}{\eta_6}\widetilde{\varepsilon}_6\left\{\eta_6 S_{i_q}tanh\left(\frac{S_{i_q}}{\chi_6}\right) - \dot{\hat{\varepsilon}}_6\right\} - \varepsilon_6^* S_{i_q}tanh\left(\frac{S_{i_q}}{\chi_6}\right) + \hat{\varepsilon}_6 S_{i_q}tanh\left(\frac{S_{i_q}}{\chi_5}\right)\right)$$
$$+\left(-c_{id}S_{id}^2 + \frac{1}{\gamma_7}\widetilde{\Theta}_7^T\left\{\gamma_7 S_{i_d}\psi_7(x_7) - \dot{\Theta}_7\right\} + |S_{i_d}|\varepsilon_7^* - S_{i_d}\hat{\varepsilon}_7 tanh\left(\frac{S_{i_d}}{\chi_7}\right)\right.$$
$$+\frac{1}{\eta_7}\widetilde{\varepsilon}_7\left\{\eta_7 S_{i_d}tanh\left(\frac{S_{i_d}}{\chi_7}\right) - \dot{\hat{\varepsilon}}_7\right\} - \varepsilon_7^* S_{i_d}tanh\left(\frac{S_{i_d}}{\chi_7}\right) + \hat{\varepsilon}_7 S_{i_d}tanh\left(\frac{S_{i_d}}{\chi_7}\right)\right) \quad (110)$$

Lemma 2 ([43]). *The hyperbolic tangent function fulfils the following condition for all given values of $x \in \Re$ and $\chi > 0$:*

$$f(x) = |x| - xtanh\left(\frac{x}{\chi}\right) \leq \zeta\chi \quad (111)$$

where $\zeta = 0.2785$.

By replacing the adaptive rules (84) through (91) into Equation (110) and using Lemma 2, we obtain the following:

$$\dot{V} \leq \sum_{j=1}^{4}\left(-c_{zj}S_{zj}^2 + \varepsilon_j^*\zeta + \frac{\sigma_j}{\gamma_j}\widetilde{\Theta}_j^T\Theta_j + \frac{\alpha_j}{\eta_j}\widetilde{\varepsilon}_j\hat{\varepsilon}_j\right) + \left(-c_\Omega S_\Omega^2 + \varepsilon_5^*\zeta + \frac{\sigma_5}{\gamma_5}\widetilde{\Theta}_5^T\Theta_5 + \frac{\alpha_5}{\eta_5}\widetilde{\varepsilon}_5\hat{\varepsilon}_5\right) +$$
$$\left(-c_{i_q}S_{i_q}^2 + \varepsilon_6^*\zeta + \frac{\sigma_6}{\gamma_6}\widetilde{\Theta}_6^T\Theta_6 + \frac{\alpha_6}{\eta_6}\widetilde{\varepsilon}_6\hat{\varepsilon}_6\right) + \left(-c_{i_d}S_{i_d}^2 + \varepsilon_7^*\zeta + \frac{\sigma_7}{\gamma_7}\widetilde{\Theta}_7^T\Theta_7 + \frac{\alpha_7}{\eta_7}\widetilde{\varepsilon}_7\hat{\varepsilon}_7\right) \quad (112)$$

The following inequalities are derived by replacing Young's inequality for the terms $\frac{\sigma_j}{\gamma_j}\widetilde{\Theta}_j^T\Theta_j$, $\frac{\sigma_5}{\gamma_5}\widetilde{\Theta}_5^T\Theta_5$, $\frac{\sigma_6}{\gamma_6}\widetilde{\Theta}_6^T\Theta_6$, $\frac{\sigma_7}{\gamma_7}\widetilde{\Theta}_7^T\Theta_7$, $\frac{\alpha_j}{\eta_j}\widetilde{\varepsilon}_j\hat{\varepsilon}_j$, $\frac{\alpha_5}{\eta_5}\widetilde{\varepsilon}_5\hat{\varepsilon}_5$, $\frac{\alpha_6}{\eta_6}\widetilde{\varepsilon}_6\hat{\varepsilon}_6$, and $\frac{\alpha_7}{\eta_7}\widetilde{\varepsilon}_7\hat{\varepsilon}_7$:

$$\begin{cases} \frac{\sigma_j}{\gamma_j}\widetilde{\Theta}_j^T\Theta_j \leq -\frac{\sigma_j}{2\gamma_j}\widetilde{\Theta}_j^T\widetilde{\Theta}_j + \frac{\sigma_j}{2\gamma_j}\widetilde{\Theta}_j^{T*}\widetilde{\Theta}_j^* \\ \frac{\sigma_5}{\gamma_5}\widetilde{\Theta}_5^T\Theta_5 \leq -\frac{\sigma_5}{2\gamma_5}\widetilde{\Theta}_5^T\widetilde{\Theta}_5 + \frac{\sigma_5}{2\gamma_5}\widetilde{\Theta}_5^{T*}\widetilde{\Theta}_5^* \\ \frac{\sigma_6}{\gamma_6}\widetilde{\Theta}_6^T\Theta_6 \leq -\frac{\sigma_6}{2\gamma_6}\widetilde{\Theta}_6^T\widetilde{\Theta}_6 + \frac{\sigma_6}{2\gamma_6}\widetilde{\Theta}_6^{T*}\widetilde{\Theta}_6^* \\ \frac{\sigma_7}{\gamma_7}\widetilde{\Theta}_7^T\Theta_7 \leq -\frac{\sigma_7}{2\gamma_7}\widetilde{\Theta}_7^T\widetilde{\Theta}_7 + \frac{\sigma_7}{2\gamma_7}\widetilde{\Theta}_7^{T*}\widetilde{\Theta}_7^* \end{cases} \quad (113)$$

$$\begin{cases} \frac{\alpha_j}{\eta_j}\widetilde{\varepsilon}_j\hat{\varepsilon}_j \leq -\frac{\alpha_j}{2\eta_j}\widetilde{\varepsilon}_j^2 + \frac{\alpha_j}{2\eta_j}\left|\varepsilon_j^*\right|^2 \\ \frac{\alpha_5}{\eta_5}\widetilde{\varepsilon}_5\hat{\varepsilon}_5 \leq -\frac{\alpha_5}{2\eta_5}\widetilde{\varepsilon}_5^2 + \frac{\alpha_5}{2\eta_5}\left|\varepsilon_5^*\right|^2 \\ \frac{\alpha_6}{\eta_6}\widetilde{\varepsilon}_6\hat{\varepsilon}_6 \leq -\frac{\alpha_6}{2\eta_6}\widetilde{\varepsilon}_6^2 + \frac{\alpha_6}{2\eta_6}\left|\varepsilon_6^*\right|^2 \\ \frac{\alpha_7}{\eta_7}\widetilde{\varepsilon}_7\hat{\varepsilon}_7 \leq -\frac{\alpha_7}{2\eta_7}\widetilde{\varepsilon}_7^2 + \frac{\alpha_7}{2\eta_7}\left|\varepsilon_7^*\right|^2 \end{cases} \quad (114)$$

Consequently, we may restructure (112) in the following manner:

$$\begin{aligned}\dot{V} \leq &\sum_{j=1}^{4}\left(-c_{zj}S_{zj}^2 - \frac{\sigma_j}{2\gamma_j}\tilde{\Theta}_j^T\tilde{\Theta}_j + \frac{\sigma_j}{2\gamma_j}\Theta_j^{T*}\Theta_j^* - \frac{\alpha_j}{2\eta_j}\tilde{\varepsilon}_j^2 + \frac{\alpha_j}{2\eta_j}|\varepsilon_j^*|^2 + \varepsilon_j^*\zeta\right) + \left(-c_\Omega S_\Omega^2 - \frac{\sigma_5}{2\gamma_5}\tilde{\Theta}_5^T\tilde{\Theta}_5 + \frac{\sigma_5}{2\gamma_5}\Theta_5^{T*}\Theta_5^*\right.\\ &\left.- \frac{\alpha_5}{2\eta_5}\tilde{\varepsilon}_5^2 + \frac{\alpha_5}{2\eta_5}|\varepsilon_5^*|^2 + \varepsilon_5^*\zeta\right)\\ &+ \left(-c_{i_q}S_{i_q}^2 - \frac{\sigma_6}{2\gamma_6}\tilde{\Theta}_6^T\tilde{\Theta}_6 + \frac{\sigma_6}{2\gamma_6}\Theta_6^{T*}\Theta_6^* - \frac{\alpha_6}{2\eta_6}\tilde{\varepsilon}_6^2 + \frac{\alpha_6}{2\eta_6}|\varepsilon_6^*|^2 + \varepsilon_6^*\zeta\right)\\ &+ \left(-c_{i_d}S_{i_d}^2 - \frac{\sigma_7}{2\gamma_7}\tilde{\Theta}_7^T\tilde{\Theta}_7 + \frac{\sigma_7}{2\gamma_7}\Theta_7^{T*}\Theta_7^* - \frac{\alpha_7}{2\eta_7}\tilde{\varepsilon}_7^2 + \frac{\alpha_7}{2\eta_7}|\varepsilon_7^*|^2 + \varepsilon_7^*\zeta\right)\end{aligned} \quad (115)$$

Let us define

$$\vartheta = \min\left\{\sigma_j, \alpha_j, 2c_{zj}, \sigma_5, \alpha_5, 2c_\Omega, \sigma_6, \alpha_6, 2c_{i_q}, \sigma_7, \alpha_7, 2c_{i_d}\right\} \quad (116)$$

Then, (115) is transformed into the following:

$$\dot{V} \leq -\vartheta V + \rho \quad (117)$$

where

$$\begin{aligned}\rho = &\sum_{j=1}^{4}\left(\frac{\sigma_j}{2\gamma_j}\Theta_j^{T*}\Theta_j^* + \frac{\alpha_j}{2\eta_j}|\varepsilon_j^*|^2 + \varepsilon_j^*\zeta\right) + \left(\frac{\sigma_5}{2\gamma_5}\Theta_5^{T*}\Theta_5^* + \frac{\alpha_5}{2\eta_5}|\varepsilon_5^*|^2 + \varepsilon_5^*\zeta\right) + \left(\frac{\sigma_6}{2\gamma_6}\Theta_6^{T*}\Theta_6^* + \frac{\alpha_6}{2\eta_6}|\varepsilon_6^*|^2 + \varepsilon_6^*\zeta\right)\\ &+ \left(\frac{\sigma_7}{2\gamma_7}\Theta_7^{T*}\Theta_7^* + \frac{\alpha_7}{2\eta_7}|\varepsilon_7^*|^2 + \varepsilon_7^*\zeta\right)\end{aligned} \quad (118)$$

We can now establish the following theorem, which demonstrates our primary finding in this study.

Theorem 1. *Consider the six-phase PMSM nonlinear system in (28). Assuming that the previously specified Assumption 1, Assumption 2, and Lemma 1 are correct, the control laws described by Equations (92) through (95), which use adaptive fuzzy logic system, in conjunction with the parameter adaption law detailed in Equations (84) through (91), guarantee that all signals inside the closed-loop system demonstrate uniformly ultimately bounded (UUB) stability. Additionally, the output tracking error is shown to converge to a narrow area in close proximity to the origin. Furthermore, the developed controller has the ability to maintain stability.*

Proof. The integral of (117) over $[0, t]$ yields the following result:

$$V(t) \leq V(0) e^{-\vartheta t} + \frac{\rho}{\vartheta} \quad (119)$$

The inequalities represented by (117) suggest that $V \geq \frac{\rho}{\vartheta}$, $\dot{V} \leq 0$. Therefore, by utilizing the Lyapunov theorem, the signals $S_\Omega, S_{i_q}, S_{i_d}, S_{zj}, \tilde{\Theta}_j, \tilde{\Theta}_5, \tilde{\Theta}_6, \tilde{\Theta}_7, \tilde{\varepsilon}_j, \tilde{\varepsilon}_5, \tilde{\varepsilon}_6, \tilde{\varepsilon}_7, V_{zj}, i_q^*, V_q$, and V_d in the closed-loop systems are bounded. Furthermore, it can be shown that, for any $\Upsilon \geq \sqrt{\frac{\rho}{\vartheta}}$, there exists a constant $T > 0$, such that $|Z_\Omega| \leq \Upsilon$, $\left|Z_{i_q}\right| \leq \Upsilon$, $\left|Z_{i_d}\right| \leq \Upsilon$, and $\left|Z_{i_{zj}}\right| \leq \Upsilon$ for all $t \geq T$.

To attain convergence of the tracking error to a small vicinity around zero and minimize $\sqrt{\frac{\rho}{\vartheta}}$ to the desired extent, it is imperative to select the design parameters $\eta_j, \eta_5, \eta_6, \eta_7, \gamma_j, \gamma_5, \gamma_6, \gamma_7, \chi_j, \chi_5, \chi_6, \chi_7, \alpha_j, \alpha_5, \alpha_6, \alpha_7, \sigma_j, \sigma_5, \sigma_6, \sigma_7, c_{zj}, c_\Omega, c_{i_q}$, and c_{i_d} judiciously. Therefore, it is evident that $\lim_{t\to\infty}|Z_\Omega| \leq \Upsilon$, $\lim_{t\to\infty}\left|Z_{i_q}\right| \leq \Upsilon$, $\lim_{t\to\infty}\left|Z_{i_d}\right| \leq \Upsilon$, and $\lim_{t\to\infty}\left|Z_{izj}\right| \leq \Upsilon$. The proof is now concluded. □

The modification terms $\alpha_j, \alpha_5, \alpha_6, \alpha_7, \sigma_j, \sigma_5, \sigma_6$, and σ_7 were introduced in the adaptive laws (84) through (91) in order to prevent parameter drift from approximation errors. After modifying adaptive laws, the time derivative of the Lyapunov function utilized for analysis turns negative when parameter estimations surpass specified limitations [44].

Once the control law was established and the system stability was theoretically confirmed, we proposed a system structure for study, as illustrated in Figure 5, incorporating the developed control law, to subject the system to a series of tests.

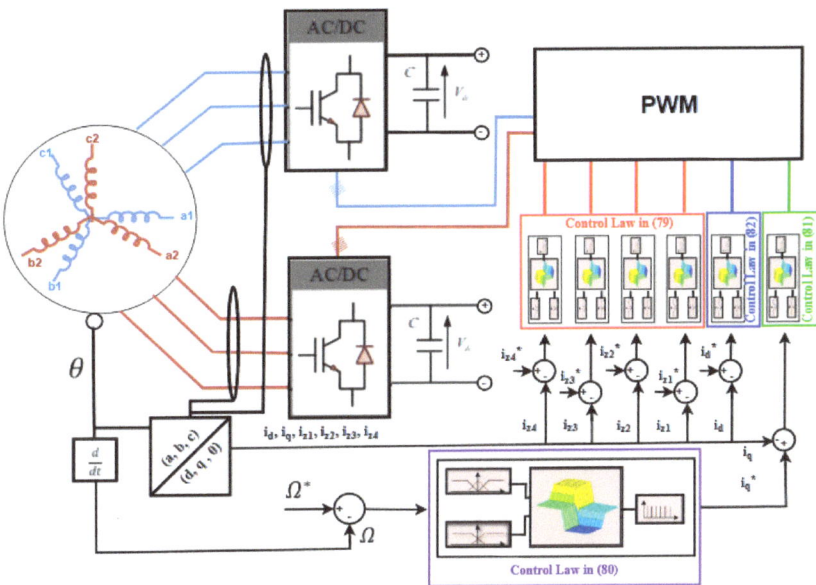

Figure 5. Control architecture of the DSPMSM.

Remark 1. *A comparative analysis is carried out in Table 1 to provide the greatest visibility and demonstrate the efficacy of the suggested control technique in relation to other relevant works.*

Table 1. Control techniques comparison.

Our Control Scheme	Other Approaches	Corresponding Papers
Uncertainty information is not required since the controller updates online to counteract the impact of uncertainty.	Information on uncertainty models is required.	[30–34]
There is no need for an approximation for disturbance since it was addressed conceptually by mathematical procedures, saving the time required for the approximation.	Disturbance was characterized as an external neutral stable system, or it was estimated. The authors considered that the time derivative of the disturbances must be limited.	[35–37]
Control gain is treated as an unknown nonlinear function.	The control gain is a straightforward constant, which restricts the scope of the systems that are taken into consideration.	[38,39]
The closed-loop system exhibits uniform ultimate bounded (UUB) stability, and the tracking error converges exponentially to the origin. This is achieved through the accurate approximation using fuzzy systems and the robust control term based on the tangent hyperbolic function, which effectively handles the residual terms from the fuzzy systems.	The closed-loop system exhibits stability, with the tracking error converging exponentially to a limited set. This behavior is attributed to the presence of residue terms resulting from the approximation.	[35,36,40]

9. Real Times Validation Results

To validate and assess the robustness of our control law, which consists of an adaptive fuzzy logic, within the framework of our research in developing control systems for the DSPMSM, we chose to use the RT-LAB platform from OPAL-RT due to its advanced features and effective integration of the hardware-in-the-loop (HIL) approach (Figure 6). By adopting this approach, we were able to combine our complex simulation models with real hardware components, allowing us to validate our designs under conditions close to reality. To demonstrate the robustness of our control, we chose to subject the system to a series of tests. Initially, we varied the control variables, namely the electromagnetic torque and the rotational speed. Subsequently, we made variations in the electrical parameters of the machine during its operation, including doubling the value of the stator resistance and halving the machine's inductance. Finally, we adjusted the mechanical parameters of the machine, such as inertia and viscous friction, by doubling their nominal values. The nominal parameters of the machine are provided in Table 2.

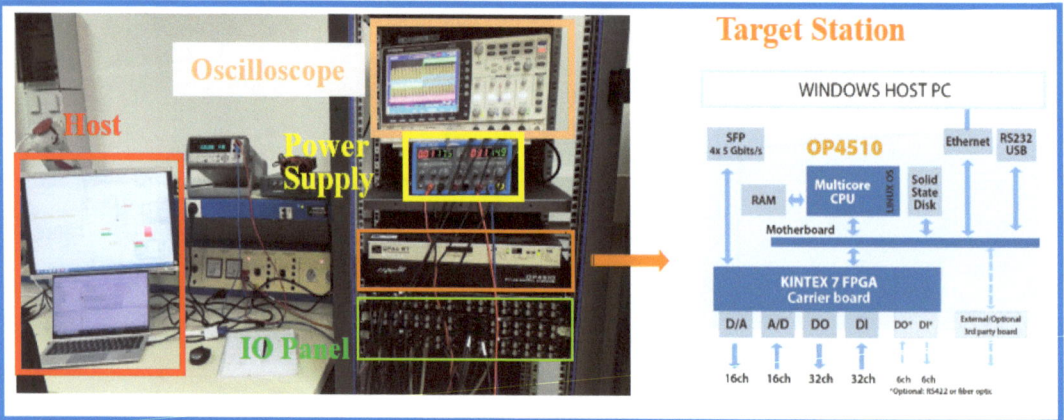

Figure 6. Test bench used for validation test.

Table 2. Machine parameters.

Parameters	Values
Stator resistance Rs (Ohm)	2
Leakage inductance l_{fs} (H)	0.562×10^{-3}
Maximum mutual inductance between two stator's windings M_{ss} (H)	3.373×10^{-3}
Rotor flux (Wb)	0.42
Moment of inertia (kg·m^2)	0.025
Numbers of pole pairs	6
Coefficient of viscose friction (N·m·s/rd)	0.01

The tests carried out to confirm the robustness of the developed control method took place in three following scenarios:

➢ Scenario 1: variation in operating references;
➢ Scenario 2: variation in electrical parameters of the DSPMSM;
➢ Scenario 3: variation in mechanical parameters of the DSPMSM.

9.1. Variation in Operating References

In this section, the references for both the torque and speed of the machine vary over time, and we will examine the behavior of the adopted control approach. At the beginning of operation, between t = 0 and t = 3 s, the applied speed reference is 300 rpm, and from 3 s onwards, the reference transitions to its nominal value of 400 rpm. Similarly, the electromagnetic torque reference is 60 Nm, and from t = 6 s onwards, the reference transitions to its nominal value of 93.5 Nm.

Figure 7a illustrates the curves of torque and velocity during a variation in their respective setpoints. The setpoint for the load torque is applied at t = 6 s, while that for velocity is set at t = 3 s. It is observed that the torque rigorously follows the setpoint with almost no oscillation. However, an undershoot is noticed during the velocity setpoint change. The integrated fuzzy logic controller quickly brings the system back to its setpoint within just 0.5 s. As for velocity, it reaches its reference after slight oscillations during startup and load torque setpoint changes, but these are promptly corrected by the controller in place, demonstrating high dynamic performance.

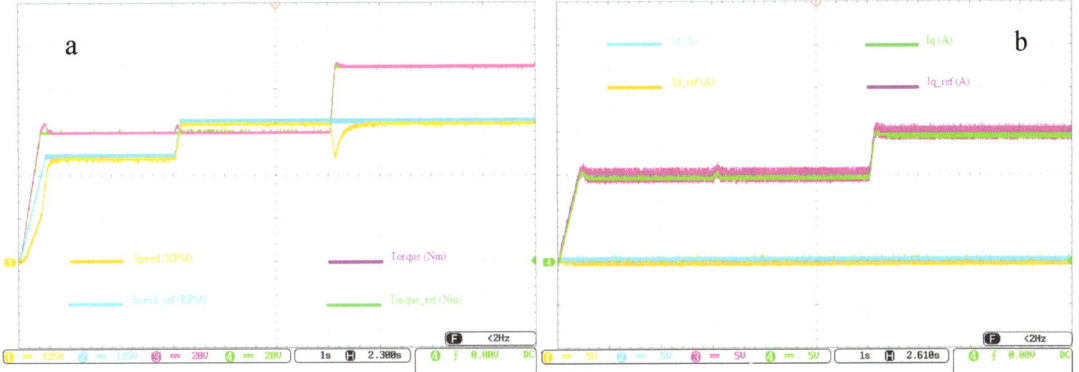

Figure 7. (**a**) Speed and torque curves, and (**b**) id and iq current curves for operating references variation.

The curves obtained during current regulation in the (d,q,o) plane are presented in Figure 7b. The id current precisely tracks its setpoint, maintained at zero in this selected control mode. Furthermore, the iq current also follows its reference, calculated based on the velocity through torque.

The Figure 8 illustrates the variation in the current in the first phase of the first star, as well as the first phase of the second star, with a zoom-in look at the current behavior at the moment when the electromagnetic torque changes from 60 Nm to 93 Nm. It should be noted that, at this moment, the current amplitudes increase from 7 A to 10 A.

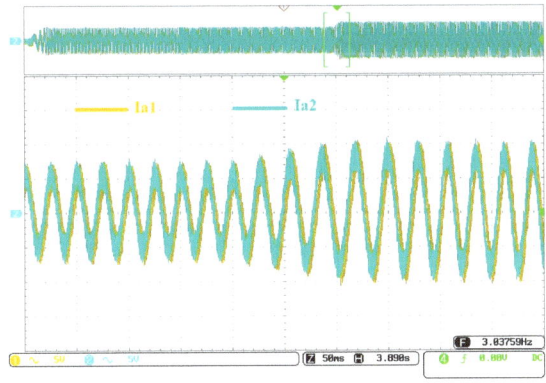

Figure 8. ia1 current curves and ia2 current curves for operating references variation.

The currents of the first phase (ia1 and ia2) for the two stator windings of the two stars of the machine are illustrated in Figure 9a. In Figure 9b, the currents of the first phase (ia1 and ia2) and those of the second phase (ib1 and ib2) of the two windings of the two stars of the machine are presented. These current curves exhibit a sinusoidal shape, with a 30° phase shift between the currents of the two stars, in accordance with the electrical and mechanical angle of the machine. However, due to the limitations of the oscilloscope used, only the currents of a few phases are represented, as it has only four acquisition channels.

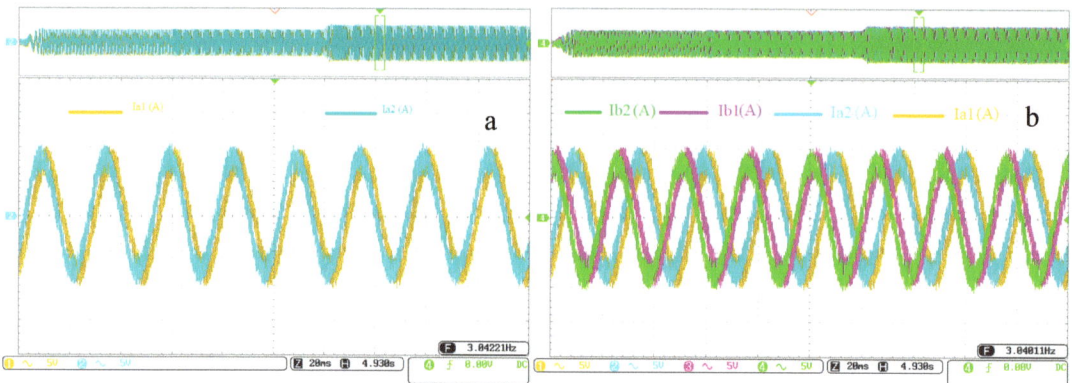

Figure 9. (**a**) ia1 current curves and ia2 current curves; (**b**) ia1, ia2, ib1, and ib2 current curves for operating references variation.

9.2. Variation in Electrical Parameters of DSPMSM

Now, we address the scenario where the electrical parameters of the machine vary (resistance and inductance). We subjected our system to the test by increasing the resistance value from its nominal value, which was 2 ohms until t = 3 s, to 4 ohms, representing a 100% increase. Then, starting from t=6 s, we reduced the machine's inductance value by 50% compared to its nominal value, which was held constant between t = 0 and t = 6 s. The curves depicting the physical quantities of the machine are shown in the following figures:

In this second case study, which presents the parametric variations in the electrical quantities of the machine, notably the resistance and inductance values, altered by several possible reasons, we can cite some of them: temperature variation within the machine; aging effects; mechanical effects, such as vibrations and shocks; and wear of the stator windings of the machine, as well as the presence of contaminants, can also influence the electrical properties of the materials of the machine, among other things.

Despite the presence of these parametric disturbances of the machine, we observe that the robustness and performance of our system are maintained thanks to the new control technique proposed in this article, which consists of an adaptive fuzzy controller, allowing us to find and readjust the parameters of the regulator to maintain the system at its optimal operating point. These performances are confirmed by the curves presented in Figures 10 and 11. In Figure 10a, the speed shows a slight increase in the overshoot value during startup, but the controller manages to bring the speed value back to its setpoint. Additionally, for the load torque and the id and iq currents, represented in Figure 10b, the setpoints are well followed and respected. The robustness of the deployed adaptive fuzzy controller is confirmed by the quality of the stator currents of the machine, which are sinusoidal in shape, like those obtained in the first case, as shown in Figure 11a,b.

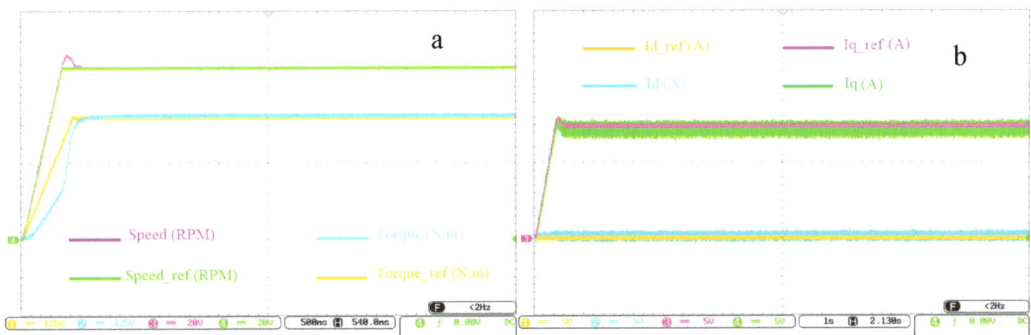

Figure 10. (**a**) Speed and torque curves, and (**b**) id and iq current curves for electrical parameters variation.

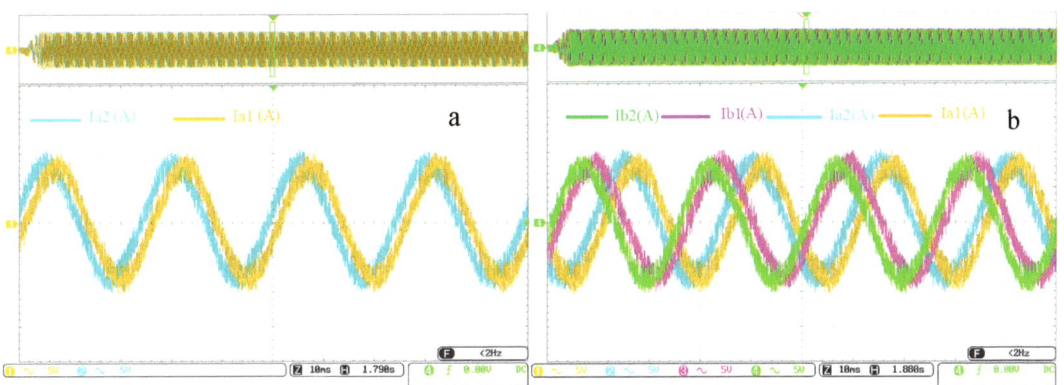

Figure 11. (**a**) ia1 current curves and ia2 current curves; and (**b**) ia1, ia2, ib1, and ib2 current curves for electrical parameters variation.

9.3. Variation in Mechanical Parameters of DSPMSM

In this new series of tests, we explore variations in the mechanical parameters of the machine, namely inertia and viscous friction. These variations may be due to various factors, such as a coupling issue leading to an increase in the system's inertia, or problems in the machine's bearings, resulting in a significant increase in friction. In this section, we examine the impact of these variations by increasing the inertia value to twice its nominal value at t = 3 s, while the friction, initially at its nominal value, is doubled at t = 6 s. The resulting curves from these tests are presented below.

The last case studied concerns the parametric variation in the mechanical quantities of the machine, which may be disturbed by factors such as the mechanical load fluctuations; wear of internal parts, such as bearings and rollers; or lubrication issues with mechanical components. Once again, the robustness of our adaptive fuzzy controller is validated by the maintenance of the machine's performance despite these disturbances.

Figure 12a shows that the load torque is not affected by the parametric variations, and the id and iq currents, depicted in Figure 12b, also perfectly follow their setpoints. However, a slight increase in the speed overshoot during startup is observed, but the setpoint is quickly reached and followed. The curves in Figure 13a represent two currents: one corresponding to the first phase of the first star, and the other to the first phase of the second star. As for the curves in Figure 13b, they illustrate the currents of the first and second phases of the first star, as well as those of the first and second phases of the second star. These currents maintain a sinusoidal shape, with a peak value close to 10 amperes.

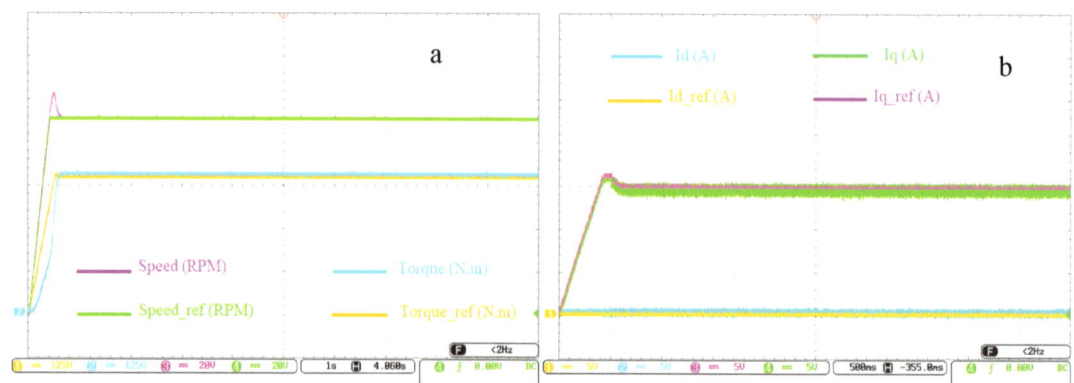

Figure 12. (**a**) Speed and torque curves, and (**b**) id and iq current curves for mechanical parameters variation.

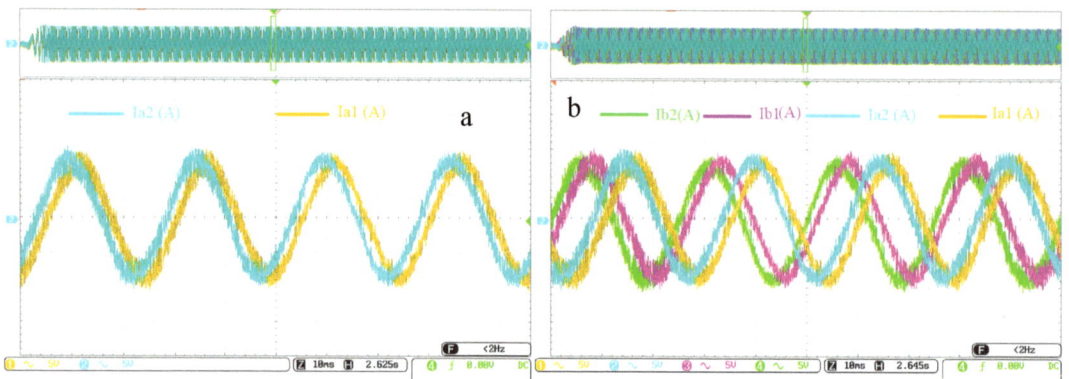

Figure 13. (**a**) ia1 current curves and ia2 current curves; and (**b**) ia1, ia2, ib1 and ib2 current curves for mechanical parameters variation.

To make a quantitative comparison between the proposed control method and the controllers proposed in Refs. [30,33], three well-known performance criteria are used. These are the integral of square error (ISE), integral of the absolute value of the error (IAE), and integral of the time multiplied by the absolute value of the error (ITAE). The obtained values for each criterion are summarized in Table 3. It is noted that the proposed control method offers the smallest values control of ISE, IAE, and ITAE as compared to the other two controllers. Hence, it is evident that the suggested controller is optimal and exhibits superior tracking of desired values compared to the other two controllers.

Table 3. Performance indices: ISE, IAE, and ITAE for speed and torque controls.

Control Method	ISE		IAE		ITAE	
	Speed Control	Torque Control	Speed Control	Torque Control	Speed Control	Torque Control
Proposed control in [33]	1.988	2.141×10^{-2}	4.387	0.719	3.302	2.821
Proposed control in [30]	1.275	9.541×10^{-3}	3.416	0.517	2.782	1.365
Proposed control method	0.813	5.771×10^{-3}	2.138	0.365	1.096	0.219

10. Conclusions

In this paper, we present a robust fuzzy adaptive control strategy for a DSPMSM, marking a significant advancement in the field of electromechanical system control amidst external disturbances and parametric uncertainties. The methodology developed in this study provides a precise and segmented representation of the system dynamics of the double-star permanent magnet synchronous machine (DSPMSM), achieved through a qualitative analysis and quantitative comparison with recent methods found in the literature. Our qualitative analysis highlighted the characteristics and advantages of our proposed approach, while the quantitative comparison demonstrated its performance and originality. Moreover, by proposing a model of the machine composed of two decoupled sub-models, the first being equivalent to that of a three-phase machine in the Park reference frame, and the second being equivalent to a fourth-order passive circuit, this facilitates the design and implementation of effective control strategies for these machines. Utilizing the Lyapunov function, we successfully developed the algorithm and adaptive parameter law, enabling the reduction of disturbances and parametric uncertainties on the DSPMSM, while maintaining tracking control efficiency and bounded stability in the global closed-loop system. Unlike active disturbance-rejection designs, our suggested technique does not rely on prior knowledge of external disturbances or a mathematical model, thus allowing it to operate optimally even in adverse conditions caused by model errors and nonlinearities. The simulation results consistently demonstrated a high tracking performance, underscoring the robustness and effectiveness of our proposed control method. In the future, our research will focus on improving the performance analysis in more complex scenarios and exploring opportunities to integrate this methodology into various domains of electrical engineering and industrial automation.

Author Contributions: Methodology, S.Z.; Validation, D.Z.; Formal analysis, A.D.; Investigation, M.F.B. All authors have read and agreed to the published version of the manuscript.

Funding: This research received no external funding.

Data Availability Statement: Data is contained within the article.

Conflicts of Interest: The authors have no conflicts of interest in this article.

References

1. Naas, B.; Nezli, L.; Naas, B.; Mahmoudi, M.O.; Elbar, M. Direct Torque Control Based Three Level Inverter-fed Double Star Permanent Magnet Synchronous Machine. *Energy Procedia* **2021**, *18*, 521–530. [CrossRef]
2. Ziane, D. *Optimisation de la Commande de la Machine Asynchrone Double Etoile en Fonctionnement Normal et Dégradé*; Thèse Université de BEJAIA: Béjaïa, Algeria, 2015; Nantes University: Nantes, France, 2021.
3. Benkhoris, M.F.; Merabtene, M.; Tabar, F.; Davat, B.; Semail, E. Approches de modélisation de la machine synchrone double étoile alimentée par des onduleurs de tension envue de la commande. *Rev. Int. Génie Électrique* **2003**, *6*. [CrossRef]
4. Naas, B.; Nezli, L.; Elbar, M.; Naas, B. Direct Torque Control of Double Star Synchronous Machine. *Int. J. Recent Trends Eng.* **2009**, *2*, 336.
5. Bojoi, R.; Lazzari, M.; Profumo, F.; Tenconi, A. Digital field-oriented control for dual three-phase induction motor drives. *IEEE Trans. Ind. Appl.* **2003**, *39*, 752–760. [CrossRef]
6. Zhang, J.-Y.; Zhou, Q.; Wang, K. Dual Three-Phase Permanent Magnet Synchronous Machines Vector Control Based on Triple Rotating Reference Frame. *Energies* **2022**, *15*, 7286. [CrossRef]
7. Hu, Y.; Zhu, Z.Q.; Odavic, M. Comparison of Two-Individual Current Control and Vector Space. *IEEE Trans. Ind. Appl.* **2017**, *53*, 4483–4492. [CrossRef]
8. Shu, M.; Ziane, D.; Oukrid, M.; Benkhoris, M.F.; Bernard, N. Dynamic modelling approach in view of vector control and behaviour analysis of multi-three-phase star Permanent Magnet Synchronous Motor drive. *Energies* **2024**, *17*, 1567. [CrossRef]
9. Laggoun, L.; Youb, L.; Belkacem, S.; Benaggoune, S.; Craciunescu, A. Direct torque control using second order Sliding mode of a double star permanent Magnet synchronous machine. In Proceedings of the 4th International Conference on Electrical Engineering and Control Applications, ICEECA 2019, Constantine, Algeria, 17–19 December 2019; pp. 139–153.
10. Wang, L.X. *Adaptive Fuzzy Systems and Control: Design and Stability Analysis*; Prentice-Hall, Inc.: Englewood Cliffs, NJ, USA, 1994.
11. Tong, S.-C.; Li, Y.-M.; Feng, G.; Li, T.-S. Observer-Based Adaptive Fuzzy Backstepping Dynamic Surface Control for a Class of MIMO Nonlinear Systems. *IEEE Trans. Syst. Man Cybern. Part B (Cybern.)* **2011**, *41*, 1124–1135. [CrossRef] [PubMed]

12. Lee, H. Robust Adaptive Fuzzy Control by Backstepping for a Class of MIMO Nonlinear Systems. *IEEE Trans. Fuzzy Syst.* **2011**, *19*, 265–275. [CrossRef]
13. Zhou, Q.; Shi, P.; Lu, J.; Xu, S. Adaptive Output-Feedback Fuzzy Tracking Control for a Class of Nonlinear Systems. *IEEE Trans. Fuzzy Syst.* **2011**, *19*, 972–982. [CrossRef]
14. Chwa, D. Fuzzy Adaptive Tracking Control of Wheeled Mobile Robots with State-Dependent Kinematic and Dynamic Disturbances. *IEEE Trans. Fuzzy Syst.* **2012**, *20*, 587–593. [CrossRef]
15. Chen, B.; Liu, X.P.; Ge, S.S.; Lin, C. Adaptive Fuzzy Control of a Class of Nonlinear Systems by Fuzzy Approximation Approach. *IEEE Trans. Fuzzy Syst.* **2012**, *20*, 1012–1021. [CrossRef]
16. Tong, S.; Li, Y. Adaptive Fuzzy Output Feedback Control of MIMO Nonlinear Systems with Unknown Dead-Zone Inputs. *IEEE Trans. Fuzzy Syst.* **2013**, *21*, 134–146. [CrossRef]
17. Liu, Z.; Wang, F.; Zhang, Y.; Chen, X.; Chen, C.P. Adaptive Fuzzy Output-Feedback Controller Design for Nonlinear Systems via Backstepping and Small-Gain Approach. *IEEE Trans. Cybern.* **2014**, *44*, 1714–1725. [CrossRef] [PubMed]
18. Zhou, Q.; Shi, P.; Xu, S.; Li, H. Adaptive Output Feedback Control for Nonlinear Time-Delay Systems by Fuzzy Approximation Approach. *IEEE Trans. Fuzzy Syst.* **2013**, *21*, 301–313. [CrossRef]
19. Wang, T.; Zhang, Y.; Qiu, J.; Gao, H. Adaptive Fuzzy Backstepping Control for A Class of Nonlinear Systems with Sampled and Delayed Measurements. *IEEE Trans. Fuzzy Syst.* **2015**, *23*, 302–312. [CrossRef]
20. Li, Y.; Tong, S.; Liu, Y.; Li, T. Adaptive Fuzzy Robust Output Feedback Control of Nonlinear Systems with Unknown Dead Zones Based on a Small-Gain Approach. *IEEE Trans. Fuzzy Syst.* **2014**, *22*, 164–176. [CrossRef]
21. Yu, J.; Chen, B.; Yu, H. Fuzzy-approximation-based adaptive control of the chaotic permanent magnet synchronous motor. *Nonlinear Dyn.* **2012**, *69*, 1479–1488. [CrossRef]
22. Li, Y.; Tong, S.; Li, T. Adaptive fuzzy backstepping control of static var compensator based on state observer. *Nonlinear Dyn.* **2013**, *73*, 133–142. [CrossRef]
23. Li, Y.; Tong, S.; Li, T. Direct adaptive fuzzy backstepping control of uncertain nonlinear systems in the presence of input saturation. *Neural Comput. Appl.* **2013**, *23*, 1207–1216. [CrossRef]
24. Li, H.; Li, Z.; Liu, J.; Zheng, X.; Yu, X.; Kaynak, O. Adaptive neural network backstepping control method for aerial manipulator based on coupling disturbance compensation. *J. Frankl. Inst.* **2024**, *361*, 106733. [CrossRef]
25. Yang, X.; Zhao, Z.; Li, Y.; Yang, G.; Zhao, J.; Liu, H. Adaptive neural network control of manipulators with uncertain kinematics and dynamics. *Eng. Appl. Artif. Intell.* **2024**, *133*, 107935. [CrossRef]
26. Zhao, N.; Zhao, D.; Liu, Y. Resilient event-triggering adaptive neural network control for networked systems under mixed cyber attacks. *Neural Netw.* **2024**, *174*, 106249. [CrossRef] [PubMed]
27. Zhao, D.; Shi, M.; Zhang, H.; Liu, Y.; Zhao, N. Event-triggering adaptive neural network output feedback control for networked systems under false data injection attacks. *Chaos Solitons Fractals* **2024**, *180*, 114584. [CrossRef]
28. Niu, S.; Wang, J.; Zhao, J.; Shen, W. Neural network-based finite-time command-filtered adaptive backstepping control of electro-hydraulic servo system with a three-stage valve. *ISA Trans.* **2024**, *144*, 419–435. [CrossRef] [PubMed]
29. Deng, D.D.; Zhao, X.W.; Lai, Q.; Liu, S. Fuzzy adaptive containment control of non-strict feedback multi-agent systems with prescribed time and accuracy under arbitrary initial conditions. *Inf. Sci.* **2024**, *663*, 120306. [CrossRef]
30. Rahmatullah, R.; Ak, A.; Serteller, N.F.O. Design of Sliding Mode Control using SVPWM Modulation Method for Speed Control of Induction Motor. *Transp. Res. Procedia* **2023**, *70*, 226–233. [CrossRef]
31. Zellouma, D.; Bekakra, Y.; Benbouhenni, H. Robust synergetic-sliding mode-based-backstepping control of induction motor with MRAS technique. *Energy Rep.* **2023**, *10*, 3665–3680. [CrossRef]
32. Rigatos, G.; Abbaszadeh, M.; Sari, B.; Siano, P.; Cuccurullo, G.; Zouari, F. Nonlinear optimal control for a gas compressor driven by an induction motor. *Results Control. Optim.* **2023**, *11*, 100226. [CrossRef]
33. Zellouma, D.; Bekakra, Y.; Benbouhenni, H. Field-oriented control based on parallel proportional–integral controllers of induction motor drive. *Energy Rep.* **2023**, *9*, 4846–4860. [CrossRef]
34. Prabhakaran, A.; Ponnusamy, T.; Janarthanan, G. Optimized fractional order PID controller with sensorless speed estimation for torque control in induction motor. *Expert Syst. Appl.* **2024**, *249*, 123574. [CrossRef]
35. Yoo, K.Y.; Kim, S.; Park, I.; Yoon, H.; Kim, H.S.; Seo, T. Disturbance torque observer-based variable impedance control for compliant stair-descending of transformable wheel mechanism. *Mech. Mach. Theory* **2024**, *194*, 105590. [CrossRef]
36. Wang, H.; Deng, J.; Zhang, L.; Bao, Q.; Mao, Y. Enhanced disturbance observer-based hybrid cascade active disturbance rejection control design for high-precise tracking system in application to aerospace satellite. *Aerosp. Sci. Technol.* **2024**, *146*, 108939. [CrossRef]
37. Fu, B.; Che, W.; Wang, Q.; Liu, Y.; Yu, H. Improved sliding-mode control for a class of disturbed systems based on a disturbance observer. *J. Frankl. Inst.* **2024**, *361*, 106699. [CrossRef]
38. Schuchert, P.; Gupta, V.; Karimi, A. Data-driven fixed-structure frequency-based H2 and H∞ controller design. *Automatica* **2024**, *160*, 111398. [CrossRef]
39. Yue, Y.L.; Sun, S.L.; Zuo, X. Discrete-time robust H∞/H2 optimal guaranteed performance control for riser recoil. *Ocean. Eng.* **2024**, *304*, 117699. [CrossRef]
40. Dong, F.; Yuan, B.; Zhao, X.; Ding, Z.; Chen, S. Adaptive robust constraint-following control for morphing quadrotor UAV with uncertainty: A segmented modeling approach. *J. Frankl. Inst.* **2024**, *361*, 106678. [CrossRef]

41. Slotine, J.J.; Li, W. *Applied Nonlinear Control*; Prentice Hall: Englewood Cliffs, NJ, USA, 1991; Volume 199.
42. Wang, L.X. Stable adaptive fuzzy control of nonlinear systems. *IEEE Trans. Fuzzy Syst.* **1993**, *1*, 146–155. [CrossRef]
43. Majid, M.Z. Robust adaptive backstepping control of uncertain fractional-order nonlinear systems with input time delay. *Math. Comput. Simul.* **2022**, *196*, 251–272.
44. Ioannou, P.; Sun, J. *Robust Adaptive Control*; Prentice Hall, Inc.: Englewood Cliffs, NJ, USA, 1996.

Disclaimer/Publisher's Note: The statements, opinions and data contained in all publications are solely those of the individual author(s) and contributor(s) and not of MDPI and/or the editor(s). MDPI and/or the editor(s) disclaim responsibility for any injury to people or property resulting from any ideas, methods, instructions or products referred to in the content.

MDPI AG
Grosspeteranlage 5
4052 Basel
Switzerland
Tel.: +41 61 683 77 34
www.mdpi.com

Mathematics Editorial Office
E-mail: mathematics@mdpi.com
www.mdpi.com/journal/mathematics

Disclaimer/Publisher's Note: The statements, opinions and data contained in all publications are solely those of the individual author(s) and contributor(s) and not of MDPI and/or the editor(s). MDPI and/or the editor(s) disclaim responsibility for any injury to people or property resulting from any ideas, methods, instructions or products referred to in the content.

www.ingramcontent.com/pod-product-compliance
Lightning Source LLC
LaVergne TN
LVHW072335090526
838202LV00019B/2426